Sampling Theory in Fourier and Signal Analysis

Advanced Topics

Sampling Theory in Fourier and Signal Analysis

Advanced Topics

Edited by

J.R. HIGGINS

Professor Emeritus
Anglia Polytechnic University, Cambridge

and

R.L. STENS

Professor of Mathematics
Rheinisch-Westfälische Technische Hochschule, Aachen

OXFORD
UNIVERSITY PRESS

OXFORD

UNIVERSITY PRESS

Great Clarendon Street, Oxford OX2 6DP

Oxford University Press is a department of the University of Oxford.
It furthers the University's objective of excellence in research, scholarship,
and education by publishing worldwide in

Oxford New York

Auckland Cape Town Dar es Salaam Hong Kong Karachi
Kuala Lumpur Madrid Melbourne Mexico City Nairobi
New Delhi Shanghai Taipei Toronto
With offices in
Argentina Austria Brazil Chile Czech Republic France Greece
Guatemala Hungary Italy Japan South Korea Poland Portugal
Singapore Switzerland Thailand Turkey Ukraine Vietnam

ISBN 0-19-853496-5

J.R Higgins and R.L. Stens take great pleasure in dedicating their work of assembling and editing this volume

to

Professor Paul Leo Butzer,

Teacher, Mentor, Friend.

PREFACE

The main theme of sampling theory is to establish connections between, on the one hand, a suitably chosen function and, on the other hand, the collection of values taken by that function on a discrete subset of its domain. The classical sampling or interpolation theorem, usually associated with the names of E.T. Whittaker, V.A. Kotel'nikov and C.E. Shannon, takes a central place in this theory; it states that for certain classes of functions there is a complete equivalence between the information contained in the whole function and that contained in its values at a discrete set of sample points. But this is only a small part of the theory; many other themes can be discerned within this general framework, and sampling theory is now a large and diverse field having connections with many parts of mathematics.

The aim of this book is to survey some of the topics in this field as they have matured in recent years, and in many cases to extend our knowledge of these topics. It is hoped that this will give the reader an idea of the rich scope of the subject, and that a contribution will thereby be made to an active area of research.

The book contains 10 chapters, each written by a different team of authors. A wide variety of topics is covered; the reader will find applications within mathematics, including connections with wavelets, approximation theory and stochastic analysis, as well as applications closer to computational and signal processing problems. Furthermore, several chapters study the theory of sampling series, and in a final chapter we see some of this theory presented in the most abstract formulation currently available.

The foundational book (with the same main title) by Higgins (1996) would serve as background to the more advanced material in the present volume.

A more detailed description of each chapter follows.

Chapter 1: *Applications of sampling theory to combinatorial analysis, Stirling numbers, special functions and the Riemann zeta function* (P.L. Butzer and M. Hauss).

The chapter treats a variety of applications of sampling theory to problems in mathematics. For example, there are applications to a fractional and infinite series form of the Vandermonde–Chu convolution formula, to new combinatorial identities, to the Gauss summation formula, to sampling of Stirling functions in terms of Stirling numbers, to Basler's problem (how to express the Riemann zeta function in terms of conjugate Bernoulli numbers), and to Hilbert versions of the Euler–Maclaurin and Poisson summation formulae. Some open problems are mentioned.

Chapter 2: *Sampling theory and the Arithmetic Fourier Transform* (W.J. Walker).

The Arithmetic Fourier Transform, or AFT, is the name now given to an algorithm for the computation of Fourier coefficients. Its history goes back to early work by Bruns, later made more rigorous by Wintner who wanted to make it "acceptable to the mathematical conscience".

Several recent and new results are developed giving sufficient conditions for the validity of the algorithm. It is interesting to find that the method involves sampling the object function at points within its interval of periodicity which are the Farey fractions of a specific order.

The AFT algorithm involves the Möbius function μ, and the famous summation formula

$$\sum_{k=1}^{\infty} \frac{\mu(k)}{k} = 0$$

(known to be equivalent to the prime number theorem) is used to demonstrate the deletion of any finite number of sample points from the algorithm. Some of the motivation here comes from certain models of the human brain.

Chapter 3: *Derivative sampling—a paradigm example of multichannel methods* (J.R. Higgins).

The process of reconstructing a function using samples taken from certain transformations of the function is called multichannel sampling. When these transformations are derivatives we speak of *derivative sampling*. In this chapter a survey is made of derivative sampling theorems, in one and in higher dimensions. The "Riesz basis method" is used to obtain some known and some new types.

Also treated here is a dual formulation of the sampling process, that of *interpolation*, in which one constructs a function whose derivatives are required to take given values at a given sequence of points. Here we look at the use to which Beurling and Selberg put such constructions in order to solve certain extremal problems in Fourier analysis.

Chapter 4: *Computational methods in linear prediction for band-limited signals based on past samples* (D.H. Mugler).

This chapter discusses linear prediction of the next few values of a band-limited function using information contained in past samples. This is in contrast to the more usual extrapolation methods which seek to determine the function in its entirety outside the region of observation, and also to the usual sampling theorems which require samples from the "future" as well as from the past. Methods developed here are universal for the class of band-limited functions and depend only on the highest frequency and a set of sample points.

Two main cases are considered; that in which an unlimited supply of samples is available and that in which there are only finitely many, the latter case receiving the chief emphasis. Further topics covered include the convergence of prediction formulas, optimal prediction and regularization methods.

Chapter 5: *Interpolation and sampling theories, and linear ordinary boundary value problems* (W.N. Everitt and G. Nasri-Roudsari).

This chapter deals with sampling expansions of Kramer type. Here, the Fourier transform is replaced by other integral transforms of a rather special type. The kernels of these transforms arise from boundary value problems associated with linear ordinary differential equations. These boundary value problems generate unbounded, self-adjoint differential operators on Hilbert function spaces, with discrete spectrum. These lead to an analytic form of Kramer's theorem, where the kernel depends analytically on the sampling parameter.

Chapter 6: *Sampling by generalized kernels* (R.L. Stens).

In this chapter the sinc-kernel in the classical sampling theorem is replaced by a different kernel function φ in order to reduce the errors occurring in practice when recovering a function from its samples. For example, if the kernel φ is chosen to have compact support then the infinite sampling series reduces to a finite one and the truncation error vanishes. Furthermore, these generalized series are more suitable when dealing with non-band-limited signals. In fact, they provide a good approximation even if the signal is continuous only. Error estimates for the aliasing error are given in terms of Jackson-type inequalities and moduli of smoothness.

Chapter 7: *Sampling theory and wavelets* (A. Fischer).

Here the ideas of the foregoing chapter are applied to the case where the kernel φ is built up from wavelets. The effectiveness of wavelet techniques in signal processing is well known; an example is their use for the compression of auditory signals or images. Here the theory of wavelets, frames and multiresolution analyses is applied to sampling expansions. In this context the Paley–Wiener space of band-limited functions, which can be regarded as a shift invariant space V_0, is replaced by a shift invariant space V_n of arbitrary level $n \in \mathbb{Z}$. If a function f belongs to one of the spaces V_n of a multiresolution analysis then one can find an exact representation of f as a series similar to the classical sampling series. For a function not belonging to one of the V_n this representation holds at least approximately.

Chapter 8: *Approximation by translates of a radial function* (N. Dyn).

In this chapter approximation schemes based on translates of radial functions are studied; these schemes provide a convenient and simple tool for global interpolation of scattered multivariate data. This is a central subject in multivariate approximation theory, and is also connected with multivariate sampling. Here the L_∞-approximation orders of such schemes are studied, first for points constituting a regular grid, and then for quasi-uniformly scattered points.

Chapter 9: *Almost sure sampling restoration of band-limited stochastic signals* (T. Pogány).

This chapter contains an overview of sampling theory published in the eastern European and former Soviet Union countries, going back to V.A. Kotel'nikov who introduced sampling principles in Russia in 1933. Much of this work is, even to this day, not at all well known in the West. Throughout this chapter Kotel'nikov's fundamental contribution is emphasized by associating his name with the classical sampling formula.

The chapter starts with some brief historial background in almost sure sampling restoration at uniformly spaced sample points, applied to scalar and vectorial weakly stationary band-limited stochastic processes and band-limited homogeneous random fields, the approach being that of oversampling. Then there are discussions of the Piranashvili–Lee extension of the Kotel'nikov formula to exponentially bounded and to (w, δ)–band-limited stochastic processes, the Gaposhkin–Klesov general theorem and related results, topics in restoration from irregularly spaced sample points, and the chapter ends with some new results generalizing the Gaposhkin condition.

Chapter 10: *Abstract harmonic analysis and the sampling theorem* (M.G. Beaty and M.M. Dodson).

This chapter gives a comprehensive introduction to sampling theory on locally compact abelian groups. Not only does the analysis provide an excellent way of seeing the group structure underlying much of Fourier analysis, but also shows how to unify many different kinds of sampling principle. Several examples involving special groups are mentioned and these emphasize the scope of the method. Reciprocity relations between the Haar measures of certain groups and sets play a prominent part in the analysis, and emerge naturally in the context of sampling theory.

After an exposition of the necessary harmonic analysis, Kluvánek's general sampling theorem and some generalizations are discussed, and then several further topics in sampling theory are brought into this general framework, including filtering, the aliasing error and multichannel sampling.

The editors J.R. Higgins and R.L. Stens take great pleasure in thanking the staff at Oxford University Press for their help and support. We are also grateful to Nancy Higgins for help with preparing the two indexes.

Cambridge   J.R.H.
Aachen   R.L.S.
February 1999

CONTENTS

LIST OF CONTRIBUTORS

M.G. Beaty
Department of Mathematics
University of Newcastle
Newcastle-upon-Tyne NE1 7RU
UK

P.L. Butzer
Lehrstuhl A für Mathematik
RWTH-Aachen
D-52056 Aachen
Germany

M.M. Dodson
Department of Mathematics
University of York
Heslington
York YO1 5DD
UK

N. Dyn
School of Mathematical Sciences
Tel Aviv University
Tel Aviv 69978
Israel

W.N. Everitt
School of Mathematics and Statistics
University of Birmingham
Birmingham B15 2TT
UK

A. Fischer
Lehrstuhl A für Mathematik
RWTH-Aachen
D-52056 Aachen
Germany

M. Hauss
Lehrstuhl A für Mathematik
RWTH-Aachen
D-52056 Aachen
Germany

J.R. Higgins
Department of Mathematics
Anglia Polytechnic University
Cambridge
UK

D.H. Mugler
Department of Mathematics
University of Akron
Akron
Ohio
USA

G. Nasri-Roudsari (née Schöttler)
Lehrstuhl A für Mathematik
RWTH-Aachen
D-52056 Aachen
Germany

T. Pogány
Sveučilište u Rijeci
Pomorski Fakultet u Rijeci
Studentska 2
Rijeka
Hrvatska

R.L. Stens
Lehrstuhl A für Mathematik
RWTH-Aachen
D-52056 Aachen
Germany

W.J. Walker
Department of Theoretical Mathematics
University of Auckland
Private Bag 92019
Aukland
New Zealand

APPLICATIONS OF SAMPLING THEORY TO COMBINATORIAL ANALYSIS, STIRLING NUMBERS, SPECIAL FUNCTIONS AND THE RIEMANN ZETA FUNCTION

1.1 Introduction

The Shannon sampling theorem together with its version for not necessarily band-limited functions is essentially equivalent (in the sense that each can be deduced from any of the others by elementary means) to at least five fundamental theorems in four different fields of mathematics, namely to the Poisson summation formula of Fourier analysis, to a particular form of Cauchy's integral formula of complex function theory, to the Euler–Maclaurin and Abel–Plana summation formulae of numerical analysis as well as to the celebrated functional equation of the Riemann zeta function of analytical number theory. Recall (Higgins 1996, pp. 90–96) and check Section 1.6 below.

Let us, for the sake of completeness, recall the classical sampling theorem. It states that any $f \in B_{\pi W}^p$, for $1 \le p < \infty$ and some $W > 0$, can be reconstructed from its sampled values $f(k/W)$, $k \in \mathbb{Z}$, in terms of

$$f(z) = \sum_{k=-\infty}^{\infty} f\left(\frac{k}{W}\right) \operatorname{sinc}(Wz - k) \qquad (1.1.1)$$

the series being absolutely and uniformly convergent on compact subsets of $\mathbb{C}$. Here $\operatorname{sinc} z := (\sin \pi z)/\pi z$, and $B_{\pi W}^p$ is the class of entire functions (on $\mathbb{C}$) of exponential type πW, namely,

$$|f(z)| \le \exp(\pi W|y|)\|f\|_C \qquad (z = x + iy \in \mathbb{C}), \qquad (1.1.2)$$

which belong to $L^p(\mathbb{R})$ when restricted to $\mathbb{R}$.

Here, we want to point out connections of the sampling theorem and its equivalents to combinatorial analysis, the theory of special functions and number theory. Firstly we will deduce a sampling representation of the binomial coefficient function $\binom{z}{\alpha}$, which will be used to establish the continuous Vandermonde formula

$$\sum_{k=0}^{\infty} \binom{z}{k}\binom{w}{\alpha - k} = \binom{z + w}{\alpha} \qquad (\alpha \in \mathbb{C}).$$

Using this formula a variety of infinite combinatorial identities can be deduced, all of them generalizing classical finite combinatorial identities. A very interesting fact is that this generalized Vandermonde formula is equivalent to Gauss's formula of the theory of special functions. In fact, Gauss's formula will be deduced from Shannon's sampling theorem and is a particular case of it.

Introducing the Stirling functions $s(\alpha, k)$ via

$$\binom{z}{\alpha} = \sum_{k=0}^{\infty} \frac{s(\alpha, k)}{\Gamma(\alpha + 1)} z^k \qquad (|z| < 1, \alpha \in \mathbb{C}),$$

a Shannon sampling theorem for these Stirling functions is also established, namely

$$\frac{s(\alpha, k)}{\Gamma(\alpha + 1)} = \sum_{j=k}^{\infty} \frac{s(j, k)}{j!} \operatorname{sinc}(\alpha - j) \qquad (\alpha \in \mathbb{C}, k \in \mathbb{N}_0),$$

emphasizing a close and natural connection between the Stirling functions and the (classical) Stirling numbers of the first kind, $s(j, k)$. The Stirling functions are, in turn, very closely related to the Riemann zeta function, as will be seen. Indeed, we plan to show that the $\zeta(m + 1)$ are not only connected with the $s(\alpha, m)$ for $\alpha \to 0$ but also for $\alpha = 1/2$. In fact, the Riemann zeta function can be expressed recursively in terms of Stirling functions. For even arguments Euler discovered a closed representation of the Riemann zeta function in terms of Bernoulli numbers,

$$\zeta(2m) = (-1)^{m+1} 2^{2m-1} \pi^{2m} \frac{B_{2m}}{(2m)!} \qquad (m \in \mathbb{N}). \tag{1.1.3}$$

This Euler formula is intimately connected with the Euler–Maclaurin summation formula and the Poisson sum formula, thus with Shannon's sampling theorem.

Here we will deduce the analogous Euler formula for odd arguments, namely

$$\zeta(2m + 1) = (-1)^m 2^{2m} \pi^{2m+1} \frac{\widetilde{B}_{2m+1}}{(2m + 1)!}, \tag{1.1.4}$$

using the conjugate Bernoulli numbers $\widetilde{B}_{2m+1}$ which are the particular values $\widetilde{B}_{2m+1}(0)$ of the (periodic) Hilbert transform of the Bernoulli polynomials $B_{2m+1}(x)$. The proof of the Euler formula for odd arguments is firstly based on new Hilbert–analogues of the Euler–Maclaurin summation formula and the Poisson summation formula.

Whereas the proof of Euler's original formula (1.1.3) makes use of the partial fraction representation of $\cot \pi z$, regarded as one of the most remarkable of such expansions, namely

$$\pi \cot \pi z = \frac{1}{z} + \sum_{k=1}^{\infty} \frac{2z}{z^2 - k^2} \qquad (z \in \mathbb{C} \setminus \mathbb{Z}),$$

that of formula (1.1.4) will also depend on a related partial fraction representation, namely

$$\pi \frac{\Omega(2\pi z)}{e^{-\pi z} - e^{\pi z}} = \sum_{k=1}^{\infty} (-1)^k \frac{k}{z^2 + k^2}$$

of the function

$$\Omega(w) = \int_{0+}^{1/2} (\sinh uw) \cot \pi u \, du \qquad (w \in \mathbb{C}),$$

introduced in (Butzer, Flocke and Hauss 1994).

The final section is devoted to sampling expansions of particular Jacobi *functions* in term of their associated Jacobi *polynomials*. The matter is considered for the Legendre functions, the Chebyshev functions of the first kind as well as those of the second kind. It is finally indicated how sampling theory may be used to derive many summation formulae.

1.2 Binomial coefficient function

A generalization of the very important binomial coefficients, the binomial coefficient *function* $\binom{z}{\alpha}$, is defined by

$$\binom{z}{\alpha} := \frac{\Gamma(z+1)}{\Gamma(\alpha+1)\Gamma(z-\alpha+1)} \qquad (\alpha \in \mathbb{C}, \ z \in \mathbb{C}\backslash\mathbb{Z}^-).$$

This function is known to have the infinite product expansion (see Nielsen 1965, p. 62)

$$\binom{z}{\alpha} = \prod_{k=1}^{\infty} \left(1 + \frac{\alpha}{k}\right)\left(1 - \frac{\alpha}{z+k}\right) \qquad (\alpha \in \mathbb{C}, \ z \in \mathbb{C}\backslash\mathbb{Z}^-),$$

which can be deduced from the Euler product of the Gamma function.

A recent result, first studied by Butzer, Hauss and Schmidt (1989), is that $\binom{z}{\alpha}$, considered as a function of $\alpha \in \mathbb{C}$, interpolates the coefficients $\binom{z}{k}$, $k \in \mathbb{N}_0$, in a natural way in the sense that one can deduce the following sampling representation. Thus, the binomial coefficient function also has a natural infinite series expansion. We shall establish this even for the derivatives of $\binom{z}{\alpha}$.

Theorem 1.1 (Sampling representation of the binomial coefficient function) *For* $\Re(z) > -1$ *and* $r \in \mathbb{N}_0$ *there holds true*

$$\left(\frac{d}{d\alpha}\right)^r \binom{z}{\alpha} = \sum_{k=0}^{\infty} \binom{z}{k}\left(\frac{d}{d\alpha}\right)^r \operatorname{sinc}(\alpha - k) \qquad (\alpha \in \mathbb{C}), \qquad (1.2.1)$$

the series being absolutely and uniformly convergent on each compact subset of $\mathbb{C}$.

Proof Defining $\Psi_w(z) := \left[\Gamma(\tfrac{1}{2}w + z)\Gamma(\tfrac{1}{2}w - z)\right]^{-1}$, $w, z \in \mathbb{C}$, one can readily show that for $\Re(z) \geq 0$ and $|z| \geq r_w$ (with $r_w := \max(|1 - \tfrac{1}{2}w|, |\tfrac{1}{2}w|) + 1$),

$$|\Psi_w(z)| = \frac{1}{\pi}\left|\sin \pi\left(\frac{w}{2} - z\right)\right| |z^{1-w}|\left(1 + \mathcal{O}_w\left(\frac{1}{|z|}\right)\right).$$

Thus, one has

$$|\Psi_w(z)| \leq \frac{1}{\pi}e^{\pi|\Im(w)|}e^{\pi|\Im(z)|}|z|^{1-\Re(w)}\left(1 + \mathcal{O}_w\left(\frac{1}{|z|}\right)\right),$$

which in view of $\Psi_w(-z) = \Psi_w(z)$ holds true for all $z \in \mathbb{C}$ with $|z| \geq r_w$. Further, if $\Re(w) > 1$, one has for all $z \in \mathbb{C}$, $|\Psi_w(z)| \leq M_w \exp(\pi|\Im(z)|)$, M_w being a constant depending on w only. Noting that

$$\binom{z}{\alpha} = \Gamma(z+1)\Psi_{z+2}\left(\alpha - \frac{z}{2}\right),$$

it is clear that $\binom{z}{\alpha}$ is analytic in $\alpha \in \mathbb{C}$ and that for all $\alpha \in \mathbb{C}$ and $\Re(z) > -1$

$$\left|\binom{z}{\alpha}\right| \leq M_z \exp(\pi|\Im(\alpha)|).$$

Thus, $f(\alpha) := \binom{z}{\alpha}$ is a function of exponential type π for $\Re(z) > -1$, since for $\Re(z) > -1$ and $p > 1/(\Re(z) + 1)$, $p \geq 1$, one can show that

$$\int_{\mathbb{R}} \left|\binom{z}{x}\right|^p dx < \infty.$$

Thus $\binom{z}{\cdot} \in B_\pi^p$, and the application of Shannon's sampling theorem completes the proof, noting that $f(-k) = 0$ for $k \in \mathbb{N}$.

Observing that, by $\sin \pi z = \pi(\Gamma(z)\Gamma(1-z))^{-1}$, $z \in \mathbb{C}$,

$$\binom{0}{\alpha} := \frac{\Gamma(1)}{\Gamma(\alpha+1)\Gamma(1-\alpha)} = \frac{\sin \pi\alpha}{\pi\alpha} = \mathrm{sinc}(\alpha),$$

the sampling series (1.2.1) in the case $r = 0$ can be rewritten as

$$\binom{z}{\alpha} = \sum_{k=0}^{\infty} \binom{z}{k}\binom{0}{\alpha - k} \qquad (\Re(z) > -1, \ \alpha \in \mathbb{C}). \qquad (1.2.2)$$

This convolution identity calls to mind the classical Vandermonde–Chu convolution formula, regarded as "perhaps the most widely used combinatorial identity" (cf. (Riordan 1979, p. 8)), namely

$$\sum_{k=0}^{n} \binom{z}{k}\binom{w}{n-k} = \binom{z+w}{n} \qquad (w, z \in \mathbb{N}, \ n \in \mathbb{N}_0). \qquad (1.2.3)$$

In fact, the latter can be generalized to a fractional and infinite series form with complex w, z, n which in turn is also an extension of (1.2.2). S.N. Bernstein (1958) in a lecture of 1913 on approximation theory emphasizes that the transition from the finite to the infinite is a development of mathematical ideas that lies deep; it can be characterized roughly in terms of the transition from equations to inequalities, from algebra to analysis. Below, the equations are even preserved.

The Vandermonde–Chu convolution formula can also be given in terms of the hypergeometric function ${}_2F_1(a, b; c; 1)$ (defined in Section 1.3), namely

$${}_2F_1(-\alpha, -z; w - \alpha + 1; 1) = \frac{\Gamma(z+w+1)\Gamma(w-\alpha+1)}{\Gamma(z+w-\alpha+1)\Gamma(w+1)} = \frac{[z+w-\alpha+1]^\alpha}{[w-\alpha+1]^\alpha},$$

where $n = \alpha \in \mathbb{N}_0$, $w - n + 1 \notin \mathbb{Z}_0^-$ and $[b]^\alpha := (b)_\alpha = \Gamma(b+\alpha)/\Gamma(\alpha)$, $\alpha + b \notin \mathbb{Z}_0^-$, is the rising factorial function. This relationship is studied more deeply in

Section 1.3. Concerning the history of (1.2.3), see especially (Askey 1975, p. 59). It is usually attributed to Vandermonde (1772), but it is already to be found in a book by Shih-Chieh Chu (1303). This is confirmed by Needham (1959, p. 138).

Now let us consider the case of the Vandermonde formula that α is no longer a natural number but a complex variable.

Theorem 1.2 (Generalized Vandermonde formula) *For $z, w \in \mathbb{C}$ with $\Re e(z + w) > -1$, $\Re e(z) > -1$, and $\Re e(w) > -1$ there holds true*

$$\sum_{k=0}^{\infty} \binom{z}{k}\binom{w}{\alpha - k} = \binom{z + w}{\alpha} \qquad (\alpha \in \mathbb{C}) \qquad (1.2.4)$$

or, expressed in the terminology of the falling factorial function $[z]_\alpha := \Gamma(z + 1)/\Gamma(z + 1 - \alpha)$,

$$\sum_{k=0}^{\infty} \binom{\alpha}{k}[z]_k[w]_{\alpha - k} = [z + w]_\alpha. \qquad (1.2.5)$$

Both series are absolutely and uniformly convergent on compact sets of $\mathbb{C}$.

Proof Firstly, let $\Re e(z), \Re e(w), \Re e(z + w) > -1$. Using Theorem 1.1 twice and the classical Vandermonde formula (see (1.2.3)), one has

$$\binom{z + w}{\alpha} = \sum_{j=0}^{\infty} \binom{z + w}{j}\binom{0}{\alpha - j}$$

$$= \sum_{k=0}^{\infty}\sum_{j=k}^{\infty} \binom{w}{j - k}\binom{0}{\alpha - j}\binom{z}{k} = \sum_{k=0}^{\infty} \binom{w}{\alpha - k}\binom{z}{k},$$

which proves formula (1.2.4). The order of summation can be interchanged, because in view of

$$\left|\binom{w}{j}\binom{0}{\alpha - j - k}\binom{z}{k}\right| = \mathcal{O}_{w,z,\alpha}\left(\frac{j^{-\Re e(w)-1}k^{-\Re e(z)-1}}{j + k}\right), \quad j, k \to \infty$$

the iterated series is absolutely convergent. For $\Re e(w) > 0$ or $\Re e(z) > 0$ this is obvious. If $\Re e(w), \Re e(z) \in (-1, 0]$, define

$$-\frac{1}{p} := \frac{\Re e(w) - \Re e(z) - 1}{2} \quad \text{and} \quad -\frac{1}{q} := \frac{\Re e(z) - \Re e(w) - 1}{2}.$$

Then, in view of $j + k \geq j^{1/p}k^{1/q}$, the iterated sum is convergent. Finally, observing

$$\left|\binom{z}{k}\binom{w}{\alpha - k}\right| = \left|-\frac{\Gamma(w + 1)}{\Gamma(-z)}\frac{\sin \pi \alpha}{\pi}\frac{\Gamma(k - z)\Gamma(k - \alpha)}{\Gamma(k + 1)\Gamma(w - \alpha + k + 1)}\right|$$

$$= \mathcal{O}_{\alpha,w,z}(k^{-\Re e(z)-\Re e(w)-2}), \quad k \to \infty,$$

the proof is complete in view of the identity theorem of complex function theory.

In view of (1.2.5) the falling factorials $[z]_\alpha$ could, in the terminology of Rota (1975, p. 72), be said to be of (infinite) convolution type. Rota naturally only considered the (classical) finite case.

Remark 1.3 The only work known to the authors when n in the Vandermonde formula (1.2.3) is *not* an integer is that by Davis (1962, p. 70 f.), who denotes just seven lines to the matter. In fact he states it formally in the form of (1.2.5) with $w = 1$, writing it as

$$\frac{\Gamma(2 + z)}{\Gamma(2 + z - \alpha)} = [z + 1]_\alpha$$

$$= \frac{1}{\Gamma(2 - \alpha)} \left[1 - \frac{\alpha(\alpha - 1)}{(\alpha - 1)(\alpha - 2)} z + \frac{\alpha(\alpha - 1)}{(\alpha - 2)(\alpha - 3)} \frac{[z]_2}{2!} \right.$$

$$\left. - \frac{\alpha(\alpha - 1)}{(\alpha - 3)(\alpha - 4)} \frac{[z]_3}{3!} + \frac{\alpha(\alpha - 1)}{(\alpha - 4)(\alpha - 5)} \frac{[z]_4}{4!} - \cdots \right].$$

There are no questions of convergence. As an application he takes $z_\alpha = \frac{1}{2}$, yielding the expansion

$$\frac{3\pi}{8} = 1 + \frac{1}{1 \cdot 3} \frac{1}{2} + \frac{1}{1 \cdot 5} \frac{1}{2^2} \frac{1}{2!} + \frac{1 \cdot 3}{5 \cdot 7} \frac{1}{2^3} \frac{1}{3!} + \frac{1 \cdot 3 \cdot 5}{7 \cdot 9} \frac{1}{2^4} \frac{1}{4!} + \cdots.$$

Let us first consider some simple applications of our Theorem 1.2. The cases $z = w = \alpha = \frac{3}{2}$ and $z = -\frac{1}{4}$, $w = \alpha = \frac{1}{4}$ yield, respectively,

$$\sum_{k=0}^{\infty} \binom{3/2}{k}^2 = \binom{3}{3/2} = \frac{32}{3\pi}.$$

$$\sum_{k=0}^{\infty} \binom{-1/4}{k} \binom{1/4}{k} = \binom{0}{1/4} = \frac{2\sqrt{2}}{\pi}.$$

Another identity can be obtained for the Catalan numbers, usually defined by $C_0 := -\frac{1}{2}$ and

$$C_n := \frac{1}{n} \binom{2n - 2}{n - 1} \qquad (n \in \mathbb{N}).$$

Lemma 1.4 (Identity for Catalan numbers) *There holds true*

$$\sum_{k=0}^{\infty} \frac{C_k^2}{16^k} = \frac{1}{4} \sum_{k=0}^{\infty} \binom{1/2}{k}^2 = \frac{1}{\pi}.$$

Proof Setting $z = w = \alpha = \frac{1}{2}$ in Theorem 1.2, one has

$$\sum_{k=0}^{\infty} \binom{1/2}{k}^2 = \binom{1}{1/2} = \frac{4}{\pi}.$$

Observing $\binom{1/2}{0}^2 = 1 = 4(-1/2)^2$ and

$$\sum_{k=1}^{\infty}\left(C_k 2^{-2k+1}\right)^2 = \sum_{k=1}^{\infty}\left(\frac{\Gamma(k-\frac{1}{2})}{2\sqrt{\pi}\,\Gamma(k+1)}\right)^2$$

$$= \sum_{k=1}^{\infty}\left(\binom{1/2}{k}\frac{\Gamma(k-\frac{1}{2})\Gamma(1-(k-\frac{1}{2}))\pi(-1)}{\pi\Gamma(-\frac{1}{2})\Gamma(1-(-\frac{1}{2}))}\right)^2$$

$$= \sum_{k=1}^{\infty}\binom{1/2}{k}^2,$$

the proof is complete.

Using Theorem 1.2, a great number of "infinite" combinatorial identities can be deduced. As generalizations of the well-known (finite) combinatorial identities of Hagen (1891) and Gould (1972) one obtains the following dozen assertions. They may perhaps all be new.

Proposition 1.5 (Infinite combinatorial identities) *For* $\alpha, \beta, \gamma, p, q, a, b \in \mathbb{C}$ *one has*

(a) *with* $\mathfrak{Re}(\alpha + \beta) > 0$ *and* $\alpha + \beta + \gamma \notin \mathbb{Z}^-$,

$$\sum_{j=0}^{\infty}(-1)^j\binom{\alpha+\beta+\gamma}{j+\beta+\gamma}\binom{j+\gamma}{\gamma} = \binom{\alpha+\beta-1}{\alpha};$$

(b) *with* $\mathfrak{Re}(p - q) > 0$ *and* $p \notin \mathbb{Z}^-$,

$$\sum_{j=0}^{\infty}(-1)^j(a+bj)\binom{p}{\alpha-j}\binom{q+j}{j} = \left(a+b\alpha\frac{q+1}{q+1-p}\right)\binom{p-q-1}{\alpha};$$

(c) *with* $\mathfrak{Re}(p + q) > 0$ *and* $p \notin \mathbb{Z}^-$,

$$\sum_{j=0}^{\infty}(a+bj)\binom{p}{\alpha-j}\binom{q}{j} = \left(a+bq\frac{\alpha}{q+p}\right)\binom{p+q}{\alpha};$$

(d) *with* $\mathfrak{Re}(p + q) > 0$ *and* $p \notin \mathbb{Z}^-$,

$$\sum_{j=0}^{\infty}(a+bj)\binom{p}{\alpha+j}\binom{q}{j} = \left(a+bq\frac{p-\alpha}{q+p}\right)\binom{p+q}{p+\alpha};$$

(e) *with* $\mathfrak{Re}(p) > 0$,

$$\sum_{j=0}^{\infty}(a+bj)\binom{p}{j}^2 = \left(a+\frac{bp}{2}\right)\binom{2p}{p};$$

(f) *with* $\Re e(\alpha - p + q) > 0$ *and* $p, \alpha - p, \alpha + q - p \notin \mathbb{Z}, q - p \notin \mathbb{N}$,

$$\sum_{j=0}^{\infty}(-1)^j\binom{p-j}{\alpha-j}\binom{q}{j} = \frac{\sin\pi(p-\alpha)}{\sin\pi p}\frac{\sin\pi(q-p)}{\sin\pi(\alpha+q-p)}\binom{p-q}{\alpha};$$

(g) *with* $\Re e(\alpha + q - p) > 0, \Re e(\alpha - p) > 0, \Re e(q) > 0, p - q \notin \mathbb{Z}^-, \alpha - p \notin \mathbb{Z}_0^-$
and $p, \alpha + q - p \notin \mathbb{Z}$,

$$\sum_{j=0}^{\infty}(-1)^j(a+bj)\binom{p-j}{\alpha-j}\binom{q}{j}$$

$$= \frac{\sin\pi(p-\alpha)}{\sin\pi p}\frac{\sin\pi(q-p)}{\sin\pi(\alpha+q-p)}\left(a + \frac{bq\alpha}{\alpha+q-p-1}\right)\binom{p-q}{\alpha};$$

(h) *with* $\Re e(\alpha - p - q) > 1, p, \alpha - p - q \notin \mathbb{Z}, p - \alpha + 1 \notin \mathbb{N}$ *and* $p + q + 1 \notin \mathbb{Z}^-$,

$$\sum_{j=0}^{\infty}\binom{p-j}{\alpha-j}\binom{q+j}{j} = \frac{\sin\pi(p-\alpha)}{\sin\pi p}\frac{\sin\pi(q+p)}{\sin\pi(p+q-\alpha)}\binom{p+q+1}{\alpha};$$

(i) *with* $\Re e(p + q) > -1$ *and* $p \notin \mathbb{Z}^-$,

$$\sum_{j=0}^{\infty}\binom{p}{\alpha+j}\binom{q}{j} = \binom{p+q}{\alpha+q};$$

(j) *with* $\Re e(\alpha - p - q) > 2$ *and* $p, \alpha - p - q \notin \mathbb{Z}, p - \alpha + 1 \notin \mathbb{N}, p + q + 1 \notin \mathbb{Z}^-$,

$$\sum_{j=0}^{\infty}(a+bj)\binom{p-j}{\alpha-j}\binom{q+j}{j}$$

$$= \frac{\sin\pi(p-\alpha)}{\sin\pi p}\frac{\sin\pi(q+p)}{\sin\pi(p+q-\alpha)}\left(a + \frac{b(q+1)\alpha}{p+q-\alpha+2}\right)\binom{p+q+1}{\alpha};$$

(k) *with* $\Re e(\alpha + q) < -1$ *and* $p, \alpha + p \notin \mathbb{Z}, \alpha + 1 \notin \mathbb{N}, p + q + 1 \notin \mathbb{Z}^-$,

$$\sum_{j=0}^{\infty}\binom{p-j}{\alpha}\binom{q+j}{q} = \frac{\sin\pi\alpha}{\sin\pi p}\frac{\sin\pi(q+p)}{\sin\pi(q+\alpha)}\binom{p+q+1}{p-\alpha};$$

(l) *with* $\Re e(\alpha) > -\frac{1}{2}$,

$$\sum_{j=0}^{\infty}2^{\alpha-\beta-2j}\binom{\alpha}{j}\binom{\alpha-j}{j+\beta} = \binom{2\alpha}{\alpha+\beta}.$$

Vicariously, let us establish part (f) and part (l). As to part (f), first observe that for $\alpha, \beta \in \mathbb{C}, \alpha + \beta \notin \mathbb{Z}, \alpha \notin \mathbb{N}$ there holds true the identity

$$\binom{\alpha + \beta - 1}{\beta} = \frac{\sin \pi \alpha}{\sin \pi (\alpha + \beta)} \binom{-\alpha}{\beta}.$$

Then one has with $\alpha = j - p$ and $\beta = \alpha - j$,

$$\sum_{j=0}^{\infty} (-1)^j \binom{p - j}{\alpha - j} \binom{q}{j} = -\frac{\sin \pi (\alpha - p)}{\sin \pi p} \sum_{j=0}^{\infty} \binom{\alpha - p - 1}{\alpha - j} \binom{q}{j}$$

$$= \frac{\sin \pi (p - \alpha)}{\sin \pi p} \binom{\alpha - p + q - 1}{\alpha},$$

which completes the proof of part (f).

In fact, this identity is a generalization of (Hagen 1891, p. 67), with $p, q \in \mathbb{N}$,

$$\sum_{j=0}^{n} (-1)^j \binom{p - j}{n - j} \binom{q}{j} = \binom{p - q}{n}.$$

Remark 1.6 It is important to observe that the correct substitutes for $(-1)^n$ in the fractional case $n = \alpha \in \mathbb{C}$ in this frame are the expressions

$$\frac{\sin \pi \alpha}{\sin \pi (\alpha + \beta)} \quad \text{and} \quad \frac{\sin \pi (\gamma - \alpha)}{\sin \pi \gamma}.$$

In fact, for $\alpha = n$ the product of these two expressions (see part (f)) reduces to $(-1)^n \cdot (-1)^n = 1$. In the frame of fractional order Bernoulli numbers (see also Butzer, Hauss and Leclerc 1992) it turned out that the substitute for $(-1)^n$ in the fractional case $n = \alpha \in \mathbb{C}$ is $(\cos(\pi\alpha) - \sin(\pi\alpha))$. In other areas of mathematics the simplest substitute for $(-1)^n$, namely $e^{i\pi\alpha}$, can be used to generalize classical results to fractional results.

Now let us prove part (l). Noting the Legendre duplication formula for the Gamma function, namely

$$\Gamma(2z+1) = 2^{2z} \Gamma(z+1) \Gamma\left(z + \frac{1}{2}\right) \Big/ \sqrt{\pi} \quad \left(z \in \mathbb{C} \setminus \left\{-\frac{1}{2}, -1, -\frac{3}{2}, -2, \ldots\right\}\right),$$

one has in view of the general Vandermonde convolution formula (1.2.4) for $\Re(\alpha) > -\frac{1}{2}, \alpha \notin \mathbb{N}_0$ and $\frac{1}{2}(\alpha + \beta - 1), \frac{1}{2}(\alpha - \beta) \notin \mathbb{Z}^-$,

$$\sum_{j=0}^{\infty} 2^{\alpha-\beta-2j} \binom{\alpha}{j} \binom{\alpha-j}{j+\beta}$$

$$= \sum_{j=0}^{\infty} 2^{\alpha-\beta-2j} \frac{\Gamma(\alpha+1)}{\Gamma(j+1)\Gamma(\alpha-j+1)} \frac{\Gamma(\alpha-j+1)}{\Gamma(j+\beta+1)\Gamma(\alpha-\beta-2j+1)}$$

$$= \sum_{j=0}^{\infty} \frac{\sqrt{\pi}\,\Gamma(\alpha+1)}{\Gamma(j+1)\Gamma\left(\frac{\alpha-\beta}{2}-j+1\right)} \frac{\Gamma\left(\frac{\alpha-\beta}{2}+1\right)}{\Gamma\left(\frac{\alpha-\beta}{2}+1\right)}$$

$$\times \frac{1}{\Gamma(j+\beta+1)\Gamma\left(\frac{\alpha-\beta}{2}-j+\frac{1}{2}\right)} \frac{\Gamma\left(\frac{\alpha+\beta}{2}+\frac{1}{2}\right)}{\Gamma\left(\frac{\alpha+\beta}{2}+\frac{1}{2}\right)}$$

$$= \frac{\sqrt{\pi}\,\Gamma(\alpha+1)}{\Gamma\left(\frac{\alpha-\beta}{2}+1\right)\Gamma\left(\frac{\alpha+\beta}{2}+\frac{1}{2}\right)} \sum_{j=0}^{\infty} \binom{\frac{\alpha-\beta}{2}}{j}\binom{\frac{\alpha+\beta}{2}-\frac{1}{2}}{\frac{\alpha-\beta}{2}-\frac{1}{2}-j}$$

$$= \frac{\sqrt{\pi}\,\Gamma(\alpha+1)}{\Gamma\left(\frac{\alpha-\beta}{2}+1\right)\Gamma\left(\frac{\alpha+\beta}{2}+\frac{1}{2}\right)} \binom{\alpha-\frac{1}{2}}{\frac{\alpha-\beta}{2}-\frac{1}{2}}$$

$$= \frac{\Gamma(2\alpha+1)}{\Gamma(\alpha+\beta+1)\Gamma(\alpha-\beta+1)} = \binom{2\alpha}{\alpha+\beta}.$$

Note that removable discontinuities occur in $\frac{1}{2}(\alpha+\beta-1), \frac{1}{2}(\alpha-\beta) \in \mathbb{Z}^-$ and in $\alpha \in \mathbb{N}_0$. In fact, for $\alpha = n \in \mathbb{N}_0$ the summands are zero for $j > n$, which gives in particular (see Problem 10332 in American Mathematical Monthly of 1994)

$$\binom{2n}{n+\beta} = \sum_{j=0}^{n} 2^{n-\beta-2j} \binom{n}{j}\binom{n-j}{j+\beta}.$$

$\square$

1.3 Gauss summation formula

Although Theorem 1.2 in this form seems so far only to have been studied by the authors (Butzer, Hauss and Schmidt 1989; Butzer and Hauss 1991), except for the Davis work mentioned, it can be shown that the generalized Vandermonde formula is equivalent to Gauss's summation formula for hypergeometric functions

$$_2F_1(a, b; c; z) := \sum_{k=0}^{\infty} \frac{[a]^k [b]^k}{[c]^k} \frac{z^k}{k!}, \tag{1.3.1}$$

where $a, b, c, z \in \mathbb{C}$ with $|z| < 1$ and $c \notin \mathbb{Z}_0^-$. The series converges (absolutely) throughout the entire unit circle if $\Re(c-a-b) > 0$, and it converges throughout the entire unit circle except at the point $z = 1$ for $-1 < \Re(c-a-b) \leq 0$.

For $\Re(c - a - b) \le -1$ the series diverges for all $|z| = 1$. Here $[a]^k$ is the rising factorial polynomial. In fact, one has the following theorem; it connects Shannon's sampling theorem with Gauss's summation formula via the generalized Vandermonde formula.

Theorem 1.7 (Vandermonde and Gauss's formulae) *The following two assertions can be deduced from each other:*

(a) *For $z, w \in \mathbb{C}$ with $\Re(z + w) > -1$ and $w \notin \mathbb{Z}^-$ there holds the generalized Vandermonde formula* (1.2.4).

(b) *For $a, b, c \in \mathbb{C}$ satisfying $\Re(c - a - b) > 0$ and $c \notin \mathbb{Z}_0^-$ there holds Gauss's summation formula*

$$
{}_2F_1(a, b; c; 1) = \frac{\Gamma(c - a - b)\Gamma(c)}{\Gamma(c - a)\Gamma(c - b)}. \tag{1.3.2}
$$

Proof Using the Gauss formula of part (b), one readily obtains for w, $w - \alpha \notin \mathbb{Z}^-$ and $\Re(w + z + 1) > 0$, by setting $a := -z$, $b := -\alpha$ and $c := w - \alpha + 1$

$$
\binom{w + z}{\alpha} = \frac{\Gamma(w + 1)}{\Gamma(w - \alpha + 1)\Gamma(\alpha + 1)} \frac{\Gamma(w - \alpha + 1)\Gamma(w + z + 1)}{\Gamma(w + 1)\Gamma(w + z - \alpha + 1)}
$$

$$
= \frac{\Gamma(w + 1)}{\Gamma(w - \alpha + 1)\Gamma(\alpha + 1)} \, {}_2F_1(-z, -\alpha; w - \alpha + 1; 1)
$$

$$
= \frac{\Gamma(w + 1)}{\Gamma(w - \alpha + 1)\Gamma(\alpha + 1)} \sum_{j=0}^{\infty} \frac{[-z]^j [-\alpha]^j}{[w - \alpha + 1]^j j!}
$$

$$
= \frac{\Gamma(w + 1)}{\Gamma(w - \alpha + 1)\Gamma(\alpha + 1)} \sum_{j=0}^{\infty} \frac{[-z]^j [-\alpha]^j \Gamma(w - \alpha + 1)}{\Gamma(w - \alpha + j + 1)j!}.
$$

Note that one has removable discontinuities in $w - \alpha \in \mathbb{Z}^-$. Observing the relations

$$
\frac{[-z]^j}{j!} = \binom{-z + j - 1}{j} = (-1)^j \binom{z}{j} \qquad (z \in \mathbb{C}, \, j \in \mathbb{N}_0),
$$

where $[z]^j$ with $[z]^0 := 1$ is the rising factorial polynomial,

$$
\frac{[-\alpha]^j}{\Gamma(\alpha + 1)} = \frac{(-1)^j}{\Gamma(\alpha - j + 1)} \qquad (\alpha \in \mathbb{C}, \, j \in \mathbb{N}_0),
$$

Vandermonde's formula follows for all those $z, \alpha, w \in \mathbb{C}$ satisfying $\Re(w + z) > -1$ and $w \notin \mathbb{Z}^-$, noting

$$
\frac{\Gamma(w + 1)}{\Gamma(w - \alpha + j + 1)\Gamma(\alpha - j + 1)} = \binom{w}{\alpha - j}.
$$

Vice versa, for $\Re e(c - a - b) > 0$, $c - b - 1 \notin \mathbb{Z}^-$, $1 - b, c \notin \mathbb{Z}_0^-$ and $a, b \notin \mathbb{Z}$, part (a) yields by setting $z = -a$, $w = c - b - 1$ and $\alpha = -b$,

$$
\begin{aligned}
{}_2F_1(a, b; c; 1) &= \sum_{j=0}^{\infty} \frac{\Gamma(a + j)\Gamma(b + j)\Gamma(c)}{\Gamma(a)\Gamma(b)\Gamma(c + j)\Gamma(j + 1)} \\
&= \frac{\Gamma(1 - b)\Gamma(c)}{\Gamma(c - b)} \sum_{j=0}^{\infty} \binom{-a}{j}\binom{c - b - 1}{-b - j} \\
&= \frac{\Gamma(1 - b)\Gamma(c)}{\Gamma(c - b)} \binom{c - a - b - 1}{c - a - 1},
\end{aligned}
$$

which turns out to be Gauss's formula. At $a \in \mathbb{Z}$ and $1 - b, b \in \mathbb{Z}_0^-$, one has removable discontinuities. Thus, the proof is complete.

Thus, Gauss's formula (1.3.2) is a particular case of Shannon's theorem (for $f(\alpha) = \binom{z}{\alpha}$).

1.4 Stirling functions

The classical Stirling numbers, introduced by James Stirling in his "Methodus Differentialis" in 1730, which are said to be "as important as Bernoulli's, or even more so" (see Jordan 1950, p. 143), play a major role in a variety of branches of mathematics, such as combinatorial theory, finite difference calculus, numerical analysis, interpolation theory and number theory. Those of the first kind, $s(n, k)$, can be defined via their exponential generating function

$$
\frac{(\log(1 + z))^k}{k!} = \sum_{n=k}^{\infty} \frac{s(n, k)}{n!} z^n \qquad (|z| < 1, \; k \in \mathbb{N}_0) \tag{1.4.1}
$$

or via their (horizontal) generating function

$$
[z]_n = \sum_{k=0}^{n} s(n, k) z^k \qquad (z \in \mathbb{C}, \; n \in \mathbb{N}_0) \tag{1.4.2}
$$

with the convention $s(n, 0) = \delta_{n,0}$ (Kronecker's delta).

This gives the natural possibility of defining "Stirling numbers of fractional order" $s(\alpha, k)$ with $\alpha \in \mathbb{C}$ and $k \in \mathbb{N}_0$. In fact, these "Stirling functions", as one may call them, may be defined via the ordinary generating function

$$
[z]_\alpha := \frac{\Gamma(z + 1)}{\Gamma(z - \alpha + 1)} = \sum_{k=0}^{\infty} s(\alpha, k) z^k \qquad (|z| < 1. \; \alpha \in \mathbb{C}), \tag{1.4.3}
$$

which can be rewritten as

$$\binom{z}{\alpha} = \sum_{k=0}^{\infty} \frac{s(\alpha, k)}{\Gamma(\alpha + 1)} z^k. \tag{1.4.4}$$

Both series are well defined for all $\alpha \in \mathbb{C}$, since $[z]_\alpha$ is holomorphic for $|z| < 1$. Noting that (1.4.1) is a Taylor expansion, these numbers could equivalently be defined by

$$s(n, k) = \frac{1}{k!} \left(\frac{d}{dx}\right)^n (\log x)^k |_{x=1} \qquad (k, n \in \mathbb{N}_0). \tag{1.4.5}$$

These functions $s(\alpha, k)$ were introduced by Butzer, Hauss and Schmidt (1989).

Of major interest is a sampling series representation of the functions $s(\alpha, k)$, actually of $s(\alpha, k)/\Gamma(\alpha + 1)$, for $\alpha \in \mathbb{C}$, in terms of the numbers $s(j, k)/j!$ for $j \in \mathbb{N}_0$, to stress the natural connection between the Stirling functions and the classical Stirling numbers.

Theorem 1.8 (Sampling theorem for Stirling functions) *For $\alpha \in \mathbb{C}$ and $k \in \mathbb{N}_0$ there hold true*

$$\frac{s(\alpha, k)}{\Gamma(\alpha + 1)} = \sum_{j=k}^{\infty} \frac{s(j, k)}{j!} \operatorname{sinc}(\alpha - j) = (-1)^k \frac{\sin \pi \alpha}{\pi} \sum_{j=k}^{\infty} \frac{|s(j, k)|}{j!(\alpha - j)}, \tag{1.4.6}$$

$$s(\alpha, k) = \sum_{j=k}^{\infty} \binom{\alpha}{j} s(j, k) s(\alpha - j, 0) \qquad (\alpha \notin \mathbb{Z}^-), \tag{1.4.7}$$

both series being absolutely and uniformly convergent on compact sets in $\alpha \in \mathbb{C}$.

Proof Substituting (1.4.2) into (1.2.1) for $r = 0$ yields for $|z| < 1$

$$\binom{z}{\alpha} = \sum_{j=0}^{\infty} \frac{1}{j!} \left(\sum_{k=0}^{j} s(j, k) z^k\right) \operatorname{sinc}(\alpha - j)$$

$$= \sum_{k=0}^{\infty} \left(\sum_{j=k}^{\infty} \frac{s(j, k)}{j!} \operatorname{sinc}(\alpha - j)\right) z^k, \tag{1.4.8}$$

the interchange of the order of summation being possible since the double sum converges absolutely and uniformly on compact subsets of $\mathbb{C}$ by (1.2.1). A comparison of the coefficients of the series (1.4.8) and (1.4.4) finally yields (1.4.6).

Formula (1.4.7) is then obvious since with $z = 0$ in (1.4.4) there holds true

$$s(\alpha, 0) = \frac{1}{\Gamma(1 - \alpha)}.$$

Formula (1.4.7) reveals that the Stirling functions of the first kind can be said to be of (infinite) binomial type in the sense of Rota (1975).

There exist a number of different representation formulae for the $s(\alpha, k)$.

Theorem 1.9 (Representation theorem) *For $\alpha \in \mathbb{C}$ and $k > \Re(\alpha)$, $k \in \mathbb{N}$ there hold*

$$s(\alpha, k) = \frac{1}{\Gamma(-\alpha)k!} \int_{0+}^{1-} \frac{(\log u)^k}{(1-u)^{\alpha+1}} \, du, \tag{1.4.9}$$

$$s(\alpha, k) = (-1)^{k+1} \frac{\sin \pi \alpha}{\pi} \sum_{j=1}^{\infty} \frac{[j]^{\alpha}}{j^{k+1}}, \tag{1.4.10}$$

$$s(\alpha, k+1) = \frac{1}{k+1} \sum_{j=0}^{k} \left(\Psi^{(k-j)}(1) - \Psi^{(k-j)}(1-\alpha) \right) \frac{s(\alpha, j)}{(k-j)!}, \tag{1.4.11}$$

where $\Psi^{(m)}$ is the nth polygamma function, i.e.,

$$\Psi^{(m)}(z) = \left(\frac{d}{dz} \right)^m \Psi(z) \qquad (z \in \mathbb{C} \backslash \mathbb{Z}_0),$$

$\Psi(z) = \Gamma'(z)/\Gamma(z)$ being the digamma function. Equation (1.4.11) is valid for all $\alpha \notin \mathbb{N}$.

Proof Let us first sketch the proof of (1.4.9). One has formally, noting

$$\int_0^1 x^{j-\alpha-1} \, dx = \frac{1}{j-\alpha}, \tag{1.4.12}$$

equation (1.4.1) and the sampling series (1.4.6),

$$\int_0^1 (\log u)^k (1-u)^{-1-\alpha} \, du = \int_0^1 (\log(1-x))^k x^{-1-\alpha} \, dx$$

$$= k! \int_0^1 \sum_{j=k}^{\infty} \frac{s(j,k)}{j!} (-1)^j x^{j-1-\alpha} \, dx$$

$$= k! \sum_{j=k}^{\infty} (-1)^j \frac{s(j,k)}{j!(j-\alpha)} = k! \Gamma(-\alpha) s(\alpha, k).$$

Since the power series (1.4.1) has radius of convergence $\rho = 1$, Abel's limit theorem justifies the interchange of summation and integration. Observe that $s(n, k) = 0$ for $n \in \{0, 1, \ldots, k-1\}$ since $1/\Gamma(-n) = 0$.

Concerning the proof of (1.4.10), firstly this sum is convergent for $\Re(\alpha) < k$ due to Gauss's criterion. Substituting $t = -\log u$ in (1.4.9) delivers

$$s(\alpha, k) = \frac{(-1)^k}{k! \Gamma(-\alpha)} \int_{0+}^{\infty} t^k (1-e^{-t})^{-1-\alpha} e^{-t} \, dt.$$

Expanding $(1 - e^{-t})^{-1-\alpha}$ into a binomial series, interchanging the order of summation and integration, possible for $\Re(\alpha) < k$, gives

$$s(\alpha, k) = \frac{(-1)^k}{\Gamma(-\alpha)k!} \sum_{j=0}^{\infty} \binom{-1-\alpha}{j} (-1)^j \int_0^{\infty} t^k e^{-(j+1)t} \, dt.$$

The definition of the Gamma function now delivers (1.4.10), recalling $[j]^{\alpha} = \Gamma(\alpha + j)/\Gamma(j)$.

Finally to the proof of (1.4.11). Setting $\varphi(x, \alpha) := \Psi(x + 1) - \Psi(x + 1 - \alpha)$, since $\varphi(\cdot, \alpha)$ is holomorphic for $|x| < \varepsilon$ with ε small, one has $\varphi(x, \alpha) = \sum_{k=0}^{\infty} \Psi_k(\alpha) x^k$ for $|x| < \varepsilon$, $\alpha \notin \mathbb{N}$, where

$$\Psi_k(\alpha) := \frac{\Psi^{(k)}(1) - \Psi^{(k)}(1 - \alpha)}{k!} \qquad (k \in \mathbb{N}_0),$$

noting that

$$\Psi_k(\alpha) = \frac{1}{k!} \left(\frac{d}{dx}\right)^k \varphi(x, \alpha) \Bigg|_{x=0}$$

Further, differentiating the series (1.4.3) for $|x| < 1$ yields

$$\frac{d}{dx}[x]_{\alpha} = \sum_{k=0}^{\infty} (k + 1)s(\alpha, k + 1)x^k.$$

But, by definition, the left-hand derivative equals

$$\frac{d}{dx} \frac{\Gamma(x + 1)}{\Gamma(x + 1 - \alpha)} = [x]_{\alpha} \varphi(x, \alpha) = \left(\sum_{k=0}^{\infty} s(\alpha, k)x^k\right) \left(\sum_{j=0}^{\infty} \Psi_j(\alpha)x^j\right) \qquad (|x| < \epsilon).$$

The integral (1.4.9) recalls to mind the Riemann–Liouville fractional derivative of order $\Re(\alpha) > 0$ of the function $f(x) := (\log x)^k / k!$, defined via a fractional integral, namely

$$\mathcal{D}_x^{\alpha} f(x) := \left(\frac{d}{dx}\right)^N \mathcal{I}^{\alpha-N} f(x) \qquad (N - 1 < \Re(\alpha) < N; \; N \in \mathbb{N}),$$

$$\mathcal{I}^{-\alpha} f(x) := \frac{1}{\Gamma(\alpha)} \int_0^x \frac{f(t)}{(x - t)^{1-\alpha}} \, dt.$$

In fact, equation (1.4.9) in this respect reads, for $\Re(\alpha) < k, k \in \mathbb{N}_0$,

$$s(\alpha, k) = \frac{1}{k!} \mathcal{D}_x^{\alpha} (\log x)^k \Bigg|_{x=1}. \tag{1.4.13}$$

This representation is the fractional order extension of (1.4.5). For a proof, which is rather long and somewhat technical, the reader is referred to (Hauss 1989). Observe that (1.4.13) is valid also for $\Re(\alpha) < 0$ since the integral (1.4.9) exists.

An important application of (1.4.11), the proof of which follows by induction and is intricate, is

Proposition 1.10 (Derivative property) *The Stirling functions $s(\alpha, k)$ are arbitrarily often continuously differentiable in α for $k \in \mathbb{N}_0$. In particular,*

$$\frac{d}{d\alpha}s(\alpha, k) = \sum_{j=0}^{k} \Psi^{(j)}(1-\alpha)\frac{s(\alpha, k-j)}{j!} \qquad (\alpha \notin \mathbb{N}).$$

Definitions (1.4.2) and (1.4.11) are useful in calculating the $s(\alpha, k)$ for smaller values of k. Thus, for $\alpha \notin \mathbb{Z}$,

$$s(\alpha, 1) = \frac{\Psi(1) - \Psi(1-\alpha)}{\Gamma(1-\alpha)} = -\frac{1}{\Gamma(-\alpha)}\sum_{j=1}^{\infty}\frac{1}{j(j-\alpha)}, \qquad (1.4.14)$$

$$s(\alpha, 2) = \frac{1}{\Gamma(-\alpha)}\sum_{k=2}^{\infty}\left(\sum_{j=1}^{k-1}\frac{1}{j}\right)\frac{1}{k(k-\alpha)}.$$

Thus, e.g.,

$$s\left(\frac{1}{2}, 1\right) = 2\log 2/\sqrt{\pi},$$

$$s\left(\frac{1}{3}, 1\right) = \frac{(9\log 3 - \sqrt{3}\pi)}{6\Gamma\left(\frac{2}{3}\right)},$$

$$s\left(\frac{1}{2}, 2\right) = \frac{\{4(\log 2)^2 - \frac{\pi^2}{3}\}}{2\sqrt{\pi}},$$

$$s\left(\frac{-3}{2}, 1\right) = \left(\frac{-4}{\sqrt{\pi}}\right)\left\{\frac{8}{9} - \frac{2}{3}\log 2\right\}.$$

To calculate the values $s(\alpha, k)$ for larger k, the recurrence formula (1.4.11) in k is handier. See especially (Butzer and Hauss 1991) for the foregoing matter.

1.5   Stirling functions and the Riemann zeta function $\zeta(s)$

C. Jordan (1950, pp. 166, 195) observed the following surprising link between the classical Stirling numbers $s(n, j)$ and the Riemann zeta function

$$\zeta(s) = \sum_{j=1}^{\infty}\frac{1}{j^s} \qquad (\mathfrak{Re}(s) > 1), \qquad (1.5.1)$$

one that apparently goes back to J. Stirling himself, namely that

$$\zeta(m+1) = \sum_{n=m}^{\infty}\frac{(-1)^{n+m}s(n, m)}{n \cdot n!} \qquad (m \in \mathbb{N}). \qquad (1.5.2)$$

As to a proof-sketch, replacing z by $-v$ in (1.4.1), dividing by v and integrating, using Abel's limit theorem and Raabe's convergence criterion, it can be shown that

$$\int_{0+}^{1-} \frac{[\log(1-v)]^m}{v} \, dv = m! \sum_{k=m}^{\infty} \frac{(-1)^k s(k,m)}{k \cdot k!}.$$

On the other hand, according to Nielsen (1965 p. 187), and Jordan (1950, p. 343),

$$\zeta(m+1) = \frac{(-1)^m}{m!} \int_{0+}^{1-} \frac{[\log(1-v)]^m}{v} \, dv.$$

Combining the results yields (1.5.2).

Our results give the following links between the Stirling functions $s(\alpha, m)$ and the Riemann zeta function $\zeta(m+1)$. The case $m = 2$ can be conjectured from (1.4.14) and (1.4.11), and observing (Lewin 1981; Nielsen 1965), namely,

$$\lim_{\alpha \to 0} \sum_{k=2}^{\infty} \left(\sum_{j=1}^{k-1} \frac{1}{j} \right) \frac{1}{k(k-\alpha)} = \sum_{k=2}^{\infty} \left(\sum_{j=1}^{k-1} \frac{1}{j} \right) \frac{1}{k^2} = \sum_{j=1}^{\infty} \frac{1}{k^3} = \zeta(3).$$

Theorem 1.11 (Stirling functions and the Riemann zeta function) *For $m \in \mathbb{N}$ there hold*

$$\lim_{\alpha \to 0} \Gamma(-\alpha) s(\alpha, m) = -\frac{\Psi^{(m)}(1)}{m!} = (-1)^m \zeta(m+1), \qquad (1.5.3)$$

$$\lim_{\alpha \to 0} \frac{d}{d\alpha} s(\alpha, m) = (-1)^{m+1} \zeta(m+1). \qquad (1.5.4)$$

Proof Concerning (1.5.3), formula (1.4.11) can be written as

$$\Gamma(-\alpha) s(\alpha, m+1) = \frac{1}{m+1} \sum_{j=0}^{m} \Gamma(1-\alpha) s(\alpha, j) \frac{\Psi_{m-j}(\alpha)}{-\alpha}.$$

The result now follows by induction. As to (1.5.4), applying L'Hospital's rule to (1.5.3) yields the proof in view of Proposition 1.10.

The $\zeta(m+1)$ are not only connected with $s(\alpha, m)$ for α near 0 but also for $\alpha = \frac{1}{2}$. In fact, the Riemann zeta function can also be expressed recursively in terms of Stirling functions, in particular in terms of the values $s(\frac{1}{2}, m)$, $m \in \mathbb{N}$, and conversely.

Theorem 1.12 (Recurrence relation for the zeta function) *For $k \in \mathbb{N}$ there holds*

$$s\left(\frac{1}{2}, k+1\right) = \frac{2\log 2 \, s(\frac{1}{2}, k)}{k+1} + \frac{1}{k+1} \sum_{j=2}^{k+1} s\left(\frac{1}{2}, k+1-j\right) (-1)^j (2 - 2^j) \zeta(j).$$

Conversely, one has for $k \in \mathbb{N}$

$$\zeta(k+1) = \frac{\sqrt{\pi}(-1)^k}{2-2^{k+1}}\left(2\log 2\, s\left(\frac{1}{2},k\right) - (k+1)\, s\left(\frac{1}{2},k+1\right)\right.$$
$$\left. + \sum_{j=2}^{k}(-1)^j(2-2^j)\, s\left(\frac{1}{2},k+1-j\right)\zeta(j)\right).$$

Proof From (1.4.11) one has in view of

$$\Psi^{(n)}(1) - \Psi^{(n)}\left(\frac{1}{2}\right) = (-1)^{n+1}n!(2-2^{n+1})\zeta(n+1) \qquad (n \in \mathbb{N}),$$

$$s\left(\frac{1}{2},k+1\right) = \frac{1}{k+1}\sum_{j=0}^{k}\frac{s(\frac{1}{2},j)}{(k-j)!}\left(\Psi^{(k-j)}(1) - \Psi^{(k-j)}\left(\frac{1}{2}\right)\right)$$

$$= \frac{s(\frac{1}{2},k)}{k+1}\left(\Psi^{(0)}(1) - \Psi^{(0)}(\tfrac{1}{2})\right)$$
$$+ \frac{1}{k+1}\sum_{j=0}^{k-1}s\left(\frac{1}{2},j\right)(-1)^{k-j+1}(2-2^{k-j+1})\zeta(k-j+1)$$

$$= \frac{s(\frac{1}{2},k)}{k+1}(-\gamma-(-\gamma-2\log 2))$$
$$+ \frac{1}{k+1}\sum_{j=2}^{k+1}s\left(\frac{1}{2},k-j+1\right)(-1)^j(2-2^j)\zeta(j).$$

The second formula can then also be obtained,

$$\frac{1}{k+1}s\left(\frac{1}{2},0\right)(-1)^{k+1}(2-2^{k+1})\zeta(k+1)$$
$$= s\left(\frac{1}{2},k+1\right) - 2\log 2\frac{s(\frac{1}{2},k)}{k+1}$$
$$- \frac{1}{k+1}\sum_{j=2}^{k}s\left(\frac{1}{2},k-j+1\right)(-1)^j(2-2^j)\zeta(j),$$

since $s(\frac{1}{2},0) = 1/\sqrt{\pi}$.

The simplest particular cases of Theorem 1.12 are

$$\zeta(3) = -\frac{2}{3}(\log 2)^3 + \frac{\pi^2\log 2}{6} + \frac{\sqrt{\pi}}{2}s\left(\frac{1}{2},3\right),$$

$$\frac{\pi^4}{90} = \zeta(4) = \frac{\sqrt{\pi}}{14}\left\{(2\log 2)s(1/2, 3) - 4s(1/2, 4) + \frac{12}{\sqrt{\pi}}(\log 2)\zeta(3)\right.$$

$$\left. - \frac{\pi^{3/2}}{6}\left(4(\log 2)^2 - \frac{\pi^2}{3}\right)\right\},$$

$$\zeta(5) = -\frac{\sqrt{\pi}}{30}\left\{(2\log 2)s\left(\frac{1}{2}, 4\right) - 5s\left(\frac{1}{2}, 5\right) - \frac{\pi^2}{3}s\left(\frac{1}{2}, 3\right)\right.$$

$$\left. + \frac{3}{\sqrt{\pi}}\left(4(\log 2)^2 - \frac{\pi^2}{3}\right)\zeta(3) - \frac{14}{45}\pi^{7/2}(\log 2)\right\}.$$

Thus to determine $\zeta(3)$ we would need to know $s(\frac{1}{2}, 3)$, for $\zeta(5)$ we need $s(\frac{1}{2}, j)$ for $j = 3, 4$ and 5. In this respect, in view of (1.4.2), (1.4.13), and (1.4.9), one has for $j \in \mathbb{N}$,

$$s\left(\frac{1}{2}, j\right) = \frac{1}{j!}\left(\frac{d}{dx}\right)^j\left(\frac{\Gamma(x+1)}{\Gamma(x+\frac{1}{2})}\right)\Bigg|_{x=0} = \frac{1}{j!}\mathcal{D}_x^{\frac{1}{2}}(\log x)^j\Bigg|_{x=1}$$

$$= \frac{1}{(-2\sqrt{\pi})j!}\int_{0+}^{1-}\frac{(\log u)^j}{(1-u)^{\frac{3}{2}}}du.$$

But the calculation of these three expressions for $j = 3, 4, \ldots$, in closed form is as yet unresolved. On the other hand, it can readily be shown that

$$\zeta(3) = \frac{1}{8}\int_{0+}^{1}(\log u)^3(1-u)^{\frac{1}{2}}du$$

$$+ \frac{112}{81} - \frac{2}{3}\left(\log 2 - \frac{4}{3}\right)^3 - \left(\log 2 - \frac{4}{3}\right)\left(\frac{20}{9} - \frac{\pi^2}{6}\right).$$

More generally, the latter integral can be replaced by $c_1\int_{0+}^{1}(\log u)^3(1-u)^{\frac{1}{2}+r}du$ for *any* $r \in \mathbb{N}_0$ provided additional terms involving π and $\log 2$ are added. Likewise the multiplicative factor $(1-u)^{-\alpha-1}$ in (1.4.9) can be replaced by $(1-u)^{\alpha+1}$ provided terms involving π and $\log 2$ are added to the integral.

This expansion brings to mind the old conjecture that $\zeta(3)$ is a function of π and $\log 2$, and so possibly irrational. Note that the integral can also be written as the Laplace integral $-\int_0^\infty u^3(1-e^{-u})^{1/2}e^{-su}du$ at $s = 1$; see also (Apéry 1979) in this respect.

Let us finally remark in connection with (1.5.2) that

$$\zeta(m+1) = \sum_{j_m=m}^{\infty}\left(\frac{1}{j_m^2}\right)\sum_{j_{m-1}=m-1}^{j_m-1}\left(\frac{1}{j_{m-1}^1}\right)\cdots\sum_{j_1=1}^{j_2-1}\frac{1}{j_1^1} \qquad (m \in \mathbb{N}).$$

1.6 Euler formulae for $\zeta(s)$ and summation formulae involving the Hilbert transform

As indicated in Section 1.5, there are problems involving closed expressions for the Riemann zeta function $\zeta(n)$. For even arguments $n = 2m$ there is the famous result first discovered by Euler in 1735/42 (see Weil 1984, pp. 184, 264, 272), stating that

$$\zeta(2m) = (-1)^{m+1} 2^{2m-1} \pi^{2m} \frac{B_{2m}}{(2m)!} \qquad (m \in \mathbb{N}). \qquad (1.6.1)$$

Here $B_n := B_n(0), n \in \mathbb{N}_0$, are the classical Bernoulli *numbers*, and $B_n(x)$ the Bernoulli *polynomials*, defined via their exponential generating function

$$\sum_{n=0}^{\infty} B_n(x) \frac{w^n}{n!} = \frac{we^{wx}}{e^w - 1} \qquad (w \in \mathbb{C}, |w| < 2\pi, x \in \mathbb{R}). \qquad (1.6.2)$$

This Euler formula is intimately connected with the Euler–Maclaurin summation formula and (a special version of) the Poisson summation formula, thus with Shannon's sampling theorem; recall (Higgins 1996, p. 90 ff.). Indeed, one way of proving formula (1.6.1) is to apply the partial fraction expansion of the cotangent function, which is essentially the "alternating" generating function of the B_{2k},

$$\frac{z}{2} \cot \frac{z}{2} = \sum_{k=0}^{\infty} \frac{(-1)^k}{(2k)!} B_{2k} z^{2k} \qquad (|z| < 2\pi). \qquad (1.6.3)$$

The expansion reads (see e.g. Koecher 1987, p. 155)

$$\pi \cot \pi z = \frac{1}{z} + \sum_{k=1}^{\infty} \frac{2z}{z^2 - k^2} = \sum_{k=-\infty}^{\infty}{}_e \frac{1}{z+k} \qquad (z \in \mathbb{C} \backslash \mathbb{Z}), \qquad (1.6.4)$$

$$\left(\sum_{k=-\infty}^{\infty}{}_e := \lim_{m \to \infty} \sum_{k=-m}^{m}, \text{ the so-called Eisenstein summation} \right), \text{ which can be written as}$$

a geometric series,

$$\pi \cot \pi z - \frac{1}{z} = -2z \sum_{k=1}^{\infty} \sum_{n=0}^{\infty} \frac{1}{k^2} \left(\frac{z^2}{k^2} \right)^n$$

$$= -2z \sum_{n=0}^{\infty} \zeta(2n+2) z^{2n}, \qquad (|z| < 1). \qquad (1.6.5)$$

Then comparing coefficients between (1.6.5) and (1.6.3) yields (1.6.1).

Now one way of deducing the expansion (1.6.4), regarded as one of the most interesting series expansions of classical analysis (see Remmert 1984, p. 224), is to apply the "second" Poison summation formula—see e.g. (Koecher 1987, p. 184)—(which can be deduced from the Euler–Maclaurin formula). So let us first recall the three Poisson summation formulae:

Proposition 1.13 (Poisson summation formulae)

(a) *Let $f \in C^{(2)}[1, n]$. Then the (first) Poisson summation formula states*

$$\sum_{k=1}^{n} f(k) - \int_{1}^{n} f(x)\,dx = \frac{1}{2}(f(n) + f(1)) + 2\sum_{k=1}^{\infty} \int_{0}^{n} f(x)\cos 2\pi kx\,dx. \qquad (1.6.6)$$

(b) *Let $f \in C^{(2)}[p, q]$ and $p < q$. Then the (second) Poisson summation formula states*

$$\sum_{k=p}^{q} f(k) - \frac{1}{2}(f(p) + f(q)) = \lim_{N \to \infty} \sum_{m=-N}^{N} \int_{p}^{q} f(x)e^{2\pi imx}\,dx. \qquad (1.6.7)$$

(c) *Let $f \in C^{(2)}(\mathbb{R})$. Then the (third) Poisson summation formula states*

$$\sum_{k \in \mathbb{Z}} f(k) = \sum_{m \in \mathbb{Z}} \int_{\mathbb{R}} f(x)e^{2\pi imx}\,dx = \sqrt{2\pi} \sum_{m \in \mathbb{Z}} \mathcal{F}[f; 2\pi m], \qquad (1.6.8)$$

if the sum $\sum_{k \in \mathbb{Z}} f(k)$ is convergent and the two integrals $\int_{\mathbb{R}} f(x)\,dx$ and $\int_{\mathbb{R}} |f^{(2)}(x)|\,dx$ exist.

Our major result of this section is to present a structural solution of the so-called "Basler problem", an open question since 1689 at the latest (known to Jacob Bernoulli, Pietro Mengoli, John Wallis, etc.), namely whether there exists a counterpart of (1.6.1) for odd arguments, thus a general closed expression for $\zeta(2m + 1)$. Our result reads

Theorem 1.14 *For $m \in \mathbb{N}$ there holds*

$$\zeta(2m + 1) = (-1)^m 2^{2m} \pi^{2m+1} \frac{\tilde{B}_{2m+1}}{(2m + 1)!}. \qquad (1.6.9)$$

Here the $\tilde{B}_n$ are the conjugate Bernoulli numbers, defined in terms of the Hilbert transform below. So it is to be expected that (1.6.9), in particular its proof, will be somehow connected with Hilbert transform versions of the Euler–Maclaurin and Poisson summation formulae.

Starting with the 1-periodic Bernoulli polynomials $\mathcal{B}_n(x)$, defined as the periodic extension of $\mathcal{B}_n(x) = B_n(x)$, $x \in (0, 1]$, one can—using Hilbert transforms—introduce 1-periodic conjugate Bernoulli "polynomials" $\mathcal{B}_n^{\sim}(x)$, $x \in \mathbb{R}$ ($x \notin \mathbb{Z}$ if $n = 1$), by

$$\mathcal{B}_n^{\sim}(x) := H_1 \mathcal{B}_n(x) \qquad (n \in \mathbb{N}). \qquad (1.6.10)$$

In general, the (periodic) Hilbert transform $H_p f$ of a p-periodic function f is defined by, provided that the principal value exists (see Butzer and Nessel 1971, pp. 334 ff.),

$$H_p f(x) := \frac{1}{p}\, PV \int_{-p/2}^{p/2} f(x - u)\cot\left(\frac{\pi u}{p}\right) du. \qquad (1.6.11)$$

These conjugate periodic functions $\mathcal{B}_n^{\sim}(x)$ can be used to define non-periodic functions $B_n^{\sim}(x)$, which can be regarded as the conjugate Bernoulli "polynomials" (see Butzer, Hauss and Leclerc 1992). The idea is to define the functions $B_n^\sim(x)$ in a form such that their properties are similar to those of the classical Bernoulli polynomials $B_n(x)$. Here we only need the case $x \in [0, 1)$.

Definition 1.15 (Conjugate Bernoulli functions) *The conjugate Bernoulli functions* $B_n^\sim(x)$, $n \in \mathbb{N}$, *are defined for* $x \in [0, 1)$ ($x \neq 0$ *if* $n = 1$) *by* $B_n^\sim(x) := \mathcal{B}_n^\sim(x)$. *The conjugate Bernoulli numbers are defined by* $B_n^\sim := B_n^\sim(0)$, $n \in \mathbb{N}\setminus\{1\}$.

Since $B_0(x) = \mathcal{B}_0(x) = 1$ for all $x \in \mathbb{R}$, one defines $B_0^\sim(x) = \mathcal{B}_0^\sim(x) := 0$ for $x \in [0, 1)$.

In order to determine $B_{2m+1}^\sim$ we need to discuss some properties of $B_n^\sim(x)$.

Proposition 1.16 (Basic properties) *For* $x \in [0, 1)$ *and* $n, r \in \mathbb{N}$ *there hold true*

$$\left(\frac{d}{dx}\right)^r B_n^\sim(x) = [n]_r B_{n-r}^\sim(x) \qquad (n > r, \ x \neq 0 \text{ if } n = r + 1),$$

$$B_n^\sim(1 - x) = (-1)^{n+1} B_n^\sim(x) \qquad (x \neq 0).$$

Here $[n]_r := n(n - 1) \cdots (n - r + 1)$. Next, some values of the conjugate Bernoulli polynomials are given. Those for $x = \frac{1}{2}$ will turn out to be particularly useful. Observe that $B_k^\sim\left(\frac{1}{2}\right) = (2^{1-k} - 1)B_k^\sim$ for $k \in \mathbb{N}_0$, with $B_{2k}^\sim\left(\frac{1}{2}\right) = 0$.

Proposition 1.17 (Values of the conjugate Bernoulli functions)
(a) *For* $x \in (0, 1)$ *one has*

$$B_1^\sim(x) = -\frac{1}{\pi} \log(2 \sin \pi x).$$

(b) *There hold*

$$B_1^\sim\left(\frac{1}{2}\right) = -\frac{\log 2}{\pi}, \qquad B_3^\sim\left(\frac{1}{2}\right) = \frac{\log 2}{4\pi} - 2 \int_{0+}^{1/2} u^3 \cot \pi u \, du,$$

$$B_5^\sim\left(\frac{1}{2}\right) = \frac{11}{8} \int_{0+}^{1/2} u \cot \pi u \, du + \frac{5}{3} \int_{0+}^{1/2} u^3 \cot \pi u \, du - 2 \int_{0+}^{1/2} u^5 \cot \pi u \, du,$$

$$B_{2k}^\sim\left(\frac{1}{2}\right) = 0 \qquad (k \in \mathbb{N}_0).$$

Proof For part (a) let $x \in (0, \frac{1}{2}]$. In view of $B_1(u) = u - \frac{1}{2}$, $u \in (0, 1]$, and observing (Gradstein and Ryshik 1981, p. 465), one has

$$B_1^\sim(x) = \lim_{\varepsilon \to 0+} \left\{ -2 \int_\varepsilon^{1/2} u \cot \pi u \, du + \int_x^{1/2} \cot \pi u \, du \right\}$$

$$= -\frac{1}{\pi} \log(2 \sin \pi x).$$

In the case $x \in (\frac{1}{2}, 1)$ observe the symmetry property of Proposition 1.16.

Part (b) immediately follows from (a) and (1.6.10), respectively. The representations for $B_3^\sim(\frac{1}{2})$ and $B_5^\sim(\frac{1}{2})$ are consequences of Theorem 1.24 (see below).

Basic is the determination of a generating function for these conjugate Bernoulli numbers, just as this function for the B_n leads to its essential properties. In all of this work one special function is involved, introduced by Butzer, Flocke and Hauss (1994).

Definition 1.18 (Omega function) *For $w \in \mathbb{C}$ and $-\frac{1}{2} \le a < b \le \frac{1}{2}$ ($a, b \neq 0$) the incomplete omega function is defined by*

$$\Omega(w; a, b) := \int_a^b e^{uw} \cot(\pi u)\, du. \tag{1.6.12}$$

If $0 \in (a, b)$, the integral is to be understood in the principal value sense. The case $a = -\frac{1}{2}$ and $b = \frac{1}{2}$ gives the (complete) omega function

$$\Omega(w) := \Omega\left(w; -\frac{1}{2}, \frac{1}{2}\right) = \int_{0+}^{\frac{1}{2}} \sinh(uw) \cot(\pi u)\, du.$$

Thus, by (1.6.11), the (complete) omega function $\Omega(w)$ is the Hilbert transform at 0 of the 1-periodic function f defined by the periodic continuation of $f(x) := e^{-xw}$, $x \in [-\frac{1}{2}, \frac{1}{2})$, $w \in \mathbb{C}$.

This function and its ν-th derivatives at $w = 0$, namely

$$\Omega_\nu(a, b) := D_w^\nu \Omega(w; a, b)|_{w=0} = \int_a^b u^\nu \cot \pi u\, du, \tag{1.6.13}$$

$$\Omega_\nu := D_w^\nu \Omega(w)|_{w=0} = PV \int_{-1/2}^{1/2} u^\nu \cot \pi u\, du \tag{1.6.14}$$

($\Omega_\nu := \Omega_\nu(-\frac{1}{2}, \frac{1}{2})$), will indeed play a major role in deducing representations and generating functions also for the $\tilde{B}_{2n+1}(x)$. Note that the incomplete omega function is the exponential generating function of the $\Omega_\nu(a, b)$, i.e.,

$$\Omega(w; a, b) = \sum_{\nu=0}^{\infty} \frac{\Omega_\nu(a, b)}{\nu!} w^\nu \qquad (w \in \mathbb{C}). \tag{1.6.15}$$

Proposition 1.19 (Representation of the $\tilde{B}_{2n+1}(\frac{1}{2})$) *For $n \in \mathbb{N}_0$ there holds*

$$\tilde{B}_{2n+1}\left(\frac{1}{2}\right) = -\sum_{j=0}^{n} \sum_{\nu=0}^{j} \binom{2n+1}{2j+1}\binom{2j+1}{2\nu+1} B_{2n-2j} \Omega_{2\nu+1} 2^{2\nu-2j}$$

$$+ \frac{2n+1}{2} \sum_{\nu=0}^{n-1} \binom{2n}{2\nu+1} \Omega_{2\nu+1} 2^{2\nu-2n+1}. \tag{1.6.16}$$

Proof By definition (1.6.10) one has

$$\tilde{B}_{2n+1}\left(\frac{1}{2}\right) = H_1 B_{2n+1}\left(\frac{1}{2}\right) = PV \int_{-1/2}^{1/2} B_{2n+1}\left(\frac{1}{2} - u\right) \cot \pi u\, du.$$

Now there holds, noting $B_{2n+1} = 0$, $n \in \mathbb{N}$, and $B_1 = -\frac{1}{2}$,

$$B_{2n+1}\left(\frac{1}{2} - u\right) = \sum_{j=0}^{n} \binom{2n+1}{2j+1} B_{2n-2j} \cdot \left(\frac{1}{2} - u\right)^{2j+1}$$
$$- \frac{1}{2}\binom{2n+1}{2n}\left(\frac{1}{2} - u\right)^{2n}$$

as well as

$$\left(\frac{1}{2} - u\right)^{j} = \sum_{v=0}^{j} \binom{j}{v}\left(\frac{1}{2}\right)^{j-v} (-1)^{v} u^{v} \qquad (j \in \mathbb{N}_0).$$

Together this yields

$$B_{\widetilde{2n+1}}\left(\frac{1}{2}\right) = \sum_{j=0}^{n} \binom{2n+1}{2j+1} B_{2n-2j} \cdot PV \int_{-1/2}^{1/2}\left(\frac{1}{2} - u\right)^{2j+1} \cot \pi u \, du$$
$$- \frac{1}{2}\binom{2n+1}{2n} PV \int_{-1/2}^{1/2}\left(\frac{1}{2} - u\right)^{2n} \cot \pi u \, du.$$

Observing that the principal value equals

$$PV \int_{-1/2}^{1/2}\left(\frac{1}{2} - u\right)^{j} \cot \pi u \, du = - \sum_{v=0}^{\lfloor\frac{j-1}{2}\rfloor} \binom{j}{2v+1}\left(\frac{1}{2}\right)^{j-2v-1} \Omega_{2v+1},$$

with $\Omega_{2v} = 0$ for $v \in \mathbb{N}_0$, the proof is complete.

Proposition 1.20 (Exponential generating function of $B_k^{\sim}(\frac{1}{2})$) *For $|w| < 2\pi$ there holds*

$$\sum_{k=0}^{\infty} B_k^{\sim}\left(\frac{1}{2}\right)\frac{w^k}{k!} = -\frac{we^{\frac{w}{2}}}{e^w - 1}\Omega(w). \qquad (1.6.17)$$

Proof For $|w| < 2\pi$ by Cauchy's product formula, (1.6.2) and (1.6.15), one has

$$\frac{we^{w/2}}{e^w - 1}\Omega(w) = \sum_{k=0}^{\infty}\left\{\sum_{j=0}^{k} \binom{k}{j} B_{k-j}\left(\frac{1}{2}\right)\Omega_j\right\} \cdot \frac{w^k}{k!}.$$

In view of

$$B_k\left(\frac{1}{2} - u\right) = \sum_{j=0}^{k} \binom{k}{j} B_{k-j}\left(\frac{1}{2}\right)(-u)^j \qquad (k \in \mathbb{N}_0)$$

the term in curly brackets can be rewritten as

$$-PV \int_{-1/2}^{1/2} B_k \left(\frac{1}{2} - u \right) \cot \pi u \, du = -PV \int_{-1/2}^{1/2} B_k \left(\frac{1}{2} - u \right) \cot \pi u \, du$$

$$= -B_k^{\sim} \left(\frac{1}{2} \right),$$

which completes the proof, noting that $B_k^{\sim}(\frac{1}{2}) = B_k^{\sim}(\frac{1}{2})$.

Now to the analogue of the Euler–Maclaurin summation formula involving the conjugate Bernoulli functions. The main difference between this formula, which shall be called the Hilbert–Euler–Maclaurin formula, and the classical Euler–Maclaurin summation formula, is the fact that the difference $\sum_{k=1}^{n} f(k) - \int_{1}^{n} f(x) \, dx$ is replaced by the Hilbert transform $F^{\sim}(x + y)$, and the Bernoulli polynomials $B_k(x)$ are replaced by the conjugate Bernoulli polynomials $B_k^{\sim}(x)$.

We will deduce the Hilbert–Euler–Maclaurin formula in two steps. First we prove a special case and then, in Theorem 1.22, the general formula.

Proposition 1.21 *Let* $f \in C^{(1)}[x, x + p]$ *with* $x \in \mathbb{R}$ *arbitrary and fixed. Further, let* F *be the* p-*periodic extension of* $f|_{[x,x+p)}$ *on* $\mathbb{R}$. *Then for all fixed* $y \in \mathbb{R} \setminus p\mathbb{Z}$ *there holds with* $\Delta_p f(x) := f(x + p) - f(x)$,

$$F^{\sim}(x + y) = -\frac{1}{\pi} \log \left(2 \left| \sin \frac{\pi y}{p} \right| \right) \Delta_p f(x) - R_1^{\sim}(x, y),$$

where

$$R_1^{\sim}(x, y) = PV \int_0^p B_1^{\sim} \left(\frac{y - u}{p} \right) f'(x + u) \, du.$$

Proof Let $x \in \mathbb{R}$ be arbitrary and $y \in \mathbb{R}$ with $y \neq kp(k \in \mathbb{Z})$. Since all integrals exist with $\epsilon > 0$ arbitrary (and small enough), one has

$$\frac{1}{p} \left\{ \int_{-p/2}^{-\epsilon} + \int_{\epsilon}^{p/2} \right\} F(x + y - u) \cot \frac{\pi u}{p} \, du$$

$$= \frac{1}{p} \left\{ \int_0^{w-\epsilon} + \int_{w+\epsilon}^p \right\} f(x + u) \cot \frac{\pi (w - u)}{p} \, du,$$

where $w \in (0, p)$ satisfies $w = y + mp$ for some $m \in \mathbb{Z}$.

Since $F \in L_p^1$ and is piecewise continuous, the limit $\epsilon \to 0+$ exists (see (Butzer and Nessel 1971, p. 335); even for all $y \in \mathbb{R}$, $y \neq kp$ with $k \in \mathbb{Z}$, so that

$$F^{\sim}(x + y) = \frac{1}{p} PV \int_0^p f(x + u) \cot \frac{\pi (y - u)}{p} \, du.$$

For $u \in (0, p)$ one has

$$\frac{d}{du} \log \left(2 \left| \sin \frac{\pi u}{p} \right| \right) = \frac{\pi}{p} \cot \frac{\pi u}{p},$$

which holds true even for all $u \in \mathbb{R} \setminus p\mathbb{Z}$. Then partial integration yields

$$\frac{1}{p} \left(\int_0^{w-\epsilon} + \int_{w+\epsilon}^p \right) f(x+u) \cot \frac{\pi(w-u)}{p} du$$

$$= -\frac{1}{\pi} \log \left(2 \left| \sin \frac{\pi w}{p} \right| \right) \Delta_p f(x) + \frac{1}{\pi} \log \left(2 \left| \sin \frac{\pi \epsilon}{p} \right| \right)$$

$$\times \{ f(x+w+\epsilon) - f(x+w-\epsilon) \}$$

$$- \left\{ \int_0^{w-\epsilon} + \int_{w+\epsilon}^p \right\} f'(x+u) \mathcal{B}_1^{\sim} \left(\frac{w-u}{p} \right) du.$$

Since $f \in C^{(1)}[x, x+p]$ and $w \in (0, p)$ one can easily show, using l'Hospital's rule, that

$$\lim_{\epsilon \to 0+} \frac{1}{\pi} \log \left(2 \left| \sin \frac{\pi \epsilon}{p} \right| \right) \{ f(x+w+\epsilon) - f(x+w-\epsilon) \} = 0,$$

which completes the proof.

Theorem 1.22 (Hilbert–Euler–Maclaurin formula) *Let* $f \in C^{(k)}[x, x+p]$ *and* $x \in \mathbb{R}$ *be arbitrary. Further, let* F *be the* p-*periodic extension of* $f|_{[x,x+p)}$. *Then, for all* $y \in \mathbb{R} \setminus p\mathbb{Z}$, *one has*

$$F^{\sim}(x+y) = -\frac{1}{\pi} \log \left(2 \left| \sin \frac{\pi y}{p} \right| \right) \Delta_p f(x)$$

$$+ \sum_{j=2}^k \frac{p^{j-1}}{j!} \mathcal{B}_j^{\sim} \left(\frac{y}{p} \right) \Delta_p f^{(j-1)}(x) - R_k^{\sim}(x, y), \qquad (1.6.18)$$

where the remainder term has the representation

$$R_k^{\sim}(x, y) = \frac{p^{k-1}}{k!} \int_0^p \mathcal{B}_k^{\sim} \left(\frac{y-u}{p} \right) f^{(k)}(x+u) du.$$

In the case $k = 1$ *the integral is to be understood as a principal value.*

If a p-periodic function $F : \mathbb{R} \to \mathbb{R}$ with $f \in C^{(k)}(\mathbb{R})$ is given, and $f(u) := F(u)$ for all $u \in [x, x+p]$ with $x \in \mathbb{R}$ arbitrary, one has the convolution identity

$$F^{\sim}(x+y) = -\frac{p^{k-1}}{k!} \int_0^p \mathcal{B}_k^{\sim} \left(\frac{y-u}{p} \right) F^{(k)}(x+u) du. \qquad (1.6.19)$$

Proof To prove Theorem 1.22 let $f \in C^{(k)}[x, x+p]$ be given and let $1 \le m < k$. Then, by partial integration, one has

$$R_m^{\sim}(x, y) := \frac{p^{m-1}}{m!} \int_0^p \mathcal{B}_m^{\sim} \left(\frac{y-u}{p} \right) f^{(m)}(x+u) du$$

$$= -\frac{p^m}{(m+1)!} \mathcal{B}_{m+1}^{\sim} \left(\frac{y}{p} \right) \Delta_p f^{(m)}(x) + R_{m+1}^{\sim}(x, y).$$

A telescopic argument yields

$$R_k^\sim(x, y) = \sum_{m=1}^{k-1} \left(R_{m+1}^\sim(x, y) - R_m^\sim(x, y) \right) + R_1^\sim(x, y)$$

$$= \sum_{m=1}^{k-1} \frac{p^m}{(m+1)!} B_{m+1}^\sim \left(\frac{y}{p} \right) \Delta_p f^{(m)}(x) + R_1^\sim(x, y).$$

Observing Proposition 1.21 the proof is complete.

As is to be expected, the Hilbert–Euler–Maclaurin formula yields certain counterparts of the Poisson summation formulae.

Theorem 1.23 (First Hilbert–Poisson formula) *Let* $f \in C^{(2)}[x, x + p]$ *with* $x \in \mathbb{R}$ *arbitrary. Further, let F be the p-periodic extension of* $f|_{[x,x+p)}$. *Then, for all* $y \in \mathbb{R} \backslash p\mathbb{Z}$, *there holds*

$$F^\sim(x + y) = \frac{2}{p} \sum_{k=1}^{\infty} \int_0^p f(x + u) \sin \left(2\pi k \frac{y - u}{p} \right) du. \tag{1.6.20}$$

Proof Partial integration yields

$$R_2^\sim(x, y) = \frac{p}{2} \int_0^p B_2^\sim \left(\frac{y - u}{p} \right) f^{(2)}(x + u) \, du$$

$$= 2p \int_0^p \frac{f^{(2)}(x + u)}{(2\pi)^2} \sum_{k=1}^{\infty} \frac{\sin 2\pi k \frac{y-u}{p}}{k^2} \, du$$

$$= \frac{p}{2\pi^2} \left(\sum_{k=1}^{\infty} \frac{\sin \frac{2\pi ky}{p}}{k^2} \right) \Delta_p f'(x) + \frac{1}{\pi} \left(\sum_{k=1}^{\infty} \frac{\cos \frac{2\pi ky}{p}}{k} \right) \Delta_p f(x)$$

$$- \frac{2}{p} \sum_{k=1}^{\infty} \int_0^p f(x + u) \sin \left(2\pi k \frac{y - u}{p} \right) du.$$

Noting that (see Butzer, Hauss and Leclerc 1992)

$$\sum_{k=1}^{\infty} \frac{\sin \frac{2\pi ky}{p}}{k^2} = \pi^2 B_2^\sim \left(\frac{y}{p} \right)$$

and further

$$\sum_{k=1}^{\infty} \frac{\cos \frac{2\pi ky}{p}}{k} = -2\pi \sum_{k=1}^{\infty} \frac{\sin(\frac{2\pi ky}{p} - \frac{\pi}{2})}{2\pi k} = \pi B_1^\sim \left(\frac{y}{p} \right),$$

one has the identity

$$R_2^\sim(x, y) = \frac{p}{2} B_2^\sim \left(\frac{y}{p} \right) \Delta_p f'(x) + B_1^\sim \left(\frac{y}{p} \right) \Delta_p f(x)$$

$$- \frac{2}{p} \sum_{k=1}^{\infty} \int_0^p f(x + u) \sin \left(2\pi k \frac{y - u}{p} \right) du.$$

In view of Theorem 1.22 the proof is complete.

The foregoing Hilbert versions of the two sum formulae are due to Hauss (1995). There also exists a counterpart of the Hilbert version of the second Poisson formula (see Hauss 1997) but so far not for the third one.

Now to the basic partial fraction expansion of the generating function of the conjugate Bernoulli numbers, thus essentially of the omega function $\Omega(2\pi z)$. It is the "Hilbert–version" of the expansion for the classical B_n, namely of (1.6.4).

Theorem 1.24 (Expansion of the omega function as a partial fraction) *For* $z \in \mathbb{C}\backslash$ $i\mathbb{Z}$ *there holds*

$$\frac{\pi \Omega(2\pi z)}{\left(e^{-\pi z} - e^{\pi z}\right)} = \sum_{k=1}^{\infty}(-1)^k \frac{k}{z^2 + k^2} = \frac{i}{2}\sum_{k=-\infty}^{\infty}{}_e \frac{(-1)^k \operatorname{sgn} k}{z + ik}. \tag{1.6.21}$$

Proof Choosing $p = 1$ and $f(z) := e^{2\pi zw}$ with $z \in [x, x+1]$ ($x \in \mathbb{R}$ arbitrary) in Theorem 1.23, since $2i \sin z = e^z - e^{-z}$, one obtains for arbitrary $y \in \mathbb{R}\backslash\mathbb{Z}$,

$$2\sum_{k=1}^{\infty} \int_0^1 e^{2\pi(x+u)w} \sin 2\pi k(y - u)\, du$$

$$= \frac{-ie^{2\pi xw}}{2\pi}\left\{ \sum_{k=1}^{\infty} \frac{1}{w - ik} e^{2\pi iky}\left(e^{2\pi w - 2\pi ik} - 1\right)\right.$$

$$\left. - \sum_{k=1}^{\infty} e^{-2\pi iky}\frac{1}{w + ik}\left(e^{2\pi w + 2\pi ik} - 1\right)\right\}$$

$$= -\frac{ie^{2\pi xw}}{2\pi}\left(e^{2\pi w} - 1\right)\sum_{k=1}^{\infty}\left(\frac{e^{2\pi iky}}{w - ik} - \frac{e^{-2\pi iky}}{w + ik}\right).$$

For $y = \frac{1}{2}$ there holds

$$F^{\sim}\left(x + \frac{1}{2}\right) = 2e^{2\pi xw}\sum_{k=1}^{\infty}(-1)^{k+1}\int_0^1 e^{2\pi uw}\sin 2\pi ku\, du$$

$$= -\frac{ie^{2\pi xw}}{2\pi}\left(e^{2\pi w} - 1\right)\sum_{k=1}^{\infty}\left(\frac{(-1)^k}{w - ik} - \frac{(-1)^k}{w + ik}\right).$$

Hence there results

$$\sum_{k=1}^{\infty}(-1)^{k+1}\int_0^1 e^{2\pi uw}\sin 2\pi ku\, du = \frac{1}{2\pi}\left(e^{2\pi w} - 1\right)\sum_{k=1}^{\infty}\frac{(-1)^k k}{w^2 + k^2}.$$

For $x = 0$ one has

$$F^{\sim}\left(\frac{1}{2}\right) = \lim_{\varepsilon \to 0+} \left(\int_{-1/2}^{-\varepsilon} + \int_{\varepsilon}^{1/2}\right) e^{2\pi\left(\frac{1}{2} - u\right)w} \cot \pi u \, du$$

$$= e^{\pi w} \, PV \int_{-1/2}^{1/2} e^{-2\pi u w} \cot \pi u \, du = e^{\pi w} \Omega(-2\pi w),$$

which completes the proof.

Thus the (new) function $\pi \Omega(2\pi z)(e^{-\pi z} - e^{\pi z})^{-1}$ with the partial fraction expansion (1.6.21) is the counterpart of the classical function $\pi \cot \pi z - 1/z$ with the expansion (1.6.4) when turning to the conjugate instance. It would be of interest to know whether these two functions could be put into direct connection via Hilbert transforms.

Observe that the series of (1.6.4) is the particular case $\nu = 1$ of the Eisenstein series (see e.g. Remmert 1984, p. 234; Weil 1984)

$$e_\nu(z) := \sum_{\substack{k=-\infty}}^{\infty} {}_e \frac{1}{(z+k)^\nu} \qquad (z \in \mathbb{C}\backslash\mathbb{Z}; \; \nu \geq 1),$$

the simplest examples of normally convergent meromorphic functions. So the series

$$H_\nu(z) := \sum_{\substack{k=-\infty}}^{\infty} {}_e \frac{(-1)^k \operatorname{sgn} k}{(z+ik)^\nu} \qquad (z \in \mathbb{C}\backslash i\mathbb{Z}; \; \nu \geq 1)$$

could be termed a *Hilbert–Eisenstein series*. These new series have properties similar to the classical Eisenstein series (see Hauss 1997). The expansion (1.6.21) is also the counterpart of the classical (alternating) partial fraction expansion

$$\frac{\pi}{2 \sin \pi z} - \frac{1}{2z} = \sum_{k=1}^{\infty} (-1)^k \frac{z}{z^2 - k^2} = \frac{1}{2} \sum_{k=-\infty}^{\infty} \frac{(-1)^k}{z+k}.$$

We are finally able to prove the closed form for $\zeta(2m+1)$, namely Theorem 1.14.

Proof On the one hand, the partial fraction expansion of the omega function for $|z| < 1$ can be rewritten as

$$\sum_{k=1}^{\infty} (-1)^k \frac{k}{k^2 + z^2} = \sum_{k=1}^{\infty} \frac{(-1)^k}{k} \sum_{n=0}^{\infty} (-1)^n \left(\frac{z}{k}\right)^{2n}$$

$$= \sum_{k=1}^{\infty} \frac{(-1)^k}{k} \sum_{n=1}^{\infty} (-1)^n \left(\frac{z}{k}\right)^{2n} + \sum_{k=1}^{\infty} \frac{(-1)^k}{k}$$

$$= \sum_{n=0}^{\infty} \left(\sum_{k=1}^{\infty} \frac{(-1)^k}{k^{2n+1}}\right) (-1)^n z^{2n},$$

the interchange of the order of summation being justified by Cauchy's theorem on double series, see (Heuser 1989, p. 258), if $|z| < 1$. On the other hand, in view of (1.6.17), one has

$$\frac{\pi \Omega (2\pi z)}{e^{-\pi z} - e^{\pi z}} = \frac{1}{2z} \sum_{n=0}^{\infty} \frac{B_n^{\sim}(1/2)}{n!} (2\pi)^n z^n.$$

Thus, comparing coefficients,

$$B_{2n}^{\sim}\left(\frac{1}{2}\right) = 0,$$

$$B_{2n+1}^{\sim}\left(\frac{1}{2}\right) = (-1)^n (2n+1)! 2^{-2n} \pi^{-2n-1} \sum_{k=1}^{\infty} \frac{(-1)^k}{k^{2n+1}} \qquad (n \in \mathbb{N}_0).$$

Finally (see Abramowitz and Stegun 1965, p. 807),

$$-\eta(2n+1) := \sum_{k=1}^{\infty} \frac{(-1)^k}{k^{2n+1}} = \left(2^{-2n} - 1\right)\zeta(2n+1) \qquad (n \in \mathbb{N}),$$

which completes the proof in view of $B_n^{\sim}(\frac{1}{2}) = (2^{1-n} - 1)B_n^{\sim}$.

Comparing (1.6.1) with (1.6.9), essentially the numbers B_{2m} of (1.6.1) are replaced by their conjugates $B_{2m+1}^{\sim}$ to obtain the new Euler-type formula for $\zeta(2m+1)$, which is the perfect structural counterpart of the classical Euler formula (1.6.1). Theorem 1.14 states that $\zeta(2m+1) = s_{2m+1} \pi^{2m+1}$, where s_{2m+1} is the factor containing $B_{2m+1}^{\sim}$. Whether or not s_{2m+1} is rational is however still unsolved. The counterpart of the factor s_{2m+1} for $\zeta(2m)$ is of course rational.

We have presented Hilbert versions of the Euler–Maclaurin and the second Poisson summation formula. What remains unresolved are Hilbert versions of the third or classical Poisson sum formula as well as of Shannon's sampling theorem. Whether the latter possible result will be related to the known sampling representations of the Hilbert transform (cf. Higgins 1996, p. 126) is then the next open question. Whether the Hilbert versions of the various summation formulae will also be equivalent to the sampling theorem in question is again open.

Finally, as to the applicability of these new results, only a possible solution to Basler's problem has been stressed here. In any case the new approach can be applied to deduce Euler-type formulae for the Dirichlet function $\mathcal{L}(n) = \sum_{k=0}^{\infty}(-1)^k(2k+1)^{-n}, n \in \mathbb{N}$. Whereas it is known that

$$\mathcal{L}(2m+1) = (-1)^m 2^{-2m-2} \pi^{2m+1} \frac{E_{2m}}{(2m)!} \qquad (m \in \mathbb{N}_0),$$

where E_n are the Euler numbers, Hauss (1997) has shown that

$$\mathcal{L}(2m) = (-1)^m 2^{-2m-1} \pi^{2m} \frac{E_{2m-1}^{\sim}}{(2m-1)!} \qquad (n \in \mathbb{N}_0),$$

where $E_n^{\sim}$ are the *conjugate Euler numbers* introduced by Butzer, Flocke and Hauss (1994). Further, as has been indicated above, the approach can be applied to Hilbert versions of Eisenstein series. Finally, since the Euler–Maclaurin and Poisson summation formulae play such an important role in analytic number theory, the Hilbert versions may also. They certainly play a new role in summing certain infinite series.

It was Hans Hamburger (1922) who in 1922 pointed out that the following five assertions are equivalent to one another:

(A) the functional equation of the Riemann zeta function:

$$\pi^{-s/2}\Gamma(s/2)\zeta(s) = \pi^{-(1-s)/2}\Gamma((1-s)/2)\zeta(1-s) \qquad (s \in \mathbb{C});$$

(B) the transformation formula of the Jacobi theta function $\theta(y) := \sum_{k=-\infty}^{\infty} e^{-\pi k^2 y}$, $y > 0$, namely,

$$\theta(y) = y^{-1/2}\theta(y^{-1});$$

(C) the partial fraction representation of the function $i \cot i\pi z$:

$$1 + 2\sum_{k=1}^{\infty} e^{-2\pi kz} = i \cot \pi i z = \frac{1}{\pi z} + \frac{2z}{\pi}\sum_{k=1}^{\infty} \frac{1}{z^2 + k^2} \qquad (z \neq ik); \quad (1.6.22)$$

(D) the Poisson summation formula in the form of equation (1.6.8) of Proposition 1.13;

(E) the Fourier expansion

$$\frac{1}{2\pi^2}\sum_{k=1}^{\infty} \frac{\cos 2\pi kx - 1}{k^2} = \frac{x^2 - x}{2} - \sum_{k=1}^{m}(x - k) \qquad (m < x < m + 1).$$

Thus the expansion of $\pi\cot \pi z$ in (1.6.5), which reduces to that of (1.6.22) by replacing z by iz there, and which was used in the proof of Euler's famous formula (1.6.1), is actually equivalent to Poisson's sum formula (1.6.8).

Concerning the proof of the equivalence of assertions (A) and (D), it was Ferrar (1939) who showed that the concrete formula (A) yields the general assertion (D), and Mordell (1929) who established the converse result, namely that (D) $\Rightarrow$ (A). Now it was indicated in the introduction to this chapter (see also Higgins 1996, p. 90; Butzer and Nasri-Roudsari 1997) that Poisson's formula (thus (1.6.8)), is equivalent to the approximate sampling theorem for not necessarily band-limited functions, namely that

$$f(t) = \lim_{W \to \infty} \sum_{k=-\infty}^{\infty} f\left(\frac{k}{W}\right) \mathrm{sinc}(Wt - k) \qquad (1.6.23)$$

uniformly for $t \in \mathbb{R}$, provided that just $f \in L^2(\mathbb{R}) \cap C(\mathbb{R})$ with $f^\wedge \in L^1(\mathbb{R})$. Whereas the Hamburger chain of assertions states that this sampling theorem must therefore be equivalent to the famous Riemann functional equation (A) of the zeta function,[1] what is missing are direct proofs of the two assertions, (A) $\Rightarrow$ {approximate sampling theorem in the form (1.6.23)} $\Rightarrow$ (A) which of course do not make use of the equivalent assertions (B), (C) and (E).

---

[1] Whereas Titchmarsh (1951) presents seven different proofs of the functional equation, he does not reproduce that due to Mordell (nor does any other textbook seem to treat it).

1.7 Sampling expansions of particular special functions

The aim of this section is to apply Kramer's lemma,[2] treated by Higgins (1996, pp. 78–87), to particular Jacobi functions, namely to the sampling of particular Jacobi *functions* in terms of their associated Jacobi *polynomials*. The matter will be considered for the Legendre functions, Chebyshev functions of the first kind as well as those of the second kind.

Let us first recall Kramer's lemma, putting it into the slightly more general form given by L. L. Campbell (1964).

Proposition 1.25 *Let $I \subset \mathbb{R}$ be some interval, ρ a weight function with $\rho(x) > 0$ a.e. in I and $L_\rho^2(I) := \{\varphi(x); \int_I |\varphi(x)|^2 \rho(x)\, dx\}$. Consider a kernel $K : I \times \mathbb{R} \to \mathbb{R}$ with $K(\cdot, t) \in L_\rho^2(I)$ for each $t \in \mathbb{R}$, and let $\{t_k\}_{k \in \mathbb{Z}} \subset \mathbb{R}$ be a countable set of reals such that $\{K(x, t_k)\}$ forms a complete orthogonal set in $L_\rho^2(I)$.*

If a function f is representable as

$$f(t) = \int_I K(x, t) g(x) \rho(x)\, dx \qquad (t \in \mathbb{R}) \tag{1.7.1}$$

for some $g \in L_\rho^2(I)$, then f can be reconstructed from its samples $f(t_k)$ via

$$f(t) = \sum_{k=-\infty}^{\infty} f(t_k) s_k(t) \qquad (t \in \mathbb{R}) \tag{1.7.2}$$

where

$$s_k(t) := \frac{\int_I K(x, t) \overline{K(x, t_k)} \rho(x)\, dx}{\int_I |K(x, t_k)|^2 \rho(x)\, dx}. \tag{1.7.3}$$

The series in equation (1.7.2) is absolutely convergent for each $t \in \mathbb{R}$.

As to the general (normalized) Jacobi function $R_{t-\gamma}^{(\alpha, \beta)}(x)$, it is defined via the Gaussian hypergeometric function

$$R_{t-\gamma}^{(\alpha, \beta)}(x) := {}_2F_1\left(-t + \gamma, t + \gamma; \alpha + 1; \frac{1-x}{2}\right),$$

normalized in the sense that $R_{t-\gamma}^{(\alpha, \beta)}(1) = 1$. Let us first consider the particular case of the (normalized) Legendre *functions* $P_{t-1/2}(x)$ for $\alpha = \beta = 0$, defined by

$$P_{t-1/2}(x) := R_{t-1/2}^{(0,0)}(x) = {}_2F_1\left(-t + \frac{1}{2}, t + \frac{1}{2}; 1; \frac{1-x}{2}\right), \tag{1.7.4}$$

---

[2] Kramer's lemma contains Shannon's theorem as a particular case (by taking the kernel $K(x, t) = e^{ixt}$ and using the fact that $\{e^{ixk/W}\}_{k \in \mathbb{Z}}$ is an orthogonal basis for $L^2(-\pi W, \pi W)$). It is still an open question whether the converse result holds. In any case, in the instance of the general Jacobi kernel as well as the Bessel kernel $K(x, t) = t^{-\nu/2}\sqrt{x} J_\nu(x\sqrt{t})$, $\nu > -1/2$, $x \in (0, 1)$, Kramer's sampling lemma is a particular case of that of Shannon, and so both are equivalent. See the comments by Butzer and Nasri-Roudsari (1997) to this effect.

and sampling these in terms of the Legendre *polynomials*

$$P_n(x) = \frac{(-1)^n}{2^n n!} \left(\frac{d}{dx} \right)^n (1 - x^2)^n = {}_2F_1 \left(-1, n + 1; 1; \frac{1-x}{2} \right).$$

The sampling theorem in question is the following.

Theorem 1.26 *Let* $t \in \mathbb{R}$, $t \neq 1/2$. *For* $x \in (-1, 1]$ *there holds*

$$\frac{P_{t-1/2}(x)}{t^2 - \frac{1}{2}} = \sum_{k=1}^{\infty} \frac{P_k(x)}{k(k+1)} (2k+1) \frac{\sin \pi (t - \frac{1}{2} - k)}{\pi (t^2 - (k + \frac{1}{2})^2)}$$
$$+ \frac{\sin \pi (t - \frac{1}{2})}{\pi (t^2 - \frac{1}{4})} \left\{ \frac{1}{t^2 - \frac{1}{4}} - \log \frac{2}{1+x} + 1 \right\}, \qquad (1.7.5)$$

the series converging absolutely and uniformly with respect to $t \in \mathbb{R}$ *as well as* $x \in [-1, 1]$ *or, equivalently,*

$$P_{t-1/2}(x) = \sum_{k=-\infty}^{\infty} P_k(x) \frac{\sin \pi (t - \frac{1}{2} - k)}{\pi (t - \frac{1}{2} - k)}, \qquad (1.7.6)$$

the series converging absolutely and uniformly in $t \in \mathbb{R}$ *and with respect to* x *(only) on any compact subset of* $(-1, 1)$.

Corollary 1.27 $P_{t-1/2}(x)$ *is continuous on* $(-1, 1]$ *and behaves as the function* $\log(1 + x)/2$ *in the limit* $x \to (-1)+$. *In particular, one has the estimate*

$$|P_{t-1/2}| \leq M_1(t, \gamma) + M_2(t, \gamma) \left| \log \frac{1+x}{2} \right| \qquad (x \in (-1, 1]; \ t \geq 0).$$

Using other means (see Butzer, Stens and Wehrens 1980) it is also possible to show that $|P_{t-1/2}| \leq 1 + M_t \log(2/(1+x))$, the log term being sharp.

In regard to a sketch of the proof in Theorem 1.26, we apply Proposition 1.25 not to the function $P_{1-1/2}(x)$ but to

$$f(t) = \left(\log \frac{2}{1+x} \right)^{-1} \frac{1 - P_{t-1/2}}{t^2 - 1/4} \qquad (t \in \mathbb{R}, t \neq 1/2),$$

with kernel function $K(x, t) = P_{t-1/2}(u) \in L^2(-1, 1), t \in R$ and

$$g(u) = 1/2 \left(\log \frac{2}{1+x} \right)^{-1} \log \left(\frac{1+u}{1-u} \frac{1-x}{1+x} \right).$$

Then, as shown by Butzer, Stens and Wehrens (1980),

$$\int_{-1}^{1} \frac{1}{2} \left(\log \frac{2}{1+x} \right)^{-1} \log \left(\frac{1+u}{1-u} \frac{1-x}{1+x} \right) P_{t-1/2}(u) \, du$$

$$= \begin{cases} \left(\log \frac{2}{1+x} \right) \frac{1 - P_{t-1/2}(x)}{t^2 - 1/4}, & t \neq \frac{1}{2} \\ 1, & t = \frac{1}{2} \end{cases}$$

and the associated (true) sampling series reads,

$$\left(\log \frac{2}{1+x} \right)^{-1} \frac{1 - P_{t-1/2}(x)}{t^2 - 1/4}$$

$$= \frac{\sin \pi (t - \frac{1}{2})}{\pi (t^2 - \frac{1}{4})} + \sum_{k=1}^{\infty} \left(\log \frac{2}{1+x} \right)^{-1} \frac{1 - P_k(x)}{k(k+1)} 2 \left(k + \frac{1}{2} \right) \frac{\sin \pi (t - \frac{1}{2} - k)}{\pi (t^2 - (k + \frac{1}{2})^2)}.$$

Rewriting this series in terms of the expansion of $P_{t-1/2}(x)$ yields, after some calculations, the form (1.7.5). Here in fact, essentially, $P_{t-1/2}(x)/(t^2 - 1/4)$ is sampled in terms of $P_k(x)/(k(k+1))$. For the full and difficult proof, actually carried out in the general Jacobi case, (see Butzer and Schoettler 1993).

Observe that the representation in equation (1.7.6) is a symmetric version of the actual Fourier–Legendre expansion of $P_{t-1/2}$, namely

$$P_{t-1/2}(x) = \sum_{k=0}^{\infty} P_k(x) \frac{(2k+1) \sin \pi (t - \frac{1}{2} - k)}{\pi (t^2 - (k + \frac{1}{2})^2)}.$$

Defining the (Fourier–)Legendre transform of a function f in $L^2(-1, 1)$ or $C[-1, 1]$ by

$$\mathcal{L}[f](m) = \frac{1}{2} \int_{-1}^{1} f(u) P_m(u) \, du \quad (m \in \mathbb{N}_0)$$

then, according to equations (1.3.1) and (1.7.4),

$$\mathcal{L}[P_{t-1/2}](m) = \frac{1}{2} \int_{-1}^{1} P_{t-1/2}(u) P_m(u) \, du$$

$$= \sum_{k=0}^{\infty} \frac{(-t + \frac{1}{2})_k (t + \frac{1}{2})_k}{k!} \int_{-1}^{1} P_m(u) \left(\frac{1-u}{2} \right)^k du$$

$$= (2m+1) \frac{\operatorname{sinc}(t - \frac{1}{2} - m)}{t + \frac{1}{2} + m}.$$

However, the expansion (1.7.6) could also be deduced from equation (1.7.5) by making use of the Fourier–Legendre expansion of $\log 2/(1 + x)$ (see for example

Gradstein and Ryshik 1981, p. 1092)

$$\log \frac{2}{1+x} = 1 + \sum_{k=1}^{\infty} P_k(x) \frac{2k+1}{k(k+1)} (-1)^k \quad (x \in (-1, 1]).$$

Next to the sampling expansion of the (normalized) Chebyshev *functions* of the first kind $T_t(x)$, with $\alpha = \beta - 1/2$, defined by

$$T_t(x) = P_t^{(-1/2,-1/2)}(x) = {}_2F_1\left(-t, t; \frac{1}{2}, \frac{1-x}{2}\right), \tag{1.7.7}$$

in terms of the classical (normalized) Chebyshev *polynomials* $T_n(x) = \cos(n \arccos x)$, $n \in \mathbb{N}$.

Theorem 1.28 *Let* $t \neq 0$. *For* $x \in [-1, 1]$ *there holds*

$$\frac{T_t(x)}{t^2} = \sum_{k=1}^{\infty} \frac{T_k(x)}{k^2} 2t \frac{\sin \pi(t-k)}{\pi(t^2-k^2)} + \frac{\sin \pi t}{\pi t} \left\{ \frac{1}{t^2} - \frac{1}{2}(\arccos x)^2 + \frac{\pi^2}{6} \right\} \tag{1.7.8}$$

or, equivalently,

$$T_t(x) = \sum_{k=-\infty}^{\infty} T_k(x) \frac{\sin \pi(t-k)}{\pi(t-k)}, \tag{1.7.9}$$

both series converging absolutely and uniformly in $t \in \mathbb{R}$ *and also with respect to* $x \in [-1, 1]$.

Corollary 1.29 $T_t(\cdot)$ *is continuous on* $[-1, 1]$ *and* $\lim_{x \to (-1)+} T_t(x)$ *exists.*

As to a sketch of the proof, one applies Proposition 1.25 not to $f(t) = T_t(x)$ but to the function $f(t) = 2(\arccos x)^{-1}[1 - T_t(x)]t^{-2}$, with kernel function $K(u, t) = T_t(u) \in L_\rho^2(-1, 1)$, for all $t \in \mathbb{R}$, with $\rho(u) = (1 - u^2)^{-1/2}$ and $g(u) = 2[\arccos x - \arccos u](\arccos x)^{-2}$. In fact, as established by Butzer and Schoettler (1993),

$$\int_{-1}^{2} \frac{2(\arccos x - \arccos u)}{(\arccos x)^2} T_t(u) \frac{du}{\sqrt{1-u^2}}$$
$$= \begin{cases} 2(\arccos x)^{-2}[1 - T_t(x)]t^{-2}, & t \neq 0 \\ 1, & t = 0. \end{cases} \tag{1.7.10}$$

Then the associated (true) sampling series reads

$$\frac{2}{(\arccos x)^2} \frac{1-T_t(x)}{t^2} = \frac{\sin \pi t}{\pi t} + \sum_{k=1}^{\infty} \frac{2}{(\arccos x)^2} \frac{1-T_k(x)}{k^2} \frac{2t \sin \pi(t-k)}{\pi(t^2-k^2)}.$$

Rewriting this series in terms of $T_t(x)$ yields after some calculations the form of equation (1.7.8), where actually $T_t(x)/t^2$ is sampled in terms of $T_k(x)/k^2$, part of the second term of the right of equation (1.7.8) being the "error" involved.

Note that the representation of (1.7.8) is a symmetric version of the Fourier–Chebyshev expansion of $T_t(x)$, namely

$$T_t(x) = \frac{\sin \pi t}{\pi t} + \sum_{k=1}^{\infty} T_k(x) \frac{\sin \pi (t-k)}{\pi (t^2 - k^2)}.$$

Defining the Fourier–Chebyshev transform of f in $L^2(-1, 1)$ or $C[-1, 1]$ by

$$C[f](m) = \frac{1}{\pi} \int_{-1}^{1} f(u) T_m(u) \frac{du}{\sqrt{1 - u^2}},$$

then according to equations (1.3.1) and (1.7.7),

$$
\begin{aligned}
C[T_t](m) &= \frac{1}{2} \int_{-1}^{1} T_t(u) T_m(u) \frac{du}{\sqrt{1 - u^2}} \\
&= \sum_{k=0}^{\infty} \frac{(-t)_k (t_k)}{\left(\frac{1}{2}\right)_k} \frac{1}{k!} \frac{1}{\pi} \int_{-1}^{1} T_m(u) \left(\frac{1-u}{2}\right)^k du \\
&= \begin{cases} 1/\pi \operatorname{sinc} t, & m = 0 \\ 2t \frac{\sin \pi (t-m)}{\pi^2 (t^2 - m^2)}, & m \in \mathbb{N}. \end{cases}
\end{aligned}
$$

Also observe that the expansion (1.7.9) could be deduced from (1.7.8) by taking into account the Fourier–Chebyshev expansion of $(\arccos x)^2/2$, namely

$$\frac{(\arccos x)^2}{2} = \frac{\pi^2}{6} + 2 \sum_{k=1}^{\infty} T_k(x) \frac{(-1)^k}{k^2} \qquad (x \in (-1, 1]).$$

Finally to the sampling series for the (normalized) Chebyshev *functions* of the second kind $U_{t-1}^*(x)$ with $\alpha = \beta = 1/2$ and $x \in (-1, 1]$, $t \in \mathbb{R}$, defined by

$$U_{t-1}^*(x) := R_{t-1}^{(1/2, 1/2)}(x) = {}_2F_1\left(-t + 1, t + 1; \frac{3}{2}; \frac{1-x}{2}\right).$$

The (normalized) Chebyshev *polynomials* of the second kind are given by $U_n^*(x) = \sin[(n + 1) \arccos x]/[(n + 1)\sqrt{1 - x^2}]$, $n \in \mathbb{N}_0$.

They are connected to the classical Chebyshev polynomials of the second kind $U_n(x)$ by $U_n^*(x) = U_n(x)/(n + 1)$, $n \in \mathbb{N}_0$.

Theorem 1.30 *Let $t \in \mathbb{R}$, $t \neq 1$. For $x \in (-1, 1]$ there holds for the normalized Chebyshev function of the second kind*

$$\frac{U_{t-1}^*(x)}{t^2 - 1} = \sum_{k=1}^{\infty} \frac{2(k+1)^2}{t} \frac{U_k^*(x)}{k(k+2)} \frac{\sin \pi(t - 1 - k)}{\pi(t^2 - (k+1)^2)}$$
$$+ \frac{\sin \pi(t-1)}{\pi(t^2-1)} \left\{ \frac{1}{t^2-1} - \left(\frac{1}{2} - \frac{x}{2\sqrt{1-x^2} \arccos x} \right) + \frac{3}{4} \right\},$$

the series converging absolutely and uniformly with respect to $t \in \mathbb{R}$ as well as to $x \in [-1, 1]$ or, equivalently,

$$t^2 U_{t-1}^*(x) = \sum_{\substack{k=-\infty, \\ k \neq 0}}^{\infty} k^2 U_k^*(x) \frac{\sin \pi(t - k)}{\pi(t - k)},$$

this series converging absolutely and uniformly in $t \in \mathbb{R}$ and uniformly with respect to x (only) on any compact subset of $(-1, 1)$.

Corollary 1.31 *$U_{t-1}^*(x)$ is continuous on $(-1, 1]$ and behaves as $(\sqrt{1-x^2})^{-1}$ in the limit $x \rightarrow (-1)+$.*

As Zayed (1993) has shown, sampling theorems may be used to derive many summation formulae in an easy way. Let us consider a further one.

Corollary 1.32 *For $t \geq 0$, $t \neq 1/2$ there holds*

$$\sum_{k=0}^{\infty} (4k+1) \frac{\Gamma(k + \frac{1}{2})}{\Gamma(k+1)} \frac{\sin \pi(t - \frac{1}{2} - k)}{\pi(t^2 - (2k + \frac{1}{2})^2)} = \frac{\pi}{\Gamma(\frac{3}{4} + \frac{t}{2})\Gamma(\frac{3}{4} - \frac{t}{2})}. \qquad (1.7.11)$$

Proof Setting $x = 0$ in Theorem 1.26, noting $P_{2n+1}(0) = 0$, as well as $P_{2n}(0) = (-1)^n \Gamma(n + 1/2)/(\Gamma(n+1)\sqrt{2})$, $n \in \mathbb{N}_0$, we have for $t \geq 0$, $t \neq 1/2$

$$\frac{P_{t-1/2}(0)}{t^2 - \frac{1}{2}} = \frac{\sin \pi(t - \frac{1}{2})}{\pi(t^2 - \frac{1}{4})} \left\{ \frac{1}{t^2 - \frac{1}{2}} + \sum_{j=1}^{\infty} \frac{\Gamma(j + \frac{1}{2})(-1)^j(4j+1)}{\Gamma(j+1)\sqrt{\pi}(t^2 - (2j + \frac{1}{2})^2)} \right\}.$$

Then (1.7.11) follows readily by observing formulae 1.515 (1) and (2) of Gradstein and Ryshik (1981, p. 45), as well as

$$P_{t-1/2}(0) = F\left(-\frac{t}{2} + \frac{1}{4}, \frac{t}{2} + \frac{1}{4}; 1; 1 \right) = \frac{\sqrt{\pi}}{\Gamma(\frac{3}{4} + \frac{t}{2})\Gamma(\frac{3}{4} - \frac{t}{2})}.$$

Further summation formulae may be derived from Theorems 1.26, 1.28 and 1.30 for special values of x or using particular properties of the special Jacobi functions.

SAMPLING THEORY AND THE ARITHMETIC FOURIER TRANSFORM

2.1 Historical development

The Arithmetic Fourier Transform (AFT) is an algorithm for the computation of Fourier coefficients which was given this name by Tufts and Sadasiv (1988). The algorithm has a long history differentiated by three distinct stages, the first of which began with the astronomers of the nineteenth century. At that time the calculation of the periodicities of the motions of planets or of tidal effects involved extremely large amounts of empirical data. The German mathematician H. Bruns (1903) published a method for the computation of the Fourier series of a periodic function by using Möbius inversion. This algorithm utilized the magnitude of the data available and was constructed using alternating sums of sample values.

The second stage in the development of the AFT was the monograph of Aurel Wintner (1945). Suppose that f is a periodic function of period 2π which is normalized so that

$$\int_0^{2\pi} f(\theta)d\theta = 0. \tag{2.1.1}$$

Then if f is sufficiently smooth it may be represented by its Fourier series

$$f(\theta) = \sum_{n=1}^{\infty}(a_n \cos n\theta + b_n \sin n\theta).$$

The arithmetic mean of f is defined for each positive integer N by

$$S(N) = \frac{1}{N}\sum_{m=0}^{N-1} f\left(\frac{2\pi m}{N}\right).$$

Then under suitable conditions on f the cosine coefficients a_n, $n \geq 1$, may be computed by the AFT algorithm

$$a_n = \sum_{k=1}^{\infty} \mu(k)S(nk). \tag{2.1.2}$$

Here μ is the Möbius function defined by

  (i) $\mu(1) = 1$,
 (ii) $\mu(j) = 0$ if there is a prime p such that $p^2|j$,
(iii) if $j = p_1, p_2, \ldots, p_l$ is the prime factorization of j and the factors are distinct then
     $\mu(j) = (-1)^l$.

The sine coefficients b_n cannot be calculated directly using the $S(N)$ defined above. However if the algorithm (2.1.2) is applied to the translated function $f(\theta + \pi/2n)$ then

$$
\begin{aligned}
b_n &= \frac{1}{\pi} \int_0^{2\pi} f(\phi) \sin n\phi \; d\phi \\
&= \frac{1}{\pi} \int_0^{2\pi} f\left(\theta + \frac{\pi}{2n}\right) \sin n\left(\theta + \frac{\pi}{2n}\right) d\theta \\
&= \frac{1}{\pi} \int_0^{2\pi} f\left(\theta + \frac{\pi}{2n}\right) \cos n\theta \; d\theta \\
&= \sum_{k=1}^{\infty} \mu(k) T_n(nk)
\end{aligned}
$$

where

$$
T_n(N) = \frac{1}{N} \sum_{m=0}^{N-1} f\left(\frac{2\pi m}{N} + \frac{\pi}{2n}\right). \tag{2.1.3}
$$

We shall restrict attention to the algorithm (2.1.2) for the cosine coefficients, recognizing that it can be used to obtain the sine coefficients. This is not the only approach. Bruns actually used alternating sums to obtain formulas for both a_n and b_n. Although the translation $\pi/2n$ in (2.1.3) is dependent on n there is less complexity here than within the Bruns alternation sum formulas. (Wintner himself came to this conclusion on page 10 of his monograph.) We shall not pursue the Bruns formula further but this is not to deny the ingenuity or significance of his work.

Wintner's monograph had two main objectives, which with hindsight get in each other's way. It is probably better to read the second half first (begin at Section 6) as this is concerned with a rigorous verification of (2.1.2) under suitable hypotheses on f. Indeed in the introduction Wintner indicates that his primary objective is to present the results of Bruns in a form "acceptable to the mathematical conscience". Again on page 16 he observes that although Bruns was aware of the Weierstrass double-series theorem it was not properly applied to the Möbius inversion. Wintner outlines how this may be remedied, referring indirectly to a paper of Hardy (1916) on Weierstrass's non-differentiable function. In Section 2.2 we present two distinct sets of sufficient conditions for the validity of (2.1.2) and the proof has been further simplified.

Most of the second half of the Wintner monograph is concerned with questions of a number-theoretic nature which arise from the rigorous development of the AFT and are not related to the algorithm itself. A notable exception is a brief reference to a paper of Harold Davenport (1937), which depends on Vinogradov-type estimates and "the heavy Diophantine machinery" therein. Surprisingly this paper effectively contains what is required to prove the AFT algorithm (2.1.2) for step-functions. It appears that Wintner was not aware of this possibility, although the result may not have seemed significant before the advent of digital processing. We shall not pursue this topic here and we refer the reader to Schiff and Walker (1993, 1994) for a proof of the step-function algorithm.

In the early part of his book Wintner's objective was to obtain a "direct or arithmetical" proof of the convergence of Fourier series. As he himself points out (page 7) he was unsuccessful. Notwithstanding this, an interesting estimate is obtained which we shall use in Section 2.3 to compare the aliasing properties of the AFT and Discrete Fourier Transform (DFT).

Finally it seems that the monograph had little impact on publication, perhaps because it was ahead of its time. Scientists did not yet have the computational power to calculate the DFT for large quantities of data. The advent of the computer led to an ever increasing appetite for even larger computations and the invention of the Fast Fourier Transform (FFT) by Cooley and Tukey in 1965.

The third stage in the evolution of the AFT began about 1985 when the algorithm was rediscovered by Schiff and Walker (1987) who were interested in sampling theory and at about the same time by Tufts and Sadasiv (1988) who were motivated by applications to signal processing. The arithmetic operations in the AFT can be performed in parallel and, apart from a small number of scalings, only multiplications by 0, +1 and −1 are required. In contrast to the FFT the AFT requires no storage of complex exponential functions and does not need complicated memory addressing.

On the other hand the disadvantage of the AFT lies in the irregular sampling inherent in the algorithm. In applications to signal processing f is assumed to be band-limited and sampled at the corresponding Nyquist rate whereas the AFT algorithm requires evaluation of f at irregularly spaced points. This problem was first addressed by Reed et al. (1990), where the authors used first-order (linear) and zero-order (nearest neighbour) interpolation to evaluate f at intermediate points. The implementation of these methods for VLSI architecture is further discussed by Wigley and Jullien (1992).

In Section 2.2 we shall see that the assumption that the Fourier coefficients a_n and b_n vanish for $n > M$ implies that the sums $S(N)$ vanish for $N > M$. If the period of the function f is 1 and the sampling is undertaken on the interval [0, 1], then given the band-limited assumption, the sample points are precisely the Farey fractions of order M. An asymptotic estimate for the number of sample points is $3M^2/\pi^2$ (Hardy and Wright 1979, p. 268). See also Tepedelenliglu (1989) and the reply Tufts (1989) for a discussion on the computational complexity of the AFT. At the present time it seems unlikely that the AFT will replace the FFT for the widespread calculation of the DFT. One reason is that the reduction of multiplications is no longer an advantage for VLSI architecture.

In a recent paper Walker (1994) has developed a modification of the algorithm (2.1.2) which with suitable sampling reduces the interpolation required. This idea is discussed in Section 2.4 and will be referred to as the "summability by primes" algorithm. It has its roots in the paper of Duffin (1957) in which summation undertaken by primes was used for the representation of Fourier integrals.

Another property of the algorithm (2.1.2) is easily observed. If $n = 1$ then

$$a_1 = \sum_{k=1}^{\infty} \mu(k) S(k)$$

$$= \sum_{k=1}^{\infty} \frac{\mu(k)}{k} \sum_{m=0}^{k-1} f\left(\frac{2\pi m}{k}\right).$$

The coefficient of $f(0)$ is

$$\sum_{k=1}^{\infty} \mu(k)/k$$

which is a convergent series with sum zero. As an immediate consequence the sample point $\theta = 0$ may be deleted from the algorithm.

The summation

$$\sum_{k=1}^{\infty} \mu(k)/k = 0$$

is equivalent to the prime number theorem and has an interesting history. It appears in *Euler—Introduction to Analysis of the Infinite* (1988, p. 235). It was also the subject of the dissertation of Edmund Landau (1899). In Landau (1958), he comments that in general number theory may be considered useless but this particular result was useful to him as it earned him his doctorate.

Notwithstanding these remarks, we shall demonstrate that the summation may be used to delete any finite set of sample points from the AFT algorithm. Following Walker (1995), we shall also show that it is possible to delete any finite set of sample points from the summability by primes algorithm.

Finally in this historical development we outline the motivation for these deletion procedures. It stems from two disparate concepts. The first is Karl Pribram's holographic model of the brain, developed in the late 1960s and early 1970s, in which perceptual experience is translated into Fourier transforms distributed across the brain in consciousness and memory. (For a popular science adaption of these ideas see Briggs and Peat (1985).) Of course this raises the question as to how such transforms may be "registered" or "computed".

The structure of the inner ear ensures that the input from the auditory system is received and processed in the frequency domain. By contrast the AFT suggests the possibility of Fourier analysis without an apparent spectral resolution. We have the complementary principle that an aggregate of arithmetic means from the time (or space) domain may be considered as an element in the frequency domain.

Although the AFT is suitable for parallel processing it is not an adequate model for a real (as opposed to artificial) neural network. This will be reflected by the second of our disparate concepts. A striking feature of the anatomy of the cortex is its variability. Each neuron is likely to have 8000 synaptic connections which apparently occur at random. Any reasonable paradigm for a real neural network should reflect this variability. Pellionisz (1989) makes a case for a fresh look at individual neurons, indeed for a single unit model that would encompass complex physiological (membrane) and anatomical (dendritic) properties as well as identify a well-defined computational function; at the same time it could be used as units in neuronal network models. In this paper and Pellionisz (1991) he comments that the morphology of a single neuron (dendritic arbour) reveals a fractal geometry.

As mentioned earlier we may consider the sample space for the AFT to be the Farey fractions. The branches of the Farey tree are self-similar to the whole tree in a certain sense. For a demonstration of this fractal property (in quite a different context), using continued fractions, see McCauley (1993, p. 170). However we have a totally rigid structure here and it does not reflect the variability apparent in the anatomy of the cortex. The results of Sections 2.5 and 2.6 go some way towards remedying this. There are large families of algorithms which compute Fourier coefficients with varying degrees of efficiency. On the one hand from the point of view of a communications engineer, the AFT is characterized by gross oversampling. On the other hand this oversampling is a manifestation of the variability inherent in the algorithm.

2.2   The AFT algorithm

Classical Möbius inversion depends on our first lemma.

Lemma 2.1 *If j is a positive integer then*

$$\sum_{d|j} \mu(d) = \begin{cases} 1 & \text{if } j = 1 \\ 0 & \text{if } j \neq 1. \end{cases}$$

Proof   Suppose $j > 1$ has l distinct prime factors. Then it suffices to consider the divisors obtained by taking combinations of these l factors, that is,

$$\sum_{d|j} \mu(d) = \sum_{k=0}^{l} \binom{l}{k} (-1)^k = (1 - 1)^l.$$

The next two lemmas are fundamental to all forms of the AFT algorithm.

Lemma 2.2 *If ω is an nth root of unity and j is a positive integer then*

$$\sum_{m=0}^{n-1} \omega^{jm} = \begin{cases} n & \text{if } j = kn \\ 0 & \text{if } j \neq kn \end{cases}$$

for some positive integer k.

Proof   If $j = kn$ the result is immediate. Otherwise

$$\sum_{m=0}^{n-1} \omega^{jm} = \frac{1 - (\omega^j)^n}{1 - \omega^j}$$
$$= 0.$$

Lemma 2.3 *Let f be a function of period 2π which is normalized so that (2.1.1) holds. Suppose that f has a Fourier series*

$$f(\theta) = \sum_{j=1}^{\infty} \left(c_j e^{ij\theta} + c_{-j} e^{-ij\theta} \right)$$

which converges everywhere to f. If for every positive integer n

$$S(n) = \frac{1}{n} \sum_{m=0}^{n-1} f\left(\frac{2\pi m}{n} \right)$$

then

$$S(n) = \sum_{k=1}^{\infty}(c_{kn} + c_{-kn}).$$ (2.2.1)

Proof By Lemma 2.2

$$S(n) = \frac{1}{n}\sum_{j=1}^{\infty}\left(c_j\sum_{m=0}^{n-1}\exp\left(ij\frac{2\pi m}{n}\right) + c_{-j}\sum_{m=0}^{n-1}\exp\left(-ij\frac{2\pi m}{n}\right)\right)$$

$$= \sum_{k=1}^{\infty}(c_{kn} + c_{-kn}).$$

At this stage we observe that if we have a band-limited assumption that c_j vanishes if $|j| > W$ then it is immediate that $S(n)$ vanishes for $n > W$. As indicated in the introduction we shall restrict our investigations to the cosine coefficients and from (2.2.1) we have

$$S(n) = \sum_{k=1}^{\infty}a_{kn}.$$ (2.2.2)

The fundamental problem is to invert the system of equations (2.2.2) to obtain an explicit formula for each a_n. The algorithm (2.1.2) can be obtained formally using Lemma 2.1 since

$$\sum_{k=1}^{\infty}\mu(k)S(nk) = \sum_{k=1}^{\infty}\mu(k)\sum_{j=1}^{\infty}a_{nkj}$$

$$= \sum_{m=1}^{\infty}a_{nm}\sum_{d|m}\mu(d)$$

$$= a_n.$$

The rearrangement of the double series is valid if

$$\sum_{k=1}^{\infty}\sum_{j=1}^{\infty}\mu(k)a_{nkj}$$

converges absolutely and the Weierstrass double series theorem can then be applied. This formal proof was known to Bruns who claimed that the absolute convergence

$$\sum_{j=1}^{\infty}|a_j| < \infty$$

was a sufficient condition for the rearrangement of the double series. This was refuted by Wintner who showed that due to the accumulation of divisors, the absolute convergence need not imply the validity of Möbius inversion. Our main theorem of this section gives two sets of sufficient conditions for the AFT algorithm (2.1.2).

Theorem 2.4 *Let f be a function of period 2π. Then the AFT algorithm (2.1.2) holds if either of the following conditions is satisfied.*

(i) $f \in Lip_\alpha[0, 2\pi]$, *where* $\frac{1}{2} < \alpha \leq 1$.

(ii) f *is of bounded variation on* $[0, 2\pi]$ *and* $f \in Lip_\alpha[0, 2\pi], 0 < \alpha \leq \frac{1}{2}$.

Here $f \in Lip_\alpha[0, 2\pi]$ is the usual condition that $|f(\theta_1) - f(\theta_2)| < C|\theta_1 - \theta_2|^\alpha$ for some constant C and all θ_1 and θ_2. Both (i) and (ii) imply that the Fourier series of f converges absolutely on $[0, 2\pi]$, (Zygmund 1959, p. 240).

Proof The Möbius inversion will be justified by the Weierstrass double series theorem if it is shown that the series

$$\sum_{m=1}^{\infty} a_{nm} \sum_{d|m} \mu(d)$$

converges absolutely. If m is a positive integer let $d(m)$ denote the number of divisors of m. Then $d(m) = O(m^\delta)$ for all $\delta > 0$. See Hardy and Wright (1979, p. 260). Hence it suffices to show the convergence of $\sum_{m=1}^{\infty} m^\epsilon |a_{nm}|$ for some $\epsilon > 0$.

Proof of (i). Since $f \in Lip_\alpha[0, 2\pi]$, by Zygmund (1959, p. 46) there exists $C > 0$ such that

$$|a_{nm}| < \frac{C}{(nm)^\alpha},$$

and hence for $\epsilon > 0$,

$$m^\epsilon |a_{nm}|^{\epsilon/\alpha} \leq \frac{C^{\epsilon/\alpha}}{n^\epsilon}.$$

It follows that

$$\sum_{m=1}^{\infty} m^\epsilon |a_{nm}| \leq \frac{C^{\epsilon/\alpha}}{n^\epsilon} \sum_{m=1}^{\infty} |a_{nm}|^{1-\epsilon/\alpha}$$

and by Zygmund (1959, p. 243), using the hypotheses of (i), the latter series is convergent if

$$1 - \frac{\epsilon}{\alpha} > \frac{2}{2\alpha + 1}.$$

The proof may be completed by choosing

$$\epsilon < \frac{\alpha(2\alpha - 1)}{2\alpha + 1}.$$

Proof of (ii). Since f is of bounded variation on $[0, 2\pi]$ there exists a constant C such that

$$|a_{nm}| < \frac{C}{nm}$$

and hence for $\epsilon > 0$,

$$m^\epsilon |a_{nm}|^\epsilon \leq \frac{C^\epsilon}{n^\epsilon}.$$

It follows that

$$\sum_{m=1}^{\infty} m^{\epsilon} |a_{nm}| \leq \frac{C^{\epsilon}}{n^{\epsilon}} \sum_{m=1}^{\infty} |a_{nm}|^{1-\epsilon}$$

and by Zygmund (1959, p. 243), using the hypotheses of (ii), the latter series is convergent if

$$1 - \epsilon > \frac{2}{\alpha + 2}.$$

The proof may be completed by choosing

$$\epsilon < \frac{\alpha}{\alpha + 2}.$$

Examples Suppose that f is a function of period 2π which satisfies either the condition (i) or the condition (ii) of Theorem 2.4. The following observations are easily proved using the AFT algorithm, but are likely to be more difficult by other methods.

(a) If

$$S(N) = \frac{1}{N} \sum_{m=0}^{N-1} f\left(\frac{2\pi m}{N}\right) = d$$

for each positive integer N, then $d = 0$ and f is an odd function.

This is immediate from the formula

$$a_1 = \sum_{k=1}^{\infty} \mu(k) S(k) = d \sum_{k=1}^{\infty} \mu(k).$$

Since the series is divergent we must have $d = 0$. It follows that each $a_n = 0$ and that f is odd.

(b) If

$$\sum_{m=0}^{N-1} f\left(\frac{2\pi m}{N}\right) = d$$

for each positive integer N, then $d = 0$ and f is an odd function.

Again by the AFT algorithm

$$a_n = \sum_{k=1}^{\infty} \mu(k) S(nk) = \frac{d}{n} \sum_{k=1}^{\infty} \frac{\mu(k)}{k}.$$

But the summation $\sum_{k=1}^{\infty} \mu(k)/k = 0$ and it follows that each $a_n = 0$. Again f is odd and hence $d = 0$.

2.3 Aliasing properties of the DFT and AFT

The objective of this section will be to compare the aliasing properties of the DFT and AFT. Following the usual convention, we compute the DFT of a function f defined on

$[0, N]$. We assume that the sample values $f(k)$, $0 \le k \le N - 1$, are obtained from a convergent Fourier series

$$f(x) = \sum_{m=-\infty}^{\infty} c_m \exp\left(\frac{i2\pi mx}{N}\right). \qquad (2.3.1)$$

The DFT is defined for $0 \le n \le N - 1$ by

$$d_n = \frac{1}{N} \sum_{k=0}^{N-1} f(k) \exp\left(\frac{-i2\pi nk}{N}\right). \qquad (2.3.2)$$

The following lemma is well known but we shall give the proof since it depends only on Lemma 2.2.

Lemma 2.5 *If the Fourier series* (2.3.1) *is convergent then for* $0 \le n \le N - 1$

$$d_n = \sum_{l=-\infty}^{\infty} c_{n+lN}. \qquad (2.3.3)$$

Proof Let

$$\omega = \exp(i2\pi/N).$$

Then

$$f(k) = \sum_{m=-\infty}^{\infty} c_m \omega^{mk}$$

and substitution into (2.3.2) gives

$$d_n = \frac{1}{N} \sum_{k=0}^{N-1} f(k)\omega^{-kn}$$

$$= \frac{1}{N} \sum_{k=0}^{N-1} \omega^{-kn} \sum_{m=-\infty}^{\infty} c_m \omega^{mk}$$

$$= \frac{1}{N} \sum_{m=-\infty}^{\infty} c_m \sum_{k=0}^{N-1} \omega^{k(m-n)}.$$

Then by Lemma 2.2

$$\sum_{k=0}^{N-1} \omega^{k(m-n)} = \begin{cases} N & \text{if } m - n = lN \\ 0 & \text{if } m - n \ne lN \end{cases}$$

for some integer l and the result follows.

Equation (2.3.3) encapsulates the aliasing properties of the DFT. For $0 \leq n \leq N/2$, d_n represents c_n and for $N/2 < n \leq N - 1$, d_n represents c_{n-N}. (Equation (2.3.2) defines d_n for every integer n and the sequence (d_n) is periodic of period N.)

For convenience we assume that N is even and that $-N/2 < q \leq N/2$. We seek to estimate the aliasing error induced by approximating c_q by d_q. This will require the following definition and lemma which will also be used to obtain an aliasing estimate for the AFT.

Definition 2.6 *The set of M real numbers* $(\gamma_m)_{1 \leq m \leq M}$ *is said to be equidistributed on* $[0, T]$ *if for each m*

$$(m - 1)\frac{T}{M} \leq \gamma_m \leq m\frac{T}{M}.$$

Lemma 2.7 *Suppose that* $g \in Lip_1[0, T]$ *with Lipschitz constant A. Then if* $(\gamma_m)_{1 \leq m \leq M}$ *is an equidistributed set on* $[0, T]$,

$$\left| \frac{1}{T} \int_0^T g(x)dx - \frac{1}{M} \sum_{m=1}^M g(\gamma_m) \right| \leq \frac{AT}{M}.$$

Proof By the mean value theorem for integrals there exists ξ_m in the interval $((m - 1)T/M, mT/M)$ such that

$$\int_{(m-1)T/M}^{mT/M} g(x)dx = g(\xi_m)\frac{T}{M}.$$

Then

$$\left| \frac{1}{T} \int_0^T g(x)dx - \frac{1}{M} \sum_{m=1}^M g(\gamma_m) \right| = \left| \frac{1}{M} \sum_{m=1}^M (g(\xi_m) - g(\gamma_m)) \right|$$

$$\leq \frac{1}{M} \sum_{m=1}^M A|\xi_m - \gamma_m|$$

$$\leq \frac{AT}{M}.$$

Returning to the aliasing problem we observe that

$$d_q = \frac{1}{N} \sum_{k=0}^{N-1} g(k)$$

where we define

$$g(x) = \exp\left(\frac{-i2\pi qx}{N}\right) f(x).$$

Also we may write

$$c_q = \frac{1}{N} \int_0^N g(x)dx.$$

If $g \in Lip_1[0, N]$ with Lipschitz constant A then by Lemma 2.7

$$|c_q - d_q| = \left| \frac{1}{N} \int_0^N g(x)dx - \frac{1}{N} \sum_{k=0}^{N-1} g(k) \right| < A. \tag{2.3.4}$$

This development is summarized in our next theorem.

Theorem 2.8 *Suppose that* $|f(x)| \leq C$ *on* $[0, N]$ *and that* $f \in Lip_1[0, N]$ *with Lipschitz constant* L. *Then for* $-N/2 < q \leq N/2$

$$|c_q - d_q| < \frac{4\pi |q| C}{N} + L.$$

Proof From equation (2.3.4) it suffices to find a Lipschitz constant for g. By applying the mean value theorem to the real and imaginary parts of the exponential

$$\left| \exp\left(\frac{-i2\pi q x_1}{N} \right) - \exp\left(\frac{-i2\pi q x_2}{N} \right) \right| \leq \frac{4\pi |q|}{N} |x_1 - x_2|.$$

It is now straightforward to show that g has Lipschitz constant $(4\pi |q| C)/N + L$.

In order to obtain an aliasing estimate for the AFT we make stronger smoothness assumptions on f as exhibited in the next lemma. The proof is along the lines of Wintner (1945, p. 4).

Lemma 2.9 *Let* f *be a function of period* T *which is normalized so that* $\int_0^T f(x)dx = 0$. *Suppose that* $f' \in Lip_1[0, T]$ *with Lipschitz constant* L. *Then*

$$\left| \frac{1}{M} \sum_{m=1}^M f\left(\frac{mT}{M} \right) \right| \leq \frac{LT^2}{2M^2}.$$

Proof For $1 \leq m \leq M$,

$$\frac{T}{M} f\left(\frac{mT}{M} \right) - \int_{(m-1)T/M}^{mT/M} f(x)dx = \int_{(m-1)T/M}^{mT/M} \left(f\left(\frac{mT}{M} \right) - f(x) \right) dx$$

$$= \int_{(m-1)T/M}^{mT/M} \left(\frac{mT}{M} - x \right) f'(\theta_m(x))dx,$$

where by the mean value theorem $\theta_m(x)$ lies in the interval $(x, mT/M)$. We now evaluate

$$\int_{(m-1)T/M}^{mT/M} \left(\frac{mT}{M} - x \right) dx = \frac{T^2}{2M^2}$$

and denote by $f'(\alpha_m)$ and $f'(\beta_m)$ the absolute minimum and absolute maximum respectively, of $f'(x)$ on $[(m-1)T/M, mT/M]$. Then since $\int_0^T f(x)dx = 0$ we have

$$\frac{T^2}{2M^2} \sum_{m=1}^M f'(\alpha_m) \leq \frac{T}{M} \sum_{m=1}^M f\left(\frac{mT}{M} \right) \leq \frac{T}{2M^2} \sum_{m=1}^M f'(\beta_m). \tag{2.3.5}$$

By the periodicity of f,

$$\int_0^T f'(t)dt = f(T) - f(0) = 0.$$

Hence if $(\gamma_m)_{1 \leq m \leq M}$ is any equidistributed set on $[0, T]$ it follows by Lemma 2.7 applied to f' that

$$\left| \sum_{m=1}^M f'(\gamma_m) \right| \leq LT.$$

The latter inequality holds for (α_m) and (β_m) and we have from (2.3.5)

$$-\frac{T^2 L}{2M^2} \leq \frac{1}{M} \sum_{m=1}^M f\left(\frac{mT}{M}\right) \leq \frac{T^2 L}{2M^2}.$$

This lemma may now be applied to obtain an aliasing estimate for the AFT algorithm (2.1.2).

Theorem 2.10 Let f be a function of period 2π which is normalized so that $\int_0^{2\pi} f(\theta)d\theta = 0$. Suppose that $f' \in \text{Lip}_1[0, 2\pi]$ with Lipschitz constant L. Then

$$\left| a_n - \sum_{k=1}^N \mu(k)S(nk) \right| \leq \frac{2L\pi^2}{n^2 N}.$$

Proof By Lemma 2.9, noting that $f(0) = f(2\pi)$,

$$|S(nk)| \leq \frac{L(2\pi)^2}{2(nk)^2}.$$

Hence we obtain the estimate

$$\begin{aligned}
\left| a_n - \sum_{k=1}^N \mu(k)S(nk) \right| &= \left| \sum_{k=N+1}^\infty \mu(k)S(nk) \right| \\
&\leq \frac{2L\pi^2}{n^2} \sum_{k=N+1}^\infty \frac{1}{k^2} \\
&< \frac{2L\pi^2}{n^2} \int_N^\infty \frac{1}{u^2} du \\
&= \frac{2L\pi^2}{n^2 N}.
\end{aligned}$$

If $S(j)$ is sampled for $j \leq B$ it makes sense to make the approximation $nN = B$ and we then have the estimate

$$\left| a_n - \sum_{k=1}^N \mu(k)S(nk) \right| \leq \frac{2L\pi^2}{nB}.$$

Under these circumstances the AFT error bound decreases with increasing n whereas the DFT shows a linear dependence on n.

2.4   The summability algorithm

The summability by primes algorithm has developed as an alternative method to the Möbius inversion of the system of equations (2.2.2). On the basis of our discussions in the first section we restrict attention to an even function which is represented by its Fourier cosine series. Also for the moment the function will be assumed to be normalized ($a_0 = 0$) although later we will show that this condition may be relaxed. A further point of interest is that the absolute convergence of the Fourier series will be a sufficient condition for the convergence of the summability algorithm. We recall that a principal concern of Wintner's monograph was that it was not a sufficient condition for the AFT algorithm.

In order to state the summability theorem we need to define the notation δ_j^q for j and q positive integers by

$\delta_j^q = 0$, if j contains a prime factor greater than the qth prime;
$\delta_j^q = 1$, otherwise.

Theorem 2.11 *Suppose that the Fourier series $f(\theta) = \sum_{n=1}^{\infty} a_n \cos n\theta$ of f converges absolutely. Then*

$$a_n = \lim_{q \to \infty} \sum_{j=1}^{\infty} \delta_j^q \mu(j) S(jn), \qquad n \geq 1,$$

where

$$S(N) = \frac{1}{N} \sum_{m=0}^{N-1} f\left(\frac{2\pi m}{N}\right).$$

Proof   We define for all positive integers n,

$$
\begin{aligned}
T_0(n) &= S(n), \\
T_1(n) &= T_0(n) - T_0(2n), \\
T_2(n) &= T_1(n) - T_1(3n), \\
T_q(n) &= T_{q-1}(n) - T_{q-1}(pn),
\end{aligned}
$$

where p is the qth prime. By the definition of the Möbius function μ it can be seen that

$$T_q(n) = \sum_{j=1}^{\infty} \delta_j^q \mu(j) S(jn). \qquad (2.4.1)$$

On the other hand by (2.2.2)

$$T_0(n) = S(n) = \sum_{k=1}^{\infty} a_{kn},$$

and we can find $T_q(n)$ in terms of the a_{kn}. In order to do this we define α_k^q for k and q positive integers by

$\alpha_k^q = 0$, if k contains at least one of the first q primes as a prime factor;
$\alpha_k^q = 1$, otherwise.

Then it can be seen that

$$T_q(n) = a_n + \sum_{k=2}^{\infty} \alpha_k^q a_{kn}. \tag{2.4.2}$$

The theorem will follow from equations (2.4.1) and (2.4.2) if we show that

$$\lim_{q \to \infty} \sum_{k=2}^{\infty} \alpha_k^q a_{kn} = 0.$$

But if p is the qth prime we have the estimate

$$\left| \sum_{k=2}^{\infty} \alpha_k^q a_{kn} \right| \leq \sum_{k=p+1}^{\infty} |a_{kn}|,$$

and the required result follows by the convergence of $\sum_{k=2}^{\infty} |a_{kn}|$.

Example   If $q = 3$ the expression for a_1 is

$$a_1 = S(1) - S(2) - S(3) + S(6) - S(5) + S(10) + S(15) - S(30).$$

The terms are written as they arise for $q = 1, 2, 3$ and the terms for which $\mu(j) = 0$ have been omitted.

If f is sampled at 30 equally spaced points then a_1 can be computed by using this formula without interpolation between sample points. With a suitable array the summability algorithm reduces the necessity for interpolation, especially for low-order coefficients.

Relaxation of the normalization condition. Suppose that $g(\theta) = a_0 + \sum_{n=1}^{\infty} a_n \cos n\theta$ is an absolutely convergent Fourier series and define

$$R(N) = \frac{1}{N} \sum_{m=0}^{N-1} g\left(\frac{2\pi m}{N} \right).$$

The normalization of g is $f(\theta) = \sum_{n=1}^{\infty} a_n \cos n\theta$ and we have the usual definition that

$$S(N) = \frac{1}{N} \sum_{m=0}^{N-1} f\left(\frac{2\pi m}{N} \right).$$

Then it is immediate that for all N,

$$R(N) = a_0 + S(N).$$

We now apply Theorem 2.11 to f and we claim that

$$a_n = \lim_{q \to \infty} \sum_{j=1}^{\infty} \delta_j^q \mu(j) S(jn) = \lim_{q \to \infty} \sum_{j=1}^{\infty} \delta_j^q \mu(j) R(jn).$$

To see this we fix q and note that terms in the summation with non-zero Möbius coefficients may be paired together. If 2 does not divide j and $\delta_j^q = 1$ then $\delta_{2j}^q = 1$ and $\mu(2j) = -\mu(j)$. We now have

$$\begin{aligned}
\mu(j)S(jn) + \mu(2j)S(2jn) &= \mu(j)(S(jn) - S(2jn)) \\
&= \mu(j)(R(jn) - R(2jn)) \\
&= \mu(j)R(jn) + \mu(2j)R(2jn).
\end{aligned}$$

The result follows since all other terms in the summation have zero Möbius coefficient.

The reduction of the interpolation and the relaxation of the normalization condition are noteworthy features of the summability algorithm. It has a "nicer" structure than the standard AFT algorithm and is more suitable as a paradigm along the lines suggested by Pellionisz (1989, 1991). We envisage a parallel neural network in which the array to calculate a single Fourier coefficient corresponds to the dendritic arbour of a single cell. The motivation for Section 2.6, which establishes the deletion properties of the summability algorithm, relates to the variability which must be inherent in such a paradigm.

The rate of convergence of the summability algorithm could well be the subject of further numerical investigation. Generally it seems comparable to but not as fast as the AFT algorithm. For a preliminary investigation see Walker (1994).

2.5   Deletion properties of the AFT algorithm

The deletion properties of the AFT algorithm can be derived from the next theorem. We give two proofs of the theorem. The first follows immediately from an estimate obtained by Harold Davenport. This estimate depends on deep inequalities of the Vinogradov type. The second proof is elementary and has the advantage that it readily generalizes to the summability algorithm.

Theorem 2.12   *For each positive integer q,*

$$\sum_{j=1}^{\infty} \frac{\mu(jq)}{jq} = 0.$$

Proof
First Proof. Let (k, j) denote the greatest common divisor of k and j and note that if $(k, j) = 1$ then $\mu(kj) = \mu(k)\mu(j)$. Then from an estimate of the partial sums (Davenport 1937, p. 319) we have

$$\sum_{\substack{k=1 \\ (k,q)=1}}^{\infty} \frac{\mu(k)}{k} = 0. \tag{2.5.1}$$

The theorem now follows since

$$\sum_{j=1}^{\infty} \frac{\mu(jq)}{jq} = \sum_{\substack{k=1 \\ (k,q)=1}}^{\infty} \frac{\mu(kq)}{kq}$$

$$= \frac{\mu(q)}{q} \sum_{\substack{k=1 \\ (k,q)=1}}^{\infty} \frac{\mu(k)}{k}$$

$$= 0.$$

Second Proof. Let q have the prime factorization $q = p_1, p_2, \ldots, p_M$. The theorem is immediate if there is a repeated factor so that we assume that the factors are distinct. We now obtain another proof of (2.5.1).

To avoid cumbersome notation we shall first consider the case $q = p_1 p_2$ and indicate how the proof is generalized. We have the following decomposition.

$$\sum_{k=1}^{\infty} \frac{\mu(k)}{k} = \sum_{\substack{k=1 \\ (k,q)=1}}^{\infty} \frac{\mu(k)}{k} + \sum_{\substack{k=1 \\ (k,q)=1}}^{\infty} \frac{\mu(p_1 k)}{p_1 k}$$

$$+ \sum_{\substack{k=1 \\ (k,q)=1}}^{\infty} \frac{\mu(p_2 k)}{p_2 k} + \sum_{\substack{k=1 \\ (k,q)=1}}^{\infty} \frac{\mu(p_1 p_2 k)}{p_1 p_2 k}$$

$$= \sum_{\substack{k=1 \\ (k,q)=1}}^{\infty} \frac{\mu(k)}{k} + \sum_{\substack{k=1 \\ (k,q)=1}}^{\infty} \frac{\mu(p_1)}{p_1} \frac{\mu(k)}{k}$$

$$+ \sum_{\substack{k=1 \\ (k,q)=1}}^{\infty} \frac{\mu(p_2)}{p_2} \frac{\mu(k)}{k} + \sum_{\substack{k=1 \\ (k,q)=1}}^{\infty} \frac{\mu(p_1 p_2)}{p_1 p_2} \frac{\mu(k)}{k}$$

$$= \left[1 - \frac{1}{p_1} - \frac{1}{p_2} + \frac{1}{p_1 p_2} \right] \sum_{\substack{k=1 \\ (k,q)=1}}^{\infty} \frac{\mu(k)}{k}$$

$$= \left[\frac{(p_1 - 1)(p_2 - 1)}{q} \right] \sum_{\substack{k=1 \\ (k,q)=1}}^{\infty} \frac{\mu(k)}{k}.$$

In general we have the decomposition

$$\sum_{k=1}^{\infty} \frac{\mu(k)}{k} = \left[\frac{1}{q} \prod_{i=1}^{M} (p_i - 1) \right] \sum_{\substack{k=1 \\ (k,q)=1}}^{\infty} \frac{\mu(k)}{k}$$

and since $\sum_{k=1}^{\infty} \mu(k)/k = 0$ it follows that

$$\sum_{\substack{k=1 \\ (k,q)=1}}^{\infty} \frac{\mu(k)}{k} = 0.$$

This completes our second proof.

The principal idea of this section may now be clarified. It is simply that the AFT algorithm (2.1.2) can be modified by the deletion of any finite subset of sample points. Suppose that as k varies in the summation (2.1.2) a **particular** sample point **first** occurs in $S(nq)$, that is for $k = q$. This sample point then appears in precisely the terms $S(nqj)$, j a positive integer, with coefficient

$$c = \sum_{j=1}^{\infty} \frac{\mu(qj)}{nqj} = 0.$$

Clearly this sample point may be deleted from the algorithm (2.1.2) by the subtraction of a convergent series. The procedure may be repeated for any finite set of sample points. Generally the rate of convergence of the algorithm slows as sample points are deleted.

2.6   Deletion properties of the summability algorithm

The deletion properties of the summability by primes algorithm depend on the next theorem which is the analogue of Theorem 2.12 with the second proof.

Theorem 2.13 *For each positive integer a*

$$\lim_{q \to \infty} \sum_{j=1}^{\infty} \delta_j^q \frac{\mu(ja)}{ja} = 0.$$

Proof   We first prove the result for the special case $a = 1$. Let $r = p_1 p_2 \ldots p_q$ be the product of the *first q* primes and let $\sum_{d|r}$ denote a summation over all divisors of r. Then by omitting the terms for which $\mu(j) = 0$, we have the decomposition

$$\sum_{j=1}^{\infty} \delta_j^q \frac{\mu(j)}{j} = \sum_{d|r} \frac{\mu(d)}{d}$$

$$= \prod_{i=1}^{q} \left(1 - \frac{1}{p_i}\right).$$

The special case $a = 1$ now follows since the infinite product diverges to zero (see Mertens' theorem, Hardy and Wright 1979, p. 351). In the general case it suffices to consider a positive integer a with a prime factorization $a = p_1 p_2 \ldots p_N$ containing N distinct primes. Let the largest of these primes be the Mth prime amongst *all* primes and let $\sum_{d|a}$ denote a summation over all divisors of a. If $q > M$, then $\delta_{kd}^q = \delta_k^q$ and we have the decomposition

$$\sum_{j=1}^{\infty} \delta_j^q \frac{\mu(j)}{j} = \sum_{d|a} \sum_{\substack{k=1 \\ (k,a)=1}}^{\infty} \delta_k^q \frac{\mu(kd)}{kd}$$

$$= \left[\sum_{d|a} \frac{\mu(d)}{d}\right] \sum_{\substack{k=1 \\ (k,a)=1}}^{\infty} \delta_k^q \frac{\mu(k)}{k}$$

$$= \left[\prod_{i=1}^{N}\left(1 - \frac{1}{p_i}\right)\right] \sum_{\substack{k=1 \\ (k,a)=1}}^{\infty} \delta_k^q \frac{\mu(k)}{k}.$$

If we now take $\lim_{q\to\infty}$, we have by the special case $a = 1$,

$$\lim_{q\to\infty} \sum_{\substack{k=1 \\ (k,a)=1}}^{\infty} \delta_j^q \frac{\mu(k)}{k} = 0.$$

The general case will now follow since

$$\lim_{q\to\infty} \sum_{j=1}^{\infty} \delta_j^q \frac{\mu(ja)}{ja} = \lim_{q\to\infty} \sum_{\substack{k=1 \\ (k,a)=1}}^{\infty} \delta_k^q \frac{\mu(ka)}{ka}$$

$$= \frac{\mu(a)}{a} \lim_{q\to\infty} \sum_{\substack{k=1 \\ (k,a)=1}}^{\infty} \delta_k^q \frac{\mu(k)}{k}$$

$$= 0.$$

Any finite set of sample points may now be deleted from the summability algorithm. The deletion proceeds as in Section 2.5.

3

DERIVATIVE SAMPLING—A PARADIGM EXAMPLE OF MULTICHANNEL METHODS

3.1 Introduction

This chapter is a short account of *derivative sampling*, the name usually given to the process of reconstructing a suitable function from a knowledge of its values and values of some of its derivatives at a discrete subset of its domain. Here, particularly in §3.4, we shall also take it to mean the dual problem of *constructing* a function which is to have given values and given derivative values on a given set (more remarks about this kind of duality can be found in Higgins 1996, p. 2, *et seq.*).

For example, the *first derivative sampling series* for a suitable function f is

$$
\begin{aligned}
f(t) = \sum_{n \in \mathbb{Z}} f\left(\frac{2n}{w}\right) \operatorname{sinc}^2 \frac{w}{2}\left(t - \frac{2n}{w}\right) \\
+ \frac{2}{\pi w} f'\left(\frac{2n}{w}\right) \operatorname{sinc}\frac{w}{2}\left(t - \frac{2n}{w}\right) \sin\frac{\pi w}{2}\left(t - \frac{2n}{w}\right),
\end{aligned} \tag{3.1.1}
$$

and was first given explicitly by Jagerman and Fogel (1956).

The phrase *multichannel sampling* means the reconstruction of an object function f from samples taken from certain transformations of f. An introduction to multichannel sampling can be found in Higgins (1996, Chapter 12). Derivative sampling seems to have received more attention than any other multichannel sampling scheme, and continues to be popular. We might ask why this should be so.

First, it must be said that the theory has not received much stimulus from applications. References to the usefulness of derivative sampling in the sciences and engineering are rare and not at all well documented in the literature. Those that there are tend towards potential rather than actual applications. On the other hand, within mathematics derivative sampling takes a natural place in, for example, Hermite interpolation (Kress 1972, Butzer and Splettstößer 1977), iterative reconstructions (Razafinjatovo, to appear), stochastic processes (see Chapter 9 of the present volume) and constructive function theory (to be discussed in §3.4).

Perhaps derivative sampling could be more fully exploited in such topics as quadrature formulae, summation of series and expansions for special functions, where ordinary sampling has proved useful (e.g., Rahman and Schmeisser 1985, 1994; Zayed 1993 and Butzer and Schöttler 1993, respectively).

After the values of a function itself, derivative values are a very natural source of information and so provide a natural extension of ordinary sampling. Perhaps even more importantly, derivative sampling has been, and continues to be, an excellent illustration of multichannel methods in general. It is hoped that this contention will be borne out as the present chapter develops.

This chapter is not intended to be a complete catalogue of all known results in the area. On the other hand, attempts have been made to find original sources for those topics that are discussed; this can be a hazardous task, however, because many derivative sampling formulae have been discovered independently by different people at different times.

The multidimensional Paley–Wiener space PW_B is, as usual, denoted by $\{f: f \in L^2(\mathbb{R}^N) \cap C(\mathbb{R}^N); \operatorname{supp} f^\wedge = B \subset \mathbb{R}^N\}$, and it should be borne in mind that norm convergence in this space implies uniform convergence over $\mathbb{R}$ (see, e.g., Higgins 1996, §6.6).

3.2 A brief survey of derivative sampling

It seems to have been Fogel (1955) who first took up Shannon's suggestion (1949, p. 12) that a function might be reconstructed from its samples together with samples of its derivatives. His discussion centres mainly on establishing the appropriate sample point spacing.

As we have noted, the "first derivative sampling series" (3.1.1) is due to Jagerman and Fogel (1956). Several results in this work are interesting, especially because of their early date. For example (*op. cit.* p. 135), if $\alpha = |\alpha|e^{i\theta}$ is a non-zero complex number, and f is an entire function in the complex $t = u + iv$ plane such that there is a constant K for which

$$\max_{u \in \mathbb{R}} \left| f(te^{i\theta}) \right| \le K \frac{e^{2\pi |\alpha v|}}{|v|}, \qquad |v| \to \infty, \tag{3.2.1}$$

then f is represented by the first derivative sampling series

$$f(z) = \left[\frac{\sin \alpha \pi z}{\alpha \pi} \right]^2 \sum_{n \in \mathbb{Z}} \left\{ \frac{f(n/\alpha)}{(t - n/\alpha)^2} + \frac{f'(n/\alpha)}{(t - n/\alpha)} \right\}, \tag{3.2.2}$$

with uniform convergence on compact subsets of $\mathbb{C}$. Thus, f can be sampled along a line (through the origin) in the complex plane.

It should be noted that when writing derivative sampling formulae, factors such as the 2 in (3.1.1), and more generally R in (3.2.3) below, can be written into the formulae where their effect is to increase the sample point spacing; alternatively they can be written into the hypotheses (as in the result of Jagerman and Fogel above) where they affect the growth properties of the object function. Both alternatives can be found in the literature.

As a special case of this result of Jagerman and Fogel, we have the following: Let α now be real and positive, and let μ denote the Borel measure on $(-\pi w, \pi w)$ generated by a function g of bounded variation on $(-\pi w, \pi w)$, with total variation V. Let f be band-limited in the sense that

$$f(t) = \int_{-\pi w}^{\pi w} e^{ixt} \, d\mu(x).$$

Then

$$|f(t)| \le V e^{\pi w |v|},$$

and

$$\max_{u \in \mathbb{R}} |f(t)| \leq K e^{\pi w |v|} \leq k \frac{e^{2\pi \alpha |v|}}{|v|}$$

if $2\alpha > w$ and $|v|$ is large. In this case the hypotheses of the previous result hold, and f is represented by (3.2.2). However, this does not quite capture the classical case where $2\alpha = w$ and $f \in PW_{\pi w}$ (this is, for example, a special case of Theorem 3.1).

The formula for reconstruction from samples of f and its first $R - 1$ derivatives takes the general form

$$f(t) = \sum_{n \in \mathbb{Z}} f\left(\frac{Rn}{w}\right) S_{0,n}(t) + \cdots + f^{(R-1)}\left(\frac{Rn}{w}\right) S_{R-1,n}(t)$$

and was first given explictly by Linden (1959) and Linden and Abramson (1960). The explicit form is

$$f(t) = \left[\frac{\sin(\pi wt/R)}{\pi w/R}\right]^R \sum_{n \in \mathbb{Z}} \sum_{k=0}^{R-1} (D_k f) \left(\frac{Rn}{w}\right) \frac{(t - Rn/w)^{k-R}}{k!} \qquad (3.2.3)$$

where

$$(D_k f)(t) := \sum_{j=0}^{k} \binom{k}{j} \left(\frac{\pi w}{R}\right)^{k-j} \left[\frac{d^{k-j}}{dt^{k-j}} \left(\frac{t}{\sin t}\right)^R\right]_{t=0} f^{(j)}(t).$$

The first three cases are as follows:

$$\left[\frac{\sin(\pi wt)}{\pi w}\right] \sum_{n \in \mathbb{Z}} (-1)^n f(n/w) \frac{1}{(t - n/w)}, \qquad (3.2.4)$$

$$\left[\frac{\sin(\pi wt/2)}{\pi w/2}\right]^2 \sum_{n \in \mathbb{Z}} \left\{ f(2n/w) \frac{1}{(t - 2n/w)^2} + f'(2n/w) \frac{1}{(t - 2n/w)} \right\}, \qquad (3.2.5)$$

$$\left[\frac{\sin(\pi wt/3)}{\pi w/3}\right]^3 \sum_{n \in \mathbb{Z}} (-1)^n \left\{ f(3n/w) \frac{1}{(t - 3n/w)^3} + f(3n/w) \frac{(\pi w/3)^2}{2(t - 3n/w)} \right.$$

$$\left. + f'(3n/w) \frac{1}{(t - 3n/w)^2} + f''(3n/w) \frac{1}{2(t - 3n/w)} \right\}. \qquad (3.2.6)$$

The first case, (3.2.4), is the ordinary cardinal series for f; the second, (3.2.5), is just (3.2.2) again for real α.

Next, three theorems are taken from the recent literature, and address important general aspects of sampling theory. Theorems 3.1 and 3.2 are very general results on representation in derivative sampling series, and its stability, at uniformly spaced points (inequalities of the type (3.2.8) and (3.2.9) (below) are usually called *stability* conditions; they assert that the size of the output f from a reconstruction from samples is bounded by the size of the input, from above and from below). These results have been selected from Grozev and Rahman (1994), who introduce the "stability" result (3.2.8) as a generalization to entire functions of earlier work on trigonometric polynomials by Marcinkiewicz.

Theorem 3.4 is a result of Gröchenig and Razafinjatovo (1996), and says that if derivative sampling is stable (in the sense that (3.2.9) holds), the sampling point density cannot be very small. This parallels a classical result of Landau for ordinary sampling (a discussion of this work, with many references to the original literature, can be found in Higgins 1996, particularly Chapter 17).

Theorem 3.1 (Grozev and Rahman) *Let $p > 1$, $f \in E_{\pi w}$ and $f(t) \to 0$ as $|t| \to \infty$. Furthermore, let*

$$\sum_{n \in \mathbb{Z}} \left| f^{(v)} \left(\frac{Rn}{w} \right) \right|^s < \infty, \qquad v = 0, \ldots, R - 1, \tag{3.2.7}$$

for some $s > 1$. Then (3.2.3) holds pointwise in $\mathbb{C}$. In particular, (3.2.3) holds if $f \in B_{\pi w}^p$, $p > 0$.

Theorem 3.2 (Grozev and Rahman) *If $f \in B_{\pi w}^p$, $p > 1$, and f is not null, then there exist constants $A_{R,p,w}$ and $B_{R,p,w}$, depending only on R, p and w, such that*

$$A_{R,p,w} \|f\|_p \leq \sum_{n \in \mathbb{Z}} \sum_{v=0}^{R-1} \left\{ \frac{R^v}{v! \, (\pi w)^v} \right\}^p \left| f^{(v)} \left(\frac{Rn}{w} \right) \right|^p \leq B_{R,p,w} \|f\|_p. \tag{3.2.8}$$

Preliminary to the next theorem some definitions will be needed. First, let $\Lambda := \{\lambda_n\}$, $n \in \mathbb{Z}$, where $\{\lambda_n\} \subset \mathbb{R}$. Now we introduce

Definition 3.3 *Let $I(x, r)$ be the interval centred at $x \in \mathbb{R}$ and of length $2r$. Let $n^+ := \max_{x \in \mathbb{R}} \sharp\{\Lambda \cap I(x, r)\}$. Then the upper density D^+ is defined by*

$$D^+(\Lambda) := \limsup_{r \to \infty} \frac{n^+}{2r}.$$

Similarly, n^- is defined as for n^+ but with "max" replaced with "min". Then the lower density is $D^-(\Lambda) := \liminf_{r \to \infty} (n^-/2r)$. The ordinary or uniform density is $D(\Lambda) := \lim_{r \to \infty} \{\Lambda \cap I(0, r)\}/2r$.

All three densities exist when Λ is *uniformly discrete*, that is, there is a positive constant $d > 0$ such that $|\lambda_n - \lambda_m| \geq d$, $n \neq m$.

Theorem 3.4 (Gröchenig and Razafinjatovo) *Let $\Omega \subset \mathbb{R}$, $m(\partial\Omega) = 0$, and let Λ be uniformly discrete. Let $k_0, k_1, \ldots, k_{R-1}$ be non-negative integers. Let there exist positive constants A and B such that*

$$A \|f\|^2 \leq \sum_{n \in \mathbb{Z}} \sum_{v=0}^{R-1} |f^{(k_v)}(\lambda_n)|^2 \leq B \|f\|^2 \tag{3.2.9}$$

for every $f \in PW_\Omega$. Then

$$D^-(\Lambda) \geq \frac{m(\Omega)}{2\pi R}.$$

With a certain amount of specialization these three theorems can be amalgamated, using features common to all. For example one can take $p = 2$, $\Omega = [-\pi w, \pi w]$, $k_v = v, v = 0, \ldots, R - 1$ and $\lambda_n = Rn/w$. Under these circumstances condition (3.2.8) implies condition (3.2.9); indeed, it is clear that we can find a pair of positive constants m, M, such that

$$0 < m \le \left\{ \frac{R^v}{v!(\pi w)^v} \right\}^2 \le M, \qquad v = 0, \ldots, R - 1.$$

Then the specialized version of (3.2.9) follows. Now Theorem 3.1 gives the sampling expansion (3.2.3) for each $f \in PW_\pi$; the condition (3.2.9) holds, and the sampling density is w/R which is minimal for stable sampling, by Theorem 3.4.

Some other aspects of derivative sampling that we will not have time and space to discuss are as follows; derivative sampling at irregular points (Hinsen 1993, p. 361; also Razafinjatovo (1994)); band-pass derivative sampling (Kempski 1995, Chapter 3); multiband derivative sampling (Beaty and Dodson 1989); aliasing errors for single and multiband derivative sampling (Beaty and Higgins 1994); sampling and approximation of Hermite type (Butzer and Splettstößer 1977, Chapter 5). Derivative sampling in a stochastic setting is discussed by Pogány in Chapter 9 of the present volume.

3.3 The Riesz basis method in its multichannel setting

Of the many kinds of proof which have been used in the derivation of sampling series, we prefer here the Riesz basis method with its connections to stability and minimal sampling rates. Here, the aim is to establish the multichannel sampling series (3.3.3). A brief outline follows; more details can be found in Higgins (1996, Chapters 12 and 13). In the derivative context the method goes back to Rawn (1989).

Let $S_1, \ldots, S_Q$ be operators acting on $L^2(\mathbb{R})$, given by $S_i := \mathcal{F}^{-1} \mathcal{M}_i \mathcal{F}$, where $\mathcal{F}$ denotes the Fourier transform and $\mathcal{M}_i$ denotes multiplication by $\overline{m_i(x)} \chi_{[-\pi,\pi]}$.

Let $[-\pi, \pi]$ be partitioned into subsets $\{B_i\}$, where $B_1 := [-\pi, -\pi + 2\pi/Q)$, $B_2 := [-\pi + 2\pi/Q, -\pi + 4\pi/Q), \ldots, B_Q := [-\pi + (Q-1)2\pi/Q, \pi]$. Let $e_n(x) := e^{-inQx}/\sqrt{2\pi/Q}, n \in \mathbb{Z}$.

Let us put $\varphi_{ni}(x) := m_i(x) e_n(x)$. If it can be proved that $\{\varphi_{ni}(x)\}, n \in \mathbb{Z}, i = 1, \ldots, Q$, is a Riesz basis for $L^2(-\pi, \pi)$, then for $f \in PW_\pi$,

$$f^\wedge(x) = \sum_{n \in \mathbb{Z}} \sum_{i=1}^{Q} a_{ni} \varphi_{n,i}^\star(x) \tag{3.3.1}$$

in the norm of $L^2(-\pi, \pi)$, where $\{\varphi_{n,i}^\star\}$ is the basis dual to $\{\varphi_{n,i}\}$, and

$$\begin{aligned} a_{ni} &= \langle f, \varphi_{n,i} \rangle = \sqrt{\frac{Q}{2\pi}} \int_{-\pi}^{\pi} f^\wedge(x) \overline{m_i(x)} e^{inQx} \, dx \\ &= \sqrt{Q} (\mathcal{F}^{-1} \mathcal{M}_i \mathcal{F} f)(nQ) = \sqrt{Q} \, (S_i f)(nQ). \end{aligned} \tag{3.3.2}$$

Then the Fourier dual of (3.3.1) is

$$f(t) = \sqrt{Q} \sum_{n \in \mathbb{Z}} \sum_{i=1}^{Q} (\mathcal{S}_i f)(nQ) S_{ni}(t), \qquad (3.3.3)$$

where the *reconstruction functions* $\{S_{ni}\}$ remain to be calculated (in the second part of the method). Convergence is in the norm of $L^2(\mathbb{R})$ and also uniform on $\mathbb{R}$. This completes the first part.

The method now continues into its second part, the calculation of $\{S_{ni}\}$. First, a standard way of establishing the Riesz basis property of sets such as $\{\varphi_{ni}\}$ is to introduce

$$T : L^2(B_1) \oplus \cdots \oplus L^2(B_Q) \mapsto L^2(-\pi, \pi),$$

given by

$$T(g_1, \ldots, g_Q) \mapsto F(x) := \sum_{j=1}^{Q} m_j(x)(g_j(x))_p,$$

where the subscript p denotes periodic extension with period $2\pi/Q$. By restricting x to B_1 and using the periodicities, this equation can be written in the form

$$\begin{pmatrix} F|_{B_1} \\ \vdots \\ F|_{B_Q} \end{pmatrix} = M \begin{pmatrix} g_1(x) \\ \vdots \\ g_Q(x) \end{pmatrix},$$

where $M = (m_{ij})$ with $m_{ij} := m_j(x + (i-1)2\pi/Q)$, $i, j = 1, \ldots, Q$.

Next, we note that an orthonormal basis for $L^2(B_1) \oplus \cdots \oplus L^2(B_Q)$ is given by $\{e_{ni}\} := \{(\theta, \ldots, e_n, \ldots, \theta)\}$, $n \in \mathbb{Z}$, $i = 1, \ldots, Q$ (here e_n appears in the ith position). Now $\{\varphi_{ni}\}$ is the image by T of the orthonormal basis e_{ni}. Then $\{\varphi_{ni}\}$ will be a Riesz basis for $L^2(-\pi, \pi)$ if T is bounded, linear, one-to-one and onto $L^2(-\pi, \pi)$. When this is the case, the dual basis $\{\varphi_{ni}^\star\}$ is the image of e_{ni} by $(T^\star)^{-1}$ (where $T^\star$ denotes the adjoint of T). Of course, $(T^\star)^{-1}$ is effected by $(\overline{M}^T)^{-1}$.

Now $S_{ni}(t)$ can be calculated from $(\varphi_{ni}^\star)^\vee(t)$.

Sampling is stable, in the sense that there is a pair of positive constants a and b such that

$$a \leq \|f\|^2 \leq \sum_{n \in \mathbb{Z}} \sum_{i=1}^{Q} |(\mathcal{S}_i f)(nQ)|^2 \leq b\|f\|^2.$$

This comes from (3.3.2) and appropriate properties of Riesz bases.

3.3.1 *One-dimensional derivative sampling*

The problem of derivative sampling is an excellent illustration of the complete multichannel method. Here we shall illustrate with first derivative sampling.

There are two channels so $Q = 2$, and for obvious reasons we take $m_1(x) \equiv 1$ and $m_2(x) = ix$. Also we take $B_1 := [-\pi, 0)$ and $B_2 := [0, \pi]$. Continuing with the method we consider

$$\mathcal{T} : (g_1, g_2) \mapsto F(x) = (g_1)_p(x) + ix(g_2)_p(x)$$

where the subscript p denotes periodic extension with period π. Taking $x \in B_1$, further calculation leads to

$$M = \begin{bmatrix} 1 & ix \\ 1 & i(x + \pi) \end{bmatrix} \tag{3.3.4}$$

and

$$(\overline{M}^T)^{-1} = \frac{1}{\pi} \begin{bmatrix} (x + \pi) & -i \\ -x & i \end{bmatrix}.$$

Because $\det(M) = i\pi \neq 0$, $\mathcal{T}$ is one-to-one and onto. It is also bounded and linear. Hence, since $\{e_n(x)\} := \{(\pi)^{-1/2} e^{-2inx}\}$ is an orthonormal basis for $L^2(-\pi, 0)$ and for $L^2(0, \pi)$ we find that

$$\{\varphi(x)\} \cup \{\psi(x)\} := \{e_n(x)\} \cup \{ixe_n(x)\}, \qquad n \in \mathbb{Z}, \tag{3.3.5}$$

is a Riesz basis for $L^2(-\pi, \pi)$.

Thus we find that the biorthonormal set for (3.3.5) is of the form $\{\varphi_n^\star(x)\} \cup \{\psi_n^\star(x)\}$ where

$$\begin{aligned} \varphi_n^\star(x) &= \left(1 - \frac{|x|}{\pi}\right) e_n(x) \\ \psi_n^\star(x) &= \frac{i}{\pi} \operatorname{sgn} x \, e_n(x), \end{aligned} \qquad x \in [-\pi, \pi]. \tag{3.3.6}$$

Now using special Fourier transforms (e.g., Higgins 1996, Appendix 1, nos. 9 and 11),

$$\varphi_n^{\star\vee} := \left(\mathcal{F}^{-1} \varphi_n^\star\right)(t) = \frac{1}{\sqrt{2}} \operatorname{sinc}^2 \frac{1}{2}(t - 2n)$$

$$\psi_n^{\star\vee} := \left(\mathcal{F}^{-1} \psi_n^\star\right)(t) = -\frac{\sqrt{2}}{\pi} \operatorname{sinc} \frac{1}{2}(t - 2n) \sin \frac{\pi}{2}(t - 2n),$$

and with $\langle f^\wedge, \varphi_n \rangle_{[-\pi, \pi]} = \sqrt{2} f(2n)$ and $\langle f^\wedge, \psi_n \rangle_{[-\pi, \pi]} = -\sqrt{2} f'(2n)$, we obtain (3.1.1) for $f \in PW_\pi$.

Likewise the stability condition (3.2.9) holds with $R = 2$ and $k_\nu = \nu$; further, from the amalgamated result following Theorem 3.4 the sampling density is minimal for stable sampling.

3.3.2 Asymmetrically distributed sampling density

It is not necessary to distribute the sampling density equally between the two channels. In fact the sampling series (3.1.1) can be modified so that the sampling density is weighted in favour of one channel at the expense of the other. Derivative sampling provides an example of this; we show that for $f \in PW_\pi$ there is a sampling series representation

$$f(t) = \sum_{n \in \mathbb{Z}} f(3n/2)S_{1,n}(t) + f'(3n)S_{2,n}(t), \qquad (3.3.7)$$

in which two thirds of the samples are assigned to the function and only one third to its derivative.

. The previous method needs some modification. We take $B_1 := [-\pi, \pi/3)$ and $B_2 := [\pi/3, \pi]$; we also need the fact that $\{(4\pi/3)^{-1/2}e^{i3nx/2}\}$ is an orthonormal basis for $L^2(B_1)$ and $\{(2\pi/3)^{-1/2}e^{i3nx}\}$ is an orthonormal basis for $L^2(B_2)$.

In analogy with the ordinary case we introduce

$$T : L^2(B_1) \oplus L^2(B_2) \mapsto L^2(-\pi, \pi)$$

such that

$$T((f, g)) = F(x) := f_p(x) + ixg_q(x),$$

where the subscript p means periodic extension with period $4\pi/3$, and q means periodic extension with period $2\pi/3$.

Let f_1 and f_2 denote f restricted to $[-\pi, -\pi/3)$ and to $[-\pi/3, \pi/3)$ respectively. Then for $x \in [-\pi, -\pi/3)$, the set of equations

$$F_1(x) = f_1(x) + ixg(x)$$

$$F_2(x + 2\pi/3) = f_2(x + 2\pi/3) + i(x + 2\pi/3)g(x)$$

$$F_3(x + 4\pi/3) = f_1(x) + i(x + 4\pi/3)g(x)$$

has coefficient matrix

$$M = \begin{bmatrix} 1 & 0 & ix \\ 0 & 1 & i(x + 2\pi/3) \\ 1 & 0 & i(x + 4\pi/3) \end{bmatrix},$$

so that $\det M = 4\pi i/3 \neq 0$.

The appropriate properties of T follow as before, and we now find that

$$\{(4\pi/3)^{-1/2}e^{i3nx/2}\} \cup \{(2\pi/3)^{-1/2}ixe^{i3nx}\}, \qquad n \in \mathbb{Z}, \qquad (3.3.8)$$

forms a Riesz basis for $L^2(-\pi, \pi)$. Using

$$(\overline{M}^T)^{-1} = \frac{3}{4\pi} \begin{bmatrix} x + 4\pi/3 & x + 2\pi/3 & -i \\ 0 & 4\pi/3 & 0 \\ -x & -(x + 2\pi/3) & i \end{bmatrix},$$

we find that the basis dual to that in (3.3.8) is of the form

$$\{(4\pi/3)^{-1/2}\mu_{1,n}(x)e^{i3nx/2}\} \cup \{(2\pi/3)^{-1/2}\mu_2(x)e^{i3nx}\}, \qquad n \in \mathbb{Z},$$

where the multipliers are given by

$$
\mu_{1,n}(x) = \begin{cases} 3(x + \pi)/2\pi, & x \in [-\pi, -\pi/3); \\ 1, & x \in [-\pi/3, \pi/3); \\ -3(x - \pi)/2\pi, & x \in [\pi/3, \pi] \end{cases}
$$

when n is even, and by

$$
\mu_{1,n}(x) = \begin{cases} 1/2, & x \in [-\pi, -\pi/3); \\ 1, & x \in [-\pi/3, \pi/3); \\ 1/2, & x \in [\pi/3, \pi] \end{cases}
$$

when n is odd; and

$$
\mu_2(x) = \begin{cases} -3i/4\pi, & x \in [-\pi, -\pi/3); \\ 0, & x \in [-\pi/3, \pi/3); \\ 3i/4\pi, & x \in [\pi/3, \pi]. \end{cases}
$$

Using special Fourier transforms it can be calculated that, in (3.3.7), we have

$$
S_{1,n}(t) = \begin{cases} \operatorname{sinc}\dfrac{2}{3}\left(t - \dfrac{3n}{2}\right) \operatorname{sinc}\dfrac{1}{3}\left(t - \dfrac{3n}{2}\right); & n \text{ even,} \\[2ex] \dfrac{3}{4}\left\{\dfrac{1}{3}\operatorname{sinc}\dfrac{1}{3}\left(t - \dfrac{3n}{2}\right) + \operatorname{sinc}\left(t - \dfrac{3n}{2}\right)\right\}; & n \text{ odd,} \end{cases}
$$

$$
S_{2,n}(t) = \dfrac{3}{2\pi}\operatorname{sinc}\dfrac{1}{3}(t - 3n)\, \sin\dfrac{2\pi}{3}(t - 3n).
$$

3.3.3   A different proof of the first derivative sampling series

A proof that is slightly different to the usual ones will now be given, largely because the method is useful in treating the gradient sampling formula in §3.5.1. We start with the observation that of the two multipliers in the basis (3.3.6), the second is, apart from a factor $-i$, the derivative of the first.[1] This can be traced back to the fact that the same is true of the dual basis (3.3.5), where it arises from simple properties of the Fourier transform.

Now it is noticeable that the first derivative sampling series (3.1.1) has a counterpart in the frequency domain. If T denotes the "tent" or "triangle" function $T(x) := (2/\sqrt{2\pi})(1 - |x|/\pi)\chi_{[-\pi,\pi]}$, it is

$$
f^{\wedge}(x) = \sum_{n \in \mathbb{Z}} e^{-i2nx}\left\{f(2n) + f'(2n)\,i\,\dfrac{d}{dx}\right\} T(x), \tag{3.3.9}
$$

convergence being in the norm of $L^2(-\pi, \pi)$.

---

[1]The present writer is grateful to Maurice Dodson for pointing this out; it is really the starting point for the present method.

We now take (3.3.9) as the foundation for a method of obtaining the first derivative sampling formula (3.1.1). We must show that (3.3.9) converges in the norm of $L^2(-\pi, \pi)$ for every $f \in PW_\pi$, for a universal function T that is not known *a priori* but must be determined by the method.

We show first that (3.3.9) holds for every member of the set

$$\{e^{-ikx}/\sqrt{2\pi}\}, \qquad k \in \mathbb{Z},$$

an orthonormal basis for $L^2(-\pi, \pi)$, if T is taken to be the tent function. First we consider the case $f^\wedge(x) = e^{-ikx}/\sqrt{2\pi}$ where k is odd, so that, using standard Fourier transforms, $f(2n) = \mathrm{sinc}(2n - k) = 0$ and $f'(2n) = -(2n - k)^{-1}$. Then (3.3.9) becomes

$$\frac{e^{-ikx}}{\sqrt{2\pi}} = -i \sum_{n\in\mathbb{Z}} e^{-i2nx} \frac{1}{2n - k} T'(x) = -i e^{-ikx} \sum_{n\in\mathbb{Z}} \frac{e^{-i(2n-k)x}}{2n - k} T'(x).$$

This equation can be rewritten in the form

$$1 = -i\sqrt{2\pi}T'(x) \sum_{l \text{ odd}} \frac{e^{-ilx}}{l}. \tag{3.3.10}$$

Now the series in (3.3.10) is the Fourier series for $-(i\pi/2)\mathrm{sgn}\, x$, $x \in (-\pi, \pi)$, and it converges in the norm of L^2. Hence (3.3.10), and consequently (3.3.9) with our current choice of $f^\wedge$, converges in the norm of L^2 if $T'(x)$ is taken to be $-(2/\pi\sqrt{2\pi})\mathrm{sgn}\, x$; that is, if $T(x)$ is taken to be the tent function with the possibility of an additive constant.

The case where k is even is very similar. For example if $k = 0$ we have $f^\wedge = 1/\sqrt{2\pi}$, and $f(t) = \mathrm{sinc}\, t$ so that $f(2n) = 1$ if $n = 0$ and $f(2n) = 0$ if $n \neq 0$; and $f'(2n) = 0$ if $n = 0$ and $f'(2n) = 1/2n$ if $n \neq 0$. Consequently, (3.3.9) becomes

$$\frac{1}{\sqrt{2\pi}} = T(x) + iT'(x) \sum_{n\in\mathbb{Z}}' \frac{e^{-i2nx}}{2n}. \tag{3.3.11}$$

This time the series is the Fourier series for $-(i\pi/2)(1 - 2x/\pi)$ for $x \in (0, \pi)$, and by periodic extension for $(i\pi/2)(1 + 2x/\pi)$ for $x \in (-\pi, 0)$. It is now easily verified that (3.3.11) is satisfied in norm if $T(x)$ is taken to be the tent function (suggested by the previous case, but now with no additive constant).

Other even values of k are treated similarly. The proof is completed by appeal to the usual density argument. First, the validity of (3.3.9) is extended to all finite linear combinations of the orthonormal basis elements $\{e^{-ikx}/\sqrt{2\pi}\}$, $k \in \mathbb{Z}$, and thence by standard methods to all of $L^2(-\pi, \pi)$ because these finite linear combinations are dense there.

By Fourier duality the ordinary derivative sampling series is now established, with convergence in the norm of PW_π. Pointwise and uniform convergence follow from standard principles (e.g., Higgins 1996, p. 57).

3.3.4 *Reconstruction from samples of a function and its third derivative*

It was suggested by Fogel (1955) that if a function f is reconstructed using samples of $f^{(k)}$, not all derivatives of order less than k need also be used. Fogel asserted that samples of f and of f''' taken at the points $\{2n\}$, $n \in \mathbb{Z}$, suffice for reconstruction of f, but gave no reconstruction method. Here we find that this example is easily incorporated into the Riesz basis method. Indeed, the analogue of (3.3.4) is easily seen to have determinant

$$\begin{vmatrix} 1 & -ix^3 \\ 1 & -i(x+\pi)^3 \end{vmatrix} = -i\pi(3x^2 + 3\pi x + \pi^2),$$

and this does not vanish for $x \in [-\pi, \pi)$, (in fact it does not vanish for any real x). Hence the Riesz basis method applies.

It is curious that Fogel chose f and f''', for which the Riesz basis method applies, rather than f and f'' for example, for which it does not. That it does not can be seen from

$$\begin{vmatrix} 1 & -x^2 \\ 1 & -(x+\pi)^2 \end{vmatrix}, \tag{3.3.12}$$

which vanishes when $x = -\pi/2$.

In the next section we find that another suggestion of Fogel can rectify this problem.

3.3.5 *Shifted points for sampling the derivative*

Fogel (1955, p. 48) mentions the possibility of shifting the sampling points associated with at least one of the channels. Such a shift is easily incorporated into the Riesz basis method by introducing an exponential multiplier in the frequency domain. For example, to rectify the problem with (3.3.12) we can choose

$$\begin{vmatrix} e^{iax} & -x^2 \\ e^{ia(x+\pi)} & -(x+\pi)^2 \end{vmatrix} = e^{iax}\{-(x+\pi)^2 + x^2 e^{ia\pi}\}$$

where $0 < a < 2$, which does not vanish for any real x, so the method applies and will lead to a reconstruction involving the samples $\{f(2n+a)\}$ and $\{f''(2n)\}$.

It should be noted that we cannot choose $a = 2$ since then the determinant can vanish when $x \in [-\pi, \pi)$.

3.4 Applications to constructive function theory

In order to construct certain functions (called $B(z)$ and $C(z)$ below) that Beurling and Selberg found useful for special approximation purposes, we are going to need the sampling formula of Tschakaloff ((3.4.3) below; this series is derived in, e.g., Higgins 1996, p. 60), and a derivative version of it (3.4.5 below). Indeed, if $f \in B_\sigma^\infty$, then

$$g(t) = \begin{cases} \dfrac{f(t) - f(0)}{t}, & t \neq 0 \\ f'(0), & t = 0; \end{cases} \tag{3.4.1}$$

belongs to PW_σ.

By applying the ordinary sampling series

$$g(t) = \sum_{n \in \mathbb{Z}} g\left(\frac{n}{w}\right) \operatorname{sinc}(wt - n) \tag{3.4.2}$$

to this function g, we obtain the sampling series

$$f(t) = \frac{\sin \pi wt}{\pi w} \left\{ f'(0) + \frac{f(0)}{t} + t \sum_{n \in \mathbb{Z}}' f(n/w) \frac{(-1)^n}{(n/w)(t - n/w)} \right\}, \qquad (3.4.3)$$

valid for $f \in B_\sigma^\infty$. An alternative form of this series is

$$f(t) = \frac{\sin \pi wt}{\pi w} \left\{ \frac{f(0)}{t} + B_f + \sum_{n \in \mathbb{Z}}' f(n/w) \frac{(-1)^n}{t - n/w} \right\}, \qquad (3.4.4)$$

where the constant B_f is given by

$$B_f := f'(0) + \sum_{n \in \mathbb{Z}}' f(n/w) \frac{(-1)^n}{n/w}.$$

To obtain a derivative counterpart to this series we can proceed in much the same way, but this time we apply the derivative sampling series (3.2.5) to g as defined in (3.4.1), with the additional condition $g'(0) = \frac{1}{2} f''(0)$. The special series

$$\sum_{n \in \mathbb{Z}}' \frac{1}{n^2} = \frac{\pi^2}{3} \quad \text{and} \quad \sum_{n \in \mathbb{Z}}' \frac{1}{(n - wt/2)} = \frac{\pi^2}{\sin^2(\pi wt/2)} - \frac{4}{w^2 t^2}$$

are needed (the second sum being a principal value of course), and then some rearrangement gives

$$f(t) = \left[\frac{\sin(\pi wt/2)}{\pi w/2} \right]^2 \left[f(0) \left(\frac{\pi^2 w^2}{12} + \frac{1}{t^2} \right) + \frac{f'(0)}{t} + \frac{f''(0)}{2} \right.$$

$$- t \sum_{n \in \mathbb{Z}}' \left\{ f(2n/w) \frac{t - 4n/w}{(2n/w)^2 (t - 2n/w)^2} \right.$$

$$\left. \left. - f'(2n/w) \frac{1}{(2n/w)(t - 2n/w)} \right\} \right]. \qquad (3.4.5)$$

An alternative formulation is:

$$f(t) = \left[\frac{\sin(\pi wt/2)}{\pi w/2} \right]^2 \left[A_f + \frac{f'(0)}{t} + \sum_{n \in \mathbb{Z}} \frac{f(2n/w)}{(t - 2n/w)^2} \right.$$

$$\left. + t \sum_{n \in \mathbb{Z}}' \frac{f'(2n/w)}{(2n/w)(t - 2n/w)} \right], \qquad (3.4.6)$$

where the constant A_f is given by

$$A_f := \frac{\pi^2 w^2 f(0)}{12} + \frac{f''(0)}{2} + \sum_{n \in \mathbb{Z}}' \frac{f(2n/w)}{(2n/w)^2}. \qquad (3.4.7)$$

3.4.1   The "amazing functions" of Beurling and Selberg

In about 1938 Beurling found the need to construct a function that would be of exponential type and would majorize sgn x for real x. Furthermore, it would have to be extremal for this problem in a special sense (details are to be found below). Beurling clearly knew at that time that formulae such as (3.2.2) could be used to construct special entire functions, but after using it to construct the function $B(z)$ (below) in connection with an inequality for almost periodic trigonometric polynomials (Vaaler 1985, (6.2)), which was but one step in a larger project, the result was found to be redundant and was never published.

In 1974 Selberg became interested in these matters in connection with his proof of a sharp form of the "large sieve inequality" of number theory. He rediscovered the formula (3.2.2) and used it to construct a function $C(z)$, closely related to $B(z)$, which made the proof very simple; because of this the functions $B(z)$ and $C(z)$ became known as "amazing functions". Again, Selberg's proof was never published.

Many other applications of $B(z)$ and $C(z)$ are to be found in the excellent survey article of Vaaler (1985). These striking applications of derivative sampling and interpolation are apparently still not as well known among sampling theorists as they might be, and certainly deserve further exposure.

In order to construct Beurling's function $B(z)$ and Selberg's function $C(z)$ we shall need an auxiliary function G which must belong to E^{π} and must be a good approximation to sgn t along the real axis $\mathbb{R}$ (we take sgn $t = |t|/t$ for $t \neq 0$, and sgn $0 = 0$). Indeed, if we take G to interpolate to sgn t at the integers, then an explicit construction for G is suggested by (3.4.3), except that the value of $G'(0)$ must be decided. It would seem reasonable to remove the oscillating term $(B_f/\pi w) \sin \pi wt$ from (3.4.4), and this can be done by choosing $G'(0)$ to eliminate the constant B_f. This means that, with $w = 1$,

$$0 = B_f = G'(0) - 2\sum_{n=1}^{\infty} \frac{(-1)^n}{n} = G'(0) - 2\ln 2,$$

and our choice is $G'(0) = \ln 4$. Then from (3.4.4) we obtain

$$G(t) := \sideset{}{'}\sum_{n\in\mathbb{Z}} \text{sgn}(n)\text{sinc}(t - n). \tag{3.4.8}$$

These choices do indeed give a suitable function G, in that the following lemma holds:

Lemma 3.5 *Let G be the function defined by (3.4.8). Then*

- $G \in E^{\pi}$;
- $G(t) - \text{sgn}\, t \in L^1(\mathbb{R})$;
- $G'(t) \in L^1(\mathbb{R})$.

Proofs of these properties are rather too lengthy for inclusion here; they can be found in Vaaler (1985, §2).

Now we are ready to investigate the functions B and C mentioned above. Let us take B first.

In the late 1930s Beurling considered the following:

Problem (Beurling) Find a function f such that

(a) $f \in E^{2\pi}$;

(b) $f(x) \geq \operatorname{sgn} x, \ x \in \mathbb{R}$;

(c) $\int_{\mathbb{R}} \{ f(x) - \operatorname{sgn} x \} \, dx \geq 1$.

This problem asks for a smooth function that majorizes $\operatorname{sgn} x$ and is a good approximation when (c) is near to equality. Beurling showed that

$$
B(z) = \left[\frac{\sin \pi z}{\pi} \right]^2 \left\{ \sum_{n=0}^{\infty} (z-n)^{-2} - \sum_{m=-\infty}^{-1} (z-m)^{-2} + 2z^{-1} \right\} \tag{3.4.9}
$$

is the unique extremal for this problem, in the sense that if we take $f = B$, then (a) and (b) are satisfied, and if f is any function satisfying (a) and (b), then (c) holds, and there is equality if and only if $f = B$.

Let us see where the solution of this problem might come from, and then prove that it does indeed turn out to be the function B of (3.4.9) discovered by Beurling. To construct a solution f it is natural to turn to (3.4.6), since we shall need to assign function values obtained from $\operatorname{sgn} x$ and set derivative values to zero, except those at the origin which must be determined. We take $w = 2$ since the first factor in (3.4.6) then gives the correct exponential type, 2π. As in the construction of G, it is again reasonable to choose the constant A_f to be zero: with our current choices this means taking $f''(0) = 0$, from (3.4.7). Now (3.4.6) becomes

$$
f(t) = \left[\frac{\sin \pi t}{\pi} \right]^2 \left\{ \sum_{n=0}^{\infty} (t-n)^{-2} - \sum_{m=-\infty}^{-1} (t-m)^{-2} + f'(0)t^{-1} \right\}. \tag{3.4.10}
$$

In the following analysis we shall find that the choice $f'(0) = 2$ is forced on us, and so f in (3.4.10) does indeed become the B of (3.4.9).

To show that f of (3.4.10) satisfies condition (a) of Beurling's problem, we note that it follows from simple convergence criteria that f is entire in the complex t-plane, and is evidently of exponential type 2π.

For condition (b) we must show that $f(t) \geq 1$ for $t \in \mathbb{R}, t > 0$, and $f(t) \geq -1$ for $t < 0$. Now when $t > 0$ and t is not an integer we can generate a factor 1 in (3.4.10) by adding the left side and subtracting the right side of the identity

$$
\left[\frac{\pi}{\sin \pi t} \right]^2 = \sum_{n \in \mathbb{Z}} \frac{1}{(t-n)^2}. \tag{3.4.11}
$$

This gives

$$
f(t) = 1 + \left[\frac{\sin \pi t}{\pi} \right]^2 \left\{ \frac{f'(0)}{t} - 2 \sum_{n=1}^{\infty} \frac{1}{(t+n)^2} \right\}. \tag{3.4.12}
$$

But

$$\sum_{n=1}^{\infty} \frac{1}{(t+n)^2} < \sum_{n=0}^{\infty} \frac{1}{(t+n)} \frac{1}{(t+n+1)}$$

$$= \sum_{n=0}^{\infty} \left\{ \frac{1}{(t+n)} - \frac{1}{(t+n+1)} \right\}$$

$$= \frac{1}{t}.$$

Therefore from (3.4.12),

$$f(t) > 1 + \left[\frac{\sin \pi t}{\pi} \right]^2 \frac{1}{t} \left(f'(0) - 2 \right) \geq 1 \quad \text{if } f'(0) \geq 2.$$

On the other hand $f(t) = 1$ when t is a positive integer, by construction. Hence $f(t) \geq 1$ when $t > 0$.

When $t < 0$ and t is not an integer we can generate a factor -1 in (3.4.10) by subtracting the left side and adding the right side of the identity (3.4.11). This time we obtain, after similar calculations,

$$f(t) > -1 + \left[\frac{\sin \pi t}{\pi} \right]^2 \left\{ \frac{1}{t^2} + \frac{1}{t} \left(f'(0) - 2 \right) \right\} \geq -1 \quad \text{if } f'(0) \leq 2.$$

Consequently we must take $f'(0) = 2$, and we find that our function f has become the function we called B in (3.4.9). We must now verify that B satisfies condition (c) of Beurling's problem. First, we will show that if g is any function satisfying (a) and (b) of the problem, and if $g(t) - \text{sgn}(t) \in L^1(\mathbb{R})$, then

$$\int_{\mathbb{R}} \{ g(t) - \text{sgn}(t) \} \, dt \geq 1.$$

Let $\psi(t) = g(t) - \text{sgn}(t) \geq 0$. Now

$$\psi^{\wedge}(v) = \frac{1}{\sqrt{2\pi}} \int_{\mathbb{R}} e^{-itv} (g(t) - \text{sgn}(t)) \, dt$$

$$= \frac{1}{\sqrt{2\pi}} \int_{\mathbb{R}} \frac{e^{-itv}}{iv} \, d(g(t) - \text{sgn}(t))$$

$$= \frac{1}{iv\sqrt{2\pi}} \left\{ \int_{\mathbb{R}} e^{-itv} g'(t) \, dt - 2 \right\}$$

$$= \frac{1}{iv} \left\{ (g')^{\wedge} - \frac{2}{\sqrt{2\pi}} \right\}.$$

Now $\psi \in L^1(\mathbb{R})$, and by Lemma 3.5, $\text{sgn } t - G(2t) \in L^1(\mathbb{R})$; hence $g(t) - G(2t) \in L^1(\mathbb{R})$. Now $g(t) - G(2t) \in E^{2\pi}$, therefore by a theorem of Plancherel and Pólya, $g'(t) - 2G'(2t) \in L^1(\mathbb{R})$. But $2G'(2t) \in L^1(\mathbb{R})$ by Lemma 3.5, hence $g' \in L^1(\mathbb{R})$.

It follows by the L^1-form of the Paley–Wiener theorem (Boas 1954, p. 106) that $\mathrm{supp}(g')^\wedge = [-2\pi, 2\pi]$. Therefore when $|v| \geq 2\pi$, $\psi^\wedge(v) = -2/iv\sqrt{2\pi}$.

We now apply Poisson's summation formula in the form

$$\sum_{q \in \mathbb{Z}} \psi(v + q) = \sqrt{2\pi} \sum_{m \in \mathbb{Z}} \psi^\wedge(2\pi m)e^{2\pi i m v}. \tag{3.4.13}$$

Since the left side is non-negative, the right side gives

$$0 \leq \sqrt{2\pi}\,\psi^\wedge(0) - \sideset{}{'}\sum_{m \in \mathbb{Z}} \frac{e^{2\pi i m v}}{i\pi m} = \psi^\wedge(0) - S(v),$$

where $S(v)$ is the saw-tooth function $S(v) := 2([v] - v + \frac{1}{2})$. Since $\max_{v \in \mathbb{R}} S(v) = 1$, we have

$$1 \leq \psi^\wedge(0) = \int_{\mathbb{R}} \psi(x)\,dx = \int_{\mathbb{R}} \{g(x) - \mathrm{sgn}(x)\}\,dx, \tag{3.4.14}$$

as we had set out to prove.

Here our discussion of Beurling's function B will be interrupted in order to introduce Selberg's function C; then the properties of these two functions will be completed together.

In 1974, in connection with his proof of the large sieve inequality, Selberg found the need for a function f that would solve the following problem:

Problem (Selberg) Find a function f such that

(a) $f \in E^{2\pi}$;

(b) $f(t) \geq \chi_I(t)$, $t \in \mathbb{R}$;

(c) $\int_{\mathbb{R}} \{f(t) - \chi_I(t)\}\,dt = 1$.

Here, I is a union of real intervals, but we shall only look at the case $I = [0, 1]$ of which the more general case is a simple extension. Any function solving this problem is not only an "approximate characteristic function of I", but is also a majorant, and is so smooth as to be band-limited (as we see from the L^1-form of the Paley–Wiener theorem).

By writing χ_I in terms of sgn functions, it is apparent that a reasonable candidate for a solution to Selberg's problem is

$$C(t) = \frac{1}{2}\{B(t) + B(1 - t)\} = \left[\frac{\sin \pi t}{\pi}\right]^2 \left\{\frac{1}{t^2} + \frac{1}{(t-1)^2} + \frac{1}{t} - \frac{1}{t-1}\right\}.$$

Evidently C satisfies condition (a) of Selberg's problem. Condition (c) is verified using two special integrals:

$$\int_{\mathbb{R}} \left\{\frac{\sin \pi t}{\pi(t - k)}\right\}^2 dt = 1, \qquad k \in \mathbb{Z},$$

and if s and s' are unequal integers,

$$\int_{\mathbb{R}} \sin^2 \pi t \left\{ \frac{1}{t-s} \cdot \frac{1}{t-s'} \right\} dt = 0.$$

Condition (b) follows from

$$C(t) - \chi_I(t) = \frac{1}{2} \{B(t) - \text{sgn}(t) + B(1-t) - \text{sgn}(1-t)\},$$

and using this together with

$$\int_{\mathbb{R}} \{B(t) - \text{sgn}(t)\} \, dt = \int_{\mathbb{R}} \{B(1-t) - \text{sgn}(1-t)\} \, dt,$$

we obtain

$$\int_{\mathbb{R}} \{B(t) - \text{sgn}(t)\} \, dt = 1.$$

It remains only to verify the uniqueness property of B. Before doing so it is interesting to note that C does not have the same uniqueness property as B, contrary to what one might expect. Neither does it have the same extremal properties; in fact, the nature of the set I plays a determining rôle. More information can be found in Vaaler (1985, p. 186).

Finally, let equality hold in (3.4.14) for some g; that is, $\psi^\wedge(0) = 1$. From (3.4.13) we find

$$\psi^\wedge(0) - \lim_{v \to 0^+} S(v) = 0,$$

so that from the left side we get $\psi(q) = 0$, $q \in \mathbb{Z}$, that is, $g(q) = \text{sgn} q = B(q)$. Furthermore,

$$g(t) - B(t) = g(t) - \text{sgn}(t) - \{B(t) - \text{sgn}(t)\} \in L^1(\mathbb{R}),$$

and $g - B \in E^{2\pi}$. Now by the L^1 theorem for representation in sampling series (e.g., Higgins 1996, p. 51) we have $g \equiv B$.

This completes our discussion of B and C. Now we shall look at one of the simpler applications of C.

3.4.2 A simple application of Selberg's function

We can obtain a more general form of Selberg's function C by taking the interval I of the previous section to be $[a, b]$ instead of $[0, 1]$. Then it is easily verified that

$$C_I(t) := \frac{1}{2} \{B(b-t) + B(t-a)\},$$

solves Selberg's problem, given in the previous section, for our more general interval I.

To illustrate the use of Selberg's function C_I, we consider the following theorem. It was given by Selberg, and independently by Montgomery and Vaughan; here we follow Selberg's method of proof (see Vaaler 1985, pp. 184–185).

Theorem 3.6 *Let f be the almost periodic trigonometrical polynomial given by*

$$f(t) = \sum_{n=1}^{N} c_n e^{i\lambda_n t}$$

where $|\lambda_n - \lambda_m| \geq \delta > 0$, $n \neq m$. Then

$$\int_a^b |f(t)|^2 \, dt = (b - a + \theta/\delta) \sum_{n=1}^{N} |c_n|^2$$

for some θ such that $|\theta| \leq 1$.

To prove this theorem let us first take $\delta = 2\pi$. Selberg noted that, since $X_I(t) \leq C_I(t)$, then

$$\int_a^b |f(t)|^2 \, dt \leq \int_{\mathbb{R}} C_I(t)|f(t)|^2 \, dt$$

$$= \sum_{n=1}^{N} \sum_{m=1}^{N} c_n \overline{c_m} \int_{\mathbb{R}} C_I(t) e^{-i(\lambda_m - \lambda_n)t} \, dt$$

$$= \sqrt{2\pi} \sum_{n=1}^{N} \sum_{m=1}^{N} c_n \overline{c_m} C_I^{\wedge}(\lambda_m - \lambda_n).$$

Now, by the Paley–Wiener theorem, supp $C_I^{\wedge}(x) \subseteq [-2\pi, 2\pi]$, so that when $m \neq n$, $C_I^{\wedge}(\lambda_m - \lambda_n) = 0$. Also, since

$$C_I^{\wedge}(x) = \frac{1}{\sqrt{2\pi}} \int_{\mathbb{R}} C_I(t) e^{-ixt} \, dt,$$

we find from condition (c) of Selberg's problem that when $m = n$, $C_I^{\wedge}(0) = (b-a+1)/\sqrt{2\pi}$. Hence

$$\int_a^b |f(t)|^2 \, dt \leq (b - a + 1) \sum_{n=1}^{N} |c_n|^2.$$

If, more generally, $|\lambda_n - \lambda_m| \geq \delta > 0$, $n \neq m$, then a re-scaling of this argument leads to

$$\int_a^b |f(t)|^2 \, dt \leq (b - a + 1/\delta) \sum_{n=1}^{N} |c_n|^2.$$

Furthermore, Selberg found another function, c_I, similar to C_I in all respects except one; c_I minorizes X_I instead of majorizing it. Using this function a lower bound can be obtained by similar methods, and then combined with the upper bound to give the conclusion of the theorem.

3.5   Multidimensional derivative sampling

Several different types of multidimensional derivative sampling series have appeared in the literature; some of these are listed below. Then we look at **I** in more detail, and then at a special case of **IV**.

I It seems that the first sampling series to involve partial derivative values were introduced virtually simultaneously by Montgomery (1964) (in the deterministic setting) and by Petersen and Middleton (1964) (in the context of stochastic fields). Let $f : \mathbb{R}^d \mapsto \mathbb{C}$, $d \in \mathbb{N}$, let the vectors v_j, $j = 1, \ldots, d$, be a basis for $\mathbb{R}^d$ and let $\mathbb{L}$ denote the subgroup of $\mathbb{R}^d$ generated by them; thus,

$$\mathbb{L} = \{l : l = s_1 v_1 + \cdots + s_d v_d\}, \qquad s_j \in \mathbb{Z}, \quad j = 1, \ldots, d.$$

Montgomery (1964, p. 438) obtained the representation

$$f(t) = \sum_{l \in \mathbb{L}} \{f(l) + (t - l) \cdot \nabla f(l)\} g(t - l), \tag{3.5.1}$$

for functions f belonging to the Paley–Wiener space PW_B, where the reconstruction function g and the vectors v_j depend on the nature of B.

This series uses function values and values of all the first partial derivatives of f.

II Shortly after this Montgomery (1965) gave a sampling series using function values and values of all the partial derivatives of f of order not exceeding K, an arbitrary fixed positive integer. This means that all partial derivatives of the form

$$\frac{\partial^{\alpha_1 + \alpha_2 + \cdots + \alpha_d}}{\partial t_1^{\alpha_1} \, \partial t_2^{\alpha_2} \, \cdots \, \partial t_d^{\alpha_d}} \tag{3.5.2}$$

are present, where $\alpha_1 + \alpha_2 + \cdots + \alpha_d \le K$, and the α's are non-negative integers.

III An even more general form was given by Horng (1977) (who gave results for dimension two only, but there seems to be no reason why analogous results in higher dimensions could not be formulated); among other general features it uses function values and values of all partial derivatives of f of order not exceeding the multi-index $\kappa = (K_1, \ldots, K_d) \in \mathbb{N}_0^d$. In other words all partials of the form (3.5.2), with $\alpha_j \le K_j$, $j = 1, \ldots, d$, are present. Remarkably, the K_j's are not necessarily constants but can vary with the sampling points.

IV Samples of f and of all its mixed partials are present; in other words all partials of the form (3.5.2), in which every α_j, $j = 1, \ldots, d$, is either 0 or 1, are present. This is the special case of III in which all K_j's equal 1.

3.5.1   *Gradient sampling*

Here we give a novel proof of the gradient sampling series of Montgomery (in **I** above); it is based on the method in §3.3.3.

Let us formally take Fourier transforms on both sides of (3.5.1) to obtain, with a change of notation (here R is a reconstruction function to be determined),

$$
\begin{aligned}
f^\wedge(x) &= \frac{1}{(2\pi)^{N/2}} \sum_{n\in\mathbb{Z}^n} \left\{ f(a_n) \int_{\mathbb{R}^N} e^{-ix\cdot t} R(t-a_n)\, dt \right. \\
&\quad \left. + \nabla f(a_n) \cdot \int_{\mathbb{R}^n} e^{-ix\cdot t}(t-a_n) R(t-a_n)\, dt \right\} \\
&= \frac{1}{(2\pi)^{N/2}} \sum_{n\in\mathbb{Z}^n} e^{-ix\cdot a_n} \left\{ f(a_n) \int_{\mathbb{R}^N} e^{-ix\cdot t} R(t)\, dt \right. \\
&\quad \left. + \nabla f(a_n) \cdot \int_{\mathbb{R}^n} e^{-ix\cdot t} t R(t)\, dt \right\} \\
&= \sum_{n\in\mathbb{Z}^n} e^{-ix\cdot a_n} [f(a_n) + i\nabla f(a_n)\cdot \nabla] R^\wedge(x). \quad (3.5.3)
\end{aligned}
$$

The analogy between (3.5.3) and (3.3.9) is immediately noticeable.

We now prove (3.5.1) by verifying that (3.5.3) holds, first for all members of a basis for $L^2(B)$, then by extension to all of $L^2(B)$. Since the method is wholly analogous to that of §3.3.3, only brief details will be given.

We consider here only the simplest case, that of dimension 2 in which the band-region B is a square centred at the origin. Extensions to higher dimensions, and to more complicated band-regions, is largely a matter of routine. Of course, $L^2(B)$ must possess a basis of complex exponentials.

When $B = [-\pi, \pi) \times [-\pi, \pi)$, $L^2(B)$ has an orthonormal basis of the form $\{e^{-ik\cdot x}/2\pi\}$, where $k = (k_1, k_2) \in \mathbb{Z}^2$ and $x = (x_1, x_2)$. The sample points are taken to be $\{(2n_1, n_2)\}$, $(n_1, n_2) \in \mathbb{Z}^2$. These have density $\frac{1}{2}$, that is, half the N–L density $m(B)/(2\pi)^2 = 1$, and so they are appropriate for two-channel sampling. There are other choices of course, but these points are convenient to work with.

We check the case where k_1 and k_2 are both odd. With these choices one verifies easily that

$$
f^\wedge = \frac{e^{-i(k_1 x_1 + k_2 x_2)}}{2\pi};
$$

$$
f(t) = \text{sinc}(t_1 - k_1)\, \text{sinc}(t_2 - k_2)
$$

$$
f(2n_1, n_2) = 0
$$

$$
f_{t_1}(2n_1, n_2) \doteq \begin{cases} -\dfrac{1}{2n_1 - k_1}; & n_2 = k_2, \\ 0; & \text{otherwise}; \end{cases}
$$

$$
f_{t_2}(2n_1, n_2) = 0.
$$

Now substitution into (3.5.3) gives

$$
\frac{e^{-ik_1 x_1} e^{-ik_2 x_2}}{2\pi} = -i e^{-ik_2 x_2} R^\wedge_{x_1}(x) \sum_{n_1\in\mathbb{Z}} \frac{e^{-i2n_1 x_1}}{2n_1 - k_1}.
$$

This summation is the same Fourier series that we encountered in (3.3.10), so the procedure that was used there can be used here. Then the equation reduces to

$$1 = -2\pi i R_{x_1}^{\wedge}(x) \left(-\frac{i\pi}{2} \operatorname{sgn} x \right),$$

or

$$R_{x_1}^{\wedge}(x) = -\frac{1}{\pi^2} \operatorname{sgn} x.$$

An integration gives

$$R^{\wedge}(x) = \frac{1}{\pi} \left(1 - \frac{|x_1|}{\pi} \right) + C(x_2).$$

Cases with other parities for k_1 and k_2 can now be checked, and they show that, in fact, one can take $C(x_2) \equiv 0$. Consequently, by taking the inverse two-dimensional Fourier transform of $R^{\wedge}(x) = \frac{1}{\pi} \left(1 - \frac{|x_1|}{\pi} \right)$ one finds that

$$R(t) = \frac{1}{\sqrt{2\pi}} \operatorname{sinc}^2 \frac{1}{2} t_1 \operatorname{sinc} t_2.$$

Hence, for $f \in PW_B$, where $B = [-\pi, \pi) \times [-\pi, \pi)$, we have the gradient sampling series

$$f(t) = \frac{1}{\sqrt{2\pi}} \sum_{n \in \mathbb{Z}^2} \left\{ f(2n_1, n_2) + (t - (2n_1, n_2)) \cdot (\nabla f)(2n_1, n_2) \right\}$$

$$\times \operatorname{sinc}^2 \frac{1}{2} (t_1 - 2n_1) \operatorname{sinc} (t_2 - n_2).$$

3.5.2 Mixed partial derivative sampling

It is likely that the mixed partial derivative scheme of **IV**, §3.5, can be treated by the Riesz basis method. Here we just look at two low-order cases, beginning with dimension two. Thus, $f \in PW_B$, $B = [-\pi, \pi] \times [-\pi, \pi]$, is to be reconstructed from the samples

$$\{f(2n)\} \cup \{f_{t_1}(2n)\} \cup \{f_{t_2}(2n)\} \cup \{f_{t_1 t_2}(2n)\}, \qquad n \in \mathbb{Z}^2.$$

Now B is decomposed into four equal subsquares, each in a different co-ordinate quadrant. The appropriate multipliers for the four channels are chosen, and then if B_1 denotes the subsquare in the first quadrant, the analogue of M in (3.3.4) has for its determinant, when $x = (x_1, x_2) \in B_1$,

$$\begin{vmatrix} 1 & ix_1 & ix_2 & -x_1 x_2 \\ 1 & i(x_1 - \pi) & ix_2 & -(x_1 - \pi)x_2 \\ 1 & i(x_1 - \pi) & i(x_2 - \pi) & -(x_1 - \pi)(x_2 - \pi) \\ 1 & ix_1 & i(x_2 - \pi) & -x_1(x_2 - \pi) \end{vmatrix} = -\pi^4.$$

Since this determinant does not vanish, the Riesz basis method is valid and we obtain

$$\sum_{n\in\mathbb{Z}^2} \sum_{i,j=0}^{1} \frac{\partial^{i+j}}{\partial t_1^i \, \partial t_2^j} f(2n) R_{ij}(t).$$

The reconstruction functions R_{ij} are calculated in Cheung (1992, p. 97). Note that the density of each of the four subsets $\{2n\}$, $n \in \mathbb{Z}^2$, of sampling points is $1/4$, and this accounts for the N–L density, which is $(2\pi)^2/(2\pi)^2 = 1$.

Let us consider very briefly the three-dimensional case. The band-region $B = [-\pi, \pi] \times [-\pi, \pi] \times [-\pi, \pi]$ is decomposed into eight subcubes, each in a different coordinate octant. The appropriate multipliers for the eight channels are chosen, and then if B_1 denotes the subcube in the first octant, the analogue of the determinant above has for its first row, and for $x = (x_1, x_2, x_3) \in B_1$,

$$1 \quad ix_1 \quad ix_2 \quad ix_3 \quad -x_1x_2 \quad -x_1x_3 \quad -x_2x_3 \quad -ix_1x_2x_3.$$

The remaining rows are compiled in analogy with the previous case, and the result has the value $-\pi^{12}$. Again the Riesz basis method applies since this determinant does not vanish.

There is every reason to suppose that this method would continue to work in higher-dimensional cases.

COMPUTATIONAL METHODS IN LINEAR PREDICTION FOR BAND-LIMITED SIGNALS BASED ON PAST SAMPLES

4.1 Introduction

An accurate prediction of the "next" value of a signal from its past samples is an important problem for many applications. Linear prediction coding, (Elias 1955; Rabiner and Schafer 1993), even predicts values several steps ahead of the last data sample. For the transmission of a block of signal samples, the LPC method is to record the first several samples and then use a linear prediction formula to estimate the values of the rest of the samples in the block (Meyer 1994, p. 33). The use of the prediction formula allows for data compression (Lucky 1968), such as for speech analysis and synthesis (Atal and Hanauer 1971). Other examples include applications in spectroscopy (Kauppinen et al. 1994), control theory, and geophysical signal processing.

This chapter concentrates on the computation of the linear prediction of the next few values of a band-limited signal based on past samples, whether the samples are equally spaced or are at arbitrary intervals. This differs from general extrapolation methods which seek to determine all values of the signal outside the region of observation. For the class of band-limited signals, formulas are developed that are independent of the particular signal but which depend only on its highest frequency and the set of sample times. That the results apply to this entire class of signals is fundamental to this chapter, and error formulas are presented that require only the additional information of the energy of the signal. The methods will be developed for deterministic signals, although the results carry over easily to stationary band-limited stochastic signals. The theory for linear prediction such as developed by Wiener (1949) and the statistical analysis presented in autoregressive (i.e. AR) or ARMA models is different from the scope of this chapter, which assumes only knowledge of the band-limitedness of the finite-energy signal.

A linear prediction formula can be viewed as a one-sided sampling theorem, in the sense that only samples from the past are used. This differs from the classic Whittaker–Shannon–Kotel'nikov sampling theorem that gives the band-limited signal value based on samples both from the past *and* the future,

$$f(t) = \sum_{n=-\infty}^{\infty} f(nT)\operatorname{sinc} \sigma (t - nT), \qquad (4.1.1)$$

where $\operatorname{sinc}(x) = \sin(\pi x)/(\pi x)$ and $\sigma = W$. Here, we assume that the band-limited signal has finite energy, and that it may be represented as

$$f(t) = \int_{-W/2}^{W/2} f^{\wedge}(\omega)e^{2\pi i \omega t}d\omega, \qquad (4.1.2)$$

for $f^{\wedge} \in L^2(R)$.

Although the sampling theorem (4.1.1) is based on equally spaced samples, there are many recent extensions and relations to it, see (Butzer and Gessinger 1995), for the case of non-uniform samples. Similarly in this chapter, prediction based on equally spaced samples will be initially developed, but the methods will be shown to extend to the case of non-uniformly spaced sample times. Both types of sampling will be considered in this chapter. The general problem considered begins with a set of data signal samples, labelled here as f_n, associated with t_n as sample times, and we review the case of an unlimited supply of past samples but concentrate on the case of a finite set where $n = 1, \dots, N$. With this formulation, the set of samples

$$(t_n, f_n)$$

which is a point in R^2, represents either equally spaced or non-uniformly spaced samples. In the case of equally spaced sample times, let T represent the common time period spacing between equally spaced sample times. For the case when prediction at the next sample instant t is desired, then $t_n = t - nT$, for $n = 1, 2, \dots$.

The following section provides an overview of convergence of prediction formulas to the correct signal value when there are an unlimited supply of past samples. In the next sections, an optimal prediction formula based on finitely many samples is considered from several different viewpoints. Difficulties in actual computations are discussed and regularization methods to suppress the effects of noise or quantization are developed for the prediction.

4.2 Asymptotic expansions

In the asymptotic case when samples from the past are unlimited, much is known about the convergence of linear prediction formulas to the true value of a band-limited signal. Suppose that an unlimited set of signal samples f_n may be obtained, giving data points (t_n, f_n) with t_n as sample times, $n = 1, 2, \dots$. For the case of equally spaced samples where $t_n = t - nT$, Splettstößer (1982) showed that prediction coefficients a_n^N exist which are independent of f, T, and W, such that

$$f(t) = \lim_{N \to \infty} \sum_{n=1}^{N} a_n^N f(t_n), \qquad (4.2.1)$$

as long as $TW < 1$, where W is as in (4.1.2) and the superscript on a_n^N indicates the dependence on the number of samples N and does not indicate exponentiation. That is, for any sampling rate exceeding the Nyquist rate, prediction coefficients exist which assure convergence of the prediction formula to the signal value.

Much of this carries over to the prediction of stationary stochastic signals. Slepian (1978) considered a stationary white noise $X(t)$ with bandwidth W (where $W < 1$) and mean zero. In a least-squares linear prediction formula based on equally spaced samples, he showed that the error η_0 satisfies

$$\lim_{N \to \infty} \frac{1}{N} \log(\eta_0) = \log(\sin(\pi TW/2))^2.$$

For large N, this means that the error η_0 is on the order of $\sin(\pi T W/2)^{2N}$, and this specifies the convergence rate as well as the result that, similar to the above, convergence results whenever $T W < 1$.

An upper bound on the error of the prediction can be derived using standard inequalities and the integral representation (4.1.2). For the general formulation of the prediction problem for data (t_n, f_n), if $f(t_n) = f_n$, the error can be bounded by

$$\left| f(t) - \sum_{n=1}^{N} a_n^N f(t_n) \right|^2 \leq \|f\|_2^2 \cdot \left\{ \int_{-W/2}^{W/2} \left| e^{2\pi i \omega t} - \sum_{n=1}^{N} a_n^N e^{2\pi i \omega t_n} \right|^2 d\omega \right\}. \quad (4.2.2)$$

For the case of a prediction at the "next" point from a set of equally spaced sample times, the error bound is simpler. For this case, the error bound can be put in the form

$$\left| f(t) - \sum_{n=1}^{N} a_n^N f(t - nT) \right|^2 \leq \|f\|_2^2 \cdot \left\{ \frac{1}{2\pi T} \int_{-\pi T W}^{\pi T W} |d_N(\omega)|^2 d\omega \right\}, \quad (4.2.3)$$

where $d_N(\omega)$ is the difference given by

$$d_N(\omega) := 1 - \sum_{n=1}^{N} a_n^N e^{-in\omega}. \quad (4.2.4)$$

The existence of predictor coefficients a_n^N that cause the error to vanish as $N \to \infty$ then follows from the density of the set $\{e^{in\omega}\}$ in the space $L^2[-\pi T W, \pi T W]$ provided that $T W < 1$.

Although the existence of appropriate prediction coefficients is known whenever $T W < 1$, formulas available for prediction are not as advanced. Wainstein and Zubakov (1962) defined the prediction coefficients by $a_n^N = (-1)^{n+1} \binom{N}{n}$, so that $d_N(\omega)$ from (4.2.4) is a binomial sum for which $|d_N(\omega)| = |1 - e^{-i\omega}|^N$. From the integral bound in (4.2.3), it then follows that the error goes to zero if $T W \leq 1/3$. Later, J. L. Brown, Jr., (1972), modified the coefficients to $a_n^N = (-1)^{n+1} a^n \binom{N}{n}$, leading to another binomial sum with $|d_N(\omega)| = |1 - a e^{-i\omega}|^N$. It turns out that one now has convergence whenever $T W < 1/2$ for each $a \leq 2\cos(\pi T W)$, with $a = \cos(\pi T W)$ being the optimal modifier to minimize the truncation error of the prediction formula. Some disadvantages of this approach are that the predictor coefficients are then dependent on T and any further information regarding additional data, such as an increase in N, requires complete recalculation of the prediction coefficients.

Mugler and Splettstößer (1986, 1987a) showed that for all of these cases, the prediction formulae could be evaluated using extensions of classical Newton series involving finite differences of the data. In particular,

$$f(t) = \lim_{N \to \infty} \sum_{n=0}^{N-1} \nabla_a^n f(t - T), \quad (4.2.5)$$

where $\nabla_a^{j+1} f(t) = \nabla_a^j f(t) - a \nabla_a^j f(t - T)$, $j = 0, 1, 2, \ldots$ and $\nabla_a^0 f(t) = f(t)$. In addition, these formulae were shown to be easily extended to the case when derivative samples are included, i.e. when both signal and derivative samples are available. These formulae adjust easily to the addition of any extra signal sample data without recomputation of the prediction coefficients. A further extension of a prediction formula involving modified binomial coefficients (4.2) was given in (Brown and Morean 1986), where two modifiers were employed to give $|d_N(\omega)| = |1 - \alpha_1 e^{-i\omega} - \alpha_2 e^{-2i\omega}|^N$. For the case that $TW \leq 1/2$, the truncation error in a prediction with N data samples may be bounded by $\|f\| \cdot (0.6863)^N$ using this formula, which may be calculated using a finite difference formulation similar to (4.2.5), as shown by Mugler and Splettstößer (1987b).

One difficulty with these prediction formulae is that they are sensitive to noise in the data samples. This is particularly clear for the initial formula described above when $a_n^N = (-1)^{n+1} \binom{N}{n}$. In this case, since the binomial coefficients grow so large, when these coefficients multiply the data signal values the noise is amplified by the magnitude of the coefficient. Somewhat less sensitivity to noise is present for the modified formula, such as in (4.2.5). In this case, the exponential factor a^n reduces the size of the coefficient when $|a| < 1$, so that the noise is not amplified as much. However, numerical trials still indicate that increasing the number of data samples, i.e. increasing N, does not keep increasing the accuracy of the prediction. Noise error, even in the form of roundoff error from finite precision arithmetic, eventually puts a limit on the accuracy that can be obtained.

In practical applications, noise is inherent in the data samples, so that prediction formulae that are not so sensitive to noise are important. In the next section, the concentration will shift from the asymptotic formulae described above to consider the case when only a finite number of samples are available on which to base the prediction. An excellent recent review that describes some other ways to generalize (4.1.1) to include prediction is (Butzer and Stens 1992). For example, they describe methods to predict even non-band-limited signals by a convolution sum of the form

$$\sum_{kT < t} f(kT)\varphi(t/T - k),$$

where the kernel φ is continuous and of compact support. They provide examples of such kernels in terms of polynomial splines and give estimates on the truncation error of the prediction.

4.3   An optimal set of prediction coefficients

In any application, prediction of a value of a band-limited signal will be based on a finite number of available samples. Suppose, as described earlier, that there is a finite set of signal samples $\{f_n\}, n = 1 \ldots, N$, corresponding to data points (t_n, f_n) with t_n as sample times. In this section, we will describe prediction coefficients that are optimal in a certain sense, but which still have some computational difficulties.

Much of the richness in the following analysis comes from two different representations for the data signal samples, f_n. As described earlier, the linear prediction formula has the form $f(t) \approx \sum_{n=1}^{N} a_n^N f_n$, where $a_n^N, n = 1 \ldots, N$, are the prediction coefficients for N samples. If f is a band-limited signal taking each of the values f_n, so that $f(t_n) = f_n$, then the integral formula (4.1.2) for f gives the representation

$$f_n = \int_{-W/2}^{W/2} f^\wedge(\omega)e^{2\pi i \omega t_n} d\omega, \quad n = 1, \ldots, N \qquad (4.3.1)$$

or by using Parseval's formula one can obtain a second representation

$$f_n = \int_{-\infty}^{\infty} f(t)\operatorname{sinc}((t - t_n)W)W dt, \quad n = 1, \dots, N \tag{4.3.2}$$

where $\operatorname{sinc}(x) = \sin(\pi x)/(\pi x)$.

In this section, we use the first representation (4.3.1) and return to (4.3.2) in the next section. Using $f(t_n) = f_n$, an upper bound of the error of the prediction for arbitrary sampling times $\{t_n\}$ was noted earlier (4.2.2) as

$$\left| f(t) - \sum_{n=1}^{N} a_n^N f(t_n) \right|^2 \leq \|f\|_2^2 \cdot \left\{ \int_{-W/2}^{W/2} \left| e^{2\pi i \omega t} - \sum_{n=1}^{N} a_n^N e^{2\pi i \omega t_n} \right|^2 d\omega \right\}.$$

There are two terms in this bound on the prediction error, so that there are two terms to minimize in order to minimize the error: (1) minimize the error by determining **optimal prediction coefficients** a_n^N such that the second integral $\int_{-W/2}^{W/2} |e^{2\pi i \omega t} - \sum_{n=1}^{N} a_n^N e^{2\pi i \omega t_n}|^2 d\omega$ is minimized, and (2) minimize the error for the unknown function f by finding f of **minimum norm** satisfying the given conditions.

We will call the prediction coefficients a_n^N which minimize the integral above the optimal prediction coefficients. They do not depend on the particular signal, but only on the time-bandwidth factor TW. For general sampling intervals, it can be shown that the coefficients which minimize the integral must satisfy the linear system

$$\sum_{n=1}^{N} \operatorname{sinc}((t_m - t_n)W)\, a_n^N = \operatorname{sinc}((t - t_m)W), \tag{4.3.3}$$

$m = 1, 2, \dots, N$. For computations involving a prediction at more than one prediction time, it is important to note that only the right side of this system depends on the prediction point t.

For sampling at arbitrary points, the system (4.3.3) can be shown to have a positive definite and symmetric matrix, while for equally spaced sampling, the system is Toeplitz, with form

$$\sum_{n=1}^{N} \operatorname{sinc}((m - n)\tau)\, a_n^N = \operatorname{sinc}(m\tau), \tag{4.3.4}$$

for $m = 1, 2, \dots, N$, where $\tau = TW$. Although the point of prediction t is not explicitly present in equation (4.3.4), the choice of sample times as $t_n = t - nT$ makes it present implicitly.

In order to emphasize the Toeplitz structure of the matrix of the system in (4.3.4), we write it out explicitly. The coefficient matrix in this system is:

$$H = \begin{bmatrix} 1.0 & \operatorname{sinc}(\tau) & \operatorname{sinc}(2\tau) & \cdots & \operatorname{sinc}((N-1)\tau) \\ \operatorname{sinc}(\tau) & 1.0 & \operatorname{sinc}(\tau) & \cdots & \operatorname{sinc}((N-2)\tau) \\ \operatorname{sinc}(2\tau) & \operatorname{sinc}(\tau) & 1.0 & \cdots & \operatorname{sinc}((N-3)\tau) \\ \cdots & \cdots & \cdots & \cdots & \cdots \\ \cdots & \cdots & \cdots & \cdots & \operatorname{sinc}(\tau) \\ \operatorname{sinc}((N-1)\tau) & \operatorname{sinc}((N-2)\tau) & \cdots & \operatorname{sinc}(\tau) & 1.0 \end{bmatrix}.$$

Table 4.1 *Initial value of N such that the condition number of H is on the order of* 10^{15}

0.10	0.20	0.30	0.40	0.50	0.60	0.70	0.80	0.90
8	11	14	18	22	29	39	60	121

Table 4.2 *First value of N for which the matrix H ceases to be positive definite*

TW value	0.10	0.20	0.30	0.40	0.50	0.60	0.70	0.80	0.90
N, with matrix H positive definite for $n < N$	10	15	17	20	26	30	57	64	126

Presenting a major computational problem for the evaluation of the optimal prediction coefficients, the matrix H may be quite ill-conditioned. For example, the condition number when $N = 6$ and $\tau = 0.10$ is $\kappa = 9.8 \times 10^{10}$. The condition number for any fixed time-bandwidth product $\tau = TW$ increases with the number of data points, N, and Table 4.1 shows this by exhibiting the first value of N for which the condition number is on the order of 10^{15}.

The matrix H may be shown to be positive definite, where that result holds theoretically for all values of N and τ. However, because of ill-conditioning, the matrix computationally does not retain this property. One numerical way to check for the property of being positive definite is to program the Cholesky decomposition of the matrix and see when that succeeds or fails. Table 4.2 lists the first value of N for which the matrix H ceases to be positive definite, according to this approach with the Cholesky decomposition applied in MATLAB.

The eigenvectors $v_i(n)$ of H are the discrete prolate spheroidal sequences, and accurate computation of these eigenvectors is important in the following. Slepian's asymptotic formulas for $v_i(n)$ are not useful for prediction here, where N must remain relatively small. However, there is a well-conditioned, symmetric, tridiagonal matrix T that has the *same* eigenvectors as does H, see Gruenbacher and Hummels (1994). That symmetric matrix can be obtained from the description

$$T(N, \tau)_{m,n} = \begin{cases} ((N + 1)/2 - m)^2 \cos(\pi \tau) & \text{if } n = m \text{ main diagonal} \\ (m - 1)(N + 1 - m)/2 & \text{if } n = m - 1 \text{ off-diagonal.} \end{cases}$$

Computing the eigenvectors of T may be done more precisely than from computing those of H directly, so the eigenvectors $v_i(n)$ in the following are computed using T.

4.4 An integral equation

An operator-theoretic approach to the prediction formula comes from the second representation described above for f_n in (4.3.2). That representation was given as

$$f_n = \int_{-\infty}^{\infty} f(t) \, \text{sinc}((t - t_n)W)W \, dt, \qquad n = 1, \ldots, N$$

In this approach, one considers the above as an integral equation to solve for $f(t)$, of basic form

$$f_n = \int_{-\infty}^{\infty} f(t)k(n,t)W dt, \qquad n = 1, \ldots, N,$$

with kernel $k(n,t) = \operatorname{sinc}((t-t_n)W)$. In this regard, suppose that the integral in the above defines the output of an integral operator on f, given by $K[n,t]f$. The bracket notation $[n,t]$ indicates that the domain depends on a function of the continuous parameter t while the range depends on a discrete parameter n, with much of these results to be found in (Wingham 1992). In general, the operator $K[n,t]$ takes f in the Paley–Wiener space with band-limit $W/2$, i.e. $f \in PW(R)$, to $y = [f_1, \ldots, f_N] \in R^N$, for

$$K : f \to y,$$

where $y = [f_1, \ldots, f_N]$.

Let $u_i(t)$ and $v_i(n)$ be right and left singular vectors of K. These satisfy

$$K u_i(t) = \sigma_i v_i(n)$$

and

$$K_a v_i(n) = \sigma_i u_i(t),$$

where K_a is the adjoint operator, and σ_i are the operator's singular values, $i = 1, \ldots, N$. For the adjoint operator,

$$K_a v_i = \sum_{n=1}^{N} v_i(n)\operatorname{sinc}((t - t_n)W), \tag{4.4.1}$$

and the kernel of K_a is $k(t,n) = \operatorname{sinc}((t - t_n)W))$.

There is a direct relation between the coefficient matrix of systems (4.3.3) and (4.3.4) and the composite operator $K \cdot K_a$. In particular, $K \cdot K_a$ is an $N \times N$ matrix, with elements $K \cdot K_a(n,m) = \int_{-\infty}^{\infty} \operatorname{sinc}((t-t_n)W)\operatorname{sinc}((t-t_m)W)dt = \operatorname{sinc}((t_n-t_m)W)$. Thus $K \cdot K_a$ is the matrix of system (4.3.3) and for the case of uniformly sampled data, the values are the same as elements of the matrix H of (4.3.4). Also, $K \cdot K_a v_i = K \cdot (\sigma_i u_i) = \sigma_i(\sigma_i v_i) = \sigma_i^2 v_i$ so that the v_i singular vectors are eigenvectors of the matrix H with eigenvalues $\lambda_i = \sigma_i^2$.

To solve $Kf = y$, where $y = [f_1, f_2, \ldots, f_N]$, apply the Moore–Penrose (generalized) inverse $K^\dagger$ to obtain

$$f(t) = K^\dagger y = \sum_{k=1}^{N} \frac{\langle y, v_k \rangle}{\sigma_k} u_k(t), \tag{4.4.2}$$

where $\langle y, v_k \rangle = \sum_{n=1}^{N} f_n v_k(n)$. It is known that this method gives the minimum norm solution, which satisfies one of the criteria for the linear prediction solution as outlined in Section 4.3. This finite sum can be rearranged to give

$$f(t) = \sum_{n=1}^{N} \left(\sum_{k=1}^{N} \frac{1}{\sigma_k} v_k(n) u_k(t) \right) f_n,$$

which is the form of the linear prediction studied here. This formula gives prediction coefficients

$$a_n^N = \sum_{k=1}^{N} \frac{1}{\sigma_k} v_k(n) u_k(t).$$

Since

$$K_a v_k = \sigma_k u_k,$$

then

$$u_k = \frac{1}{\sigma_k} K_a v_k,$$

and thus

$$a_n^N = \sum_{k=1}^{N} \frac{1}{\sigma_k^2} v_k(n) K_a v_k, \qquad (4.4.3)$$

for $n = 1, \ldots, N$. But $\sigma_k^2 = \lambda_k$ are the singular values of the matrix H, and v_k are eigenvectors of H. Thus the computation of a_n^N in (4.4.3) can all be done using known quantities from H.

One of the advantages of the prediction formula (4.4.2) is that it can be employed for samples that need not be equally spaced. In addition, as will be developed in Section 4.7, regularization methods can be used with this formulation to suppress the effects of noise. As noted by Wingham (1992), the singular vectors $v_k(n)$ reduce to the discrete prolate spheroidal sequences in the case of equally spaced samples. Discussion of general extrapolation algorithms in this case that also includes noise factors is given by Marks (1991, pp. 257ff).

4.5   Size of the optimal prediction coefficients

Even if the computational problems in solving (4.3.4) could be overcome so that the optimal prediction coefficients, a_n^N, could be computed accurately, the values of these coefficients can be so large as to make the prediction be sensitive to noise in the data. This sensitivity follows from the linear prediction formula, with factors that include multiples $a_n^N f_n$ where any noise in a signal sample will be magnified by the value of the corresponding prediction coefficient.

To see the magnitude of these coefficients, Table 4.3 exhibits the prediction coefficients a_n^N for $N = 2, \ldots, 9$, and $TW = 0.20$, computed using as accurate methods as available. In the table, the rows start with $N = 2$ and go to $N = 9$, the columns start with a_1^N on the left and end with a_N^N on the right.

In Section 4.2, the development of asymptotic prediction formulae was traced, and it was noted that formulae depending on binomial coefficients were some of the first proposed for the class of band-limited signals. For this case of the optimal prediction coefficients, the values of the coefficients may still be related to the binomial coefficients. In particular, if the a_n^N coefficients of Table 4.3 are divided by the corresponding binomial coefficients, namely $\binom{N}{n}$, one surprisingly obtains a regular, monotonic pattern exhibited in Table 4.4. In particular, pick any column, and note that the values are monotonically decreasing in magnitude in that column.

Table 4.3 *Values of optimal prediction coefficients, a_n^N for $N = 2, \ldots, 9, n = 1, \ldots, N$,* and $TW = 0.20$

1.8219	−0.9476							
2.7217	−2.6777	0.9496						
3.6241	−5.2222	3.5360	−0.9503					
4.5274	−8.5832	8.4999	−4.3951	0.9505				
5.4311	−12.7617	16.5807	−12.5553	5.2547	−0.9507			
6.3350	−17.7579	28.5183	−28.3202	17.3886	−6.1145	0.9508		
7.2391	−23.5720	45.0524	−55.2487	44.5054	−22.9999	6.9745	−0.9509	
8.1432	−30.2035	66.9212	−97.5657	97.0375	−65.8369	29.3874	−7.8340	0.9508
9.0419	−37.6088	94.7013	−159.8035	188.7729	−158.0744	92.6558	−36.3897	8.6500 −0.9455

Table 4.4 *The optimal prediction coefficients of Table 4.3 divided by corresponding binomial coefficients*

0.9110	−0.9476							
0.9072	−0.8926	0.9496						
0.9060	−0.8704	0.8840	−0.9503					
0.9055	−0.8583	0.8500	−0.8790	0.9505				
0.9052	−0.8508	0.8290	−0.8370	0.8758	−0.9507			
0.9050	−0.8456	0.8148	−0.8091	0.8280	−0.8735	0.9508		
0.9049	−0.8419	0.8045	−0.7893	0.7947	−0.8214	0.8718	−0.9509	
0.9048	−0.8390	0.7967	−0.7743	0.7701	−0.7838	0.8163	−0.8704	0.9508
0.9042	−0.8358	0.7892	−0.7610	0.7491	−0.7527	0.7721	−0.8087	0.8650 −0.9455

It was conjectured by Mugler (1992) that this monotonic pattern continues for larger values of N.

4.6 Alternative computational methods

The svd-based method for the computation of the linear prediction (4.4.2) has the advantage of being adjustable to noise or errors due to quantization of the signal samples, as will be seen in Section 4.7. Other numerical methods are discussed in this section for comparison.

The Durbin algorithm for solving Yule–Walker equations (Golub and Van Loan 1989, p. 185) can be applied to solve the system (4.3.4). This method often involves terminology borrowed from transmission line theory, see (Makhoul 1975, p. 566). An excellent recent review of related algorithms is by Strobach (1991), where the ladder form of the prediction error filter is discussed. The terminology includes the "reflection coefficients" for the nth order prediction as k_n, and the value of $\frac{1}{2\pi} \int_{-\pi}^{\pi} |d_n(\lambda\omega)|^2 d\omega$ from the error bound (4.2.3) obtained when using the optimal coefficients as "error" E_n. For the coefficient matrix H from (4.3.4), (Mugler 1992) this method is programmed to initialize the error $E_0 = 1$, $k_1 = -\text{sinc}(\tau)$ and $a_1^1 = k_1$, and then to recursively compute the other reflection coefficients k_n by

$$k_n = -\{\text{sinc}(n\tau) + \sum_{j=1}^{n-1} a_j^{n-1}\text{sinc}((n-j)\tau)\}/E_{n-1}, \qquad (4.6.1)$$

for $n = 1, \ldots, N$. In each case, the new error bound is found by

$$E_n = (1 - k_n^2)E_{n-1}, \qquad (4.6.2)$$

and the next-order prediction coefficients are given by $a_n^n = k_n$ and for $1 \leq j \leq n - 1$,

$$a_j^n = a_j^{n-1} + k_n \cdot a_{n-j}^{n-1}. \qquad (4.6.3)$$

The recursion formula (4.6.3) allows for easy computation of the prediction coefficients, given the value of the time-bandwidth product $\tau = TW$. In order to obtain the solution for N prediction coefficients, this algorithm computes the solution for all predictions less than N as a part of the calculation. Note that (4.6.3) generalizes the usual binomial coefficient calculation,

$$\binom{n-1}{j} + \binom{n-1}{j+1} = \binom{n-1}{j} + \binom{n-1}{n-j-2} = \binom{n}{j+1}$$

and connects with the conjecture in Table 4.4. Final prediction coefficients have a change in sign from the recursion above.

The numerical difficulty in this calculation of the prediction coefficients comes in equation (4.6.1) with the division by E_{n-1}. Since the error decreases to a small value, division by it may cause the loss of significant digits in the calculation of the reflection coefficients. It is known that E_n is non-increasing in n, i.e. that $E_n \leq E_{n-1}$. Since $E_n \geq 0$, this means that k_n is non-decreasing and $0 \leq k_n \leq 1$. Table 4.5 presents some of the error bounds and computed reflection coefficient values.

There are other methods available for the solution of the system (4.3.4) for the case of uniformly sampled data, as there has been much recent interest in computational methods for Toeplitz systems. For example, in (Chan and Strang 1989), methods are

Table 4.5 E_n *values and reflection coefficients for* $TW = 0.10$

n	E_n	k_n
1	$3.2 \cdot 10^{-2}$	0.986853
2	$8.5 \cdot 10^{-4}$	−0.987333
3	$2.1 \cdot 10^{-5}$	0.987491
4	$5.3 \cdot 10^{-7}$	−0.987563
5	$1.3 \cdot 10^{-8}$	0.987601
6	$3.2 \cdot 10^{-10}$	−0.987579
7	$8.0 \cdot 10^{-12}$	0.978971

developed for solving positive definite Toeplitz systems by the use of a circulant matrix as preconditioner coupled with the conjugate gradient method. Other computational methods were developed for discrete signals by Feichtinger et al. (1995). This concerned reconstruction of band-limited sequences, where a Toeplitz system was developed after formulating the reconstruction problem in terms of trigonometric polynomials. Methods introduced there involve preconditioning and the conjugate gradient method. However, the system of (4.3.3) is only Toeplitz for the prediction problem for the case of samples from equally spaced time intervals. The general case of sampling at arbitrary times was noted earlier as resulting in a matrix (4.3.3) that is symmetric but not necessarily Toeplitz. Further, it is not clear how robust these methods are relative to noise in the samples.

4.7 Suppressing the effects of noise on the prediction

The svd-based prediction formula of (4.4.2) may be adapted to suppress errors due to noise or quantization effects in the signal samples. Regularization methods are currently being applied to this type of general problem (Wingham 1992; Nagy 1995), and two methods will be discussed here: (i) Tikhonov and iterated Tikhonov regularization and (ii) truncated singular value expansion.

The size of the noise or quantization error will be measured in the norm on real N-dimensional space. The true signal samples form a point in N-space, $y = [f_1, f_2, \ldots, f_N]$, and we assume that y_δ represents the noisy samples. In the following we assume, as in (Engl 1993) that

$$\|y - y_\delta\| \le \delta,$$

where the norm is the standard Euclidean norm in N dimensions. This makes δ a bound on the error.

Tikhonov regularization employs a regularization parameter, α, so that formula (4.4.2), i.e. $f(t) = \sum_{k=1}^{N} \{\langle y_\delta, v_k\rangle/\sigma_k\} u_k(t)$, is modified to

$$f_\alpha(t) = \sum_{k=1}^{N} \frac{\sigma_k}{\sigma_k^2 + \alpha} \langle y_\delta, v_k\rangle u_k(t). \tag{4.7.1}$$

The practical consequence of this regularization is a weighting of the expansion in the singular functions $u_k(t)$. In particular, if $\alpha \ll \sigma_k$ then $\sigma_k/(\sigma_k^2 + \alpha^2) \approx 1/\sigma_k$ and no substantial change in the coefficient of $u_k(t)$ results. However, if $\sigma_k \ll \alpha$, then $\sigma_k/(\sigma_k^2 + \alpha^2) \approx 0$, so that the particular $u_k(t)$ is effectively no longer a part of the reconstruction formula.

Tikhonov regularization is discussed by Engl (1993) where it is noted that this method has the following variational characterization: f_α is the unique minimizer of the functional

$$f \to \|Kf - y_\delta\|^2 + \alpha\|f\|^2,$$

where K is the integral operator of Section 4.4.

Important to the application of Tikhonov regularization is the choice of the value of the α regularization parameter, which can be done in various ways. An *a priori* choice for α, discussed by Engl (1993), is given by α which satisfies the conditions $\lim_{\delta\to 0} \delta^2/\alpha(\delta) = 0$ and $\lim_{\delta\to 0} \alpha(\delta) \to 0$. For example, $\alpha = \delta^{1.9}$ qualifies and has given satisfactory computational results in numerical trials of prediction. A choice of regularization parameter for band-limited extrapolation that is dependent on statistical properties of the signal is discussed by Wingham (1992).

In general, the Morozov discrepancy principle is seen as an *a posteriori* choice for determining α, but may here be based on the data samples and computed before the signal value prediction. The main equation which determines α is

$$\|Kf_\alpha - y_\delta\| = \delta, \tag{4.7.2}$$

where y_δ are the noisy data samples. The norm here is the Euclidean norm in R^N, so that this equation may be written by using the prediction formula evaluated at the data points t_j:

$$\sqrt{\sum_{j=1}^{N}(f_\alpha(t_j) - y_\delta(t_j))^2} = \delta, \tag{4.7.3}$$

where $f_\alpha(t)$ is given by (4.7.1), and equation (4.7.3) is to be solved for α. The output values of the Kf_α operator, $f_\alpha(t_j)$, $j = 1, \ldots, N$, are the values of the approximation *at the sample times* and may be computed from (4.7.1) prior to that of a prediction. Since the noisy samples $y_\delta(t_j)$ are also known, equation (4.7.2) may be solved for the value of the regularization parameter α prior to computing the prediction value.

In Tables 4.6 and 4.7 some computational results are presented concerning the choice of regularization parameter α determined as described above by the Morozov discrepancy principle. In each case, the value of α was computed that solves equation (4.7.3) computationally. The tables list values for α of several magnitudes on either side of the computational solution, with the remainder term in each table being the value of the difference, $(\sum_{j=1}^{N}(f_\alpha(t_j) - y_\delta(t_j))^2)^{1/2} - \delta$. In the first table, the value of α was computed for a prediction for $f(t) = \mathrm{sinc}^2(t/2)$ with $N = 6$ samples, $TW = 0.10$, and noise error bounded by $\delta = 1.0e - 04$. The value of α in this case was $\alpha = 1.3465e - 07$.

Notice that from Table 4.6 the minimum value of the error in the prediction, i.e. $2.1e - 05$, does occur at the value of the regularization parameter α where the Morozov

Table 4.6 *Various values of regularization parameter, corresponding errors in the prediction, and the remainder in the Morozov equation (4.7.2). $N = 6$, $TW = 0.10$, and noise level 10^{-4}*

α	prediction error	Morozov remainder
$1.3e - 09$	$5.0e - 04$	$4.0e - 05$
$1.3e - 08$	$4.4e - 04$	$3.5e - 05$
$1.3e - 07$	$2.1e - 05$	$2.3e - 11$
$1.3e - 06$	$4.5e - 04$	$4.0e - 05$
$1.3e - 05$	$1.5e - 03$	$2.8e - 04$

Table 4.7 *Values of regularization parameter, corresponding errors in the prediction, with $N = 8$, noise level 10^{-6} and the remainder in the Morozov equation* (4.7.2). $N = 8$, $TW = 0.10$, *and noise level* 10^{-6}

α	prediction error	Morozov remainder
$9.3e - 17$	$1.1e - 04$	$8.7e - 07$
$9.3e - 16$	$7.7e - 05$	$6.4e - 07$
$9.3e - 15$	$4.7e - 06$	$4.3e - 09$
$9.3e - 14$	$1.9e - 05$	$2.5e - 07$
$9.3e - 13$	$1.9e - 05$	$3.0e - 07$

remainder is smallest, as specified from the Morozov discrepancy principle. A similar result is shown in Table 4.7 for the case of $N = 8$ samples, $TW = 0.10$, and noise error bounded by $\delta = 1.0e - 06$. The value of α in this case was $\alpha = 9.313e - 15$.

The method of Tikhonov regularization can be carried further under a process called iterated Tikhonov regularization (Engl 1993). An iteration method is described by initializing $f_{\alpha,0} = 0$, and calculating successive iterates $f_{\alpha,1}, f_{\alpha,2}, \ldots,$ by

$$(K_a \cdot K + \alpha I) f_{\alpha,j} = K_a y_\delta + \alpha f_{\alpha,j-1}$$

for $j = 1, \ldots$. This iteration is equivalent to the formula

$$f_{\alpha,j}(t) = \sum_{k=1}^{N} \frac{(\sigma_k^2 + \alpha)^j - \alpha^j}{\sigma_k^2(\sigma_k^2 + \alpha)^j} \sigma_k < y_\delta, v_k > u_k(t). \qquad (4.7.4)$$

The iteration formula (4.7.4) reduces to the previous formula (4.7.1) when $j = 1$. To see this, note that in this case

$$\frac{(\sigma_k^2 + \alpha)^j - \alpha^j}{\sigma_k^2(\sigma_k^2 + \alpha)^j} = \frac{(\sigma_k^2 + \alpha) - \alpha}{\sigma_k^2(\sigma_k^2 + \alpha)} = \frac{1}{\sigma_k^2 + \alpha}$$

which is the coefficient in the Tikhonov regularization formula (4.7.1).

The error from the iterated method tends to vary less with the regularization parameter α than does the error in the non-iterated case. Unfortunately, it seems difficult to determine the optimal number of iterations, j. In Table 4.8 are some examples of predictions with iterated Tikhonov regularization with $j = 5$ iterations and regularization parameter $\alpha = \delta^{1.9}$ for the $TW = 0.10$ case.

Although Tikhonov regularization with parameter determined by the Morozov discrepancy principle is the primary technique discussed here, the method of the truncated singular value expansion would also be appropriate to suppress the error due to noisy samples. This method employs a cut-off parameter, α, so that the prediction formula

$$f(t) = \sum_{k=1}^{N} \frac{\langle y_\delta, v_k \rangle}{\sigma_k} u_k(t),$$

Table 4.8 *Different noise levels and the error in iterated Tikhonov regularization compared to the error in the uncorrected prediction*

Noise 1.0e − 3	N	uncorrected	iterated Tikhonov
	8	1.1e − 01	1.5e − 03
	9	2.3e − 02	1.9e − 03
	10	4.4e − 02	1.1e − 04
Noise 1.0e − 4	N	uncorrected	iterated Tikhonov
	8	1.1e − 02	1.2e − 04
	9	2.3e − 03	1.6e − 04
	10	4.4e − 03	5.0e − 06
Noise 1.0e − 5	N	uncorrected	iterated Tikhonov
	8	1.1e − 03	1.7e − 05
	9	2.3e − 04	1.3e − 05
	10	4.4e − 04	1.7e − 05

Table 4.9 *The size of the modified prediction coefficients from (4.7.5). Compare to Table 4.3*

1.9543	−0.9869						
2.9287	−2.9164	0.9873					
3.9037	−5.7964	3.8794	−0.9875				
4.8789	−9.6275	9.6037	−4.8426	0.9876			
5.8542	−14.4101	19.0883	−14.3508	5.8059	−0.9876		
4.9955	−9.3842	6.7206	2.0272	−6.5031	3.9906	−0.8465	
5.9532	−14.1600	15.6143	−4.0706	−9.0364	10.7723	−4.9369	0.8641
5.2689	−10.2676	7.1611	2.9729	−5.8194	−1.3927	6.0374	−3.7287 0.7681

is modified by the index to

$$f_\alpha(t) = \sum_{k=1, \sigma_k^2 \geq \alpha}^{N} \frac{\langle y_\delta, v_k \rangle}{\sigma_k} u_k(t), \qquad (4.7.5)$$

which simply removes the small singular values.

With either of the methods of Tikhonov regularization or truncated singular value expansion to suppress the effect of noisy samples, the size of the prediction coefficients is not as large as the unmodified case discussed in Section 4.5. Table 4.9 shows the size of the prediction coefficients for the case of $TW = 0.10$ when the truncated singular value expansion formula (4.7.5) is applied. The noise standard deviation in this case was 10^{-6}. Compared to Table 4.3, these coefficients are much smaller than in the unmodified case, and this helps in suppressing noise error.

The use of these methods to improve accuracy can result in dramatic improvements of several magnitudes. This was indicated in Table 4.8. As another example, in a prediction of $f(t) = \text{sinc}^2(t/2)$ where $N = 11$ equally spaced samples include zero-mean Gaussian

Table 4.10 *Table of error values under different noise levels, N; the number of samples is the vertical scale, noise level is the horizontal scale*

N	$1.0e - 04$	$1.0e - 06$	$1.0e - 08$	$1.0e - 12$
4	$-2.5e - 03$	$-2.5e - 03$	$-2.5e - 03$	$-2.5e - 03$
5	$-5.4e - 04$	$-3.8e - 04$	$-3.7e - 04$	$-3.7e - 04$
6	$-2.8e - 04$	$-1.1e - 04$	$-1.0e - 04$	$-1.0e - 04$
7	$9.0e - 04$	$-5.0e - 05$	$-1.6e - 05$	$-1.6e - 05$
8	$-6.2e - 04$	$-9.9e - 05$	$-6.0e - 06$	$-5.1e - 06$
9	$-5.4e - 04$	$2.3e - 05$	$-3.2e - 06$	$-8.5e - 07$
10	$2.0e - 04$	$-2.2e - 05$	$-5.5e - 06$	$-3.1e - 07$
11	$-3.5e - 04$	$-6.2e - 06$	$-1.3e - 07$	$-7.1e - 08$
12	$-2.1e - 04$	$6.9e - 06$	$7.3e - 08$	$-9.4e - 08$
13	$1.5e - 05$	$2.2e - 05$	$-6.9e - 07$	$-4.8e - 08$
14	$3.8e - 04$	$5.5e - 06$	$-1.7e - 07$	$-7.5e - 08$
15	$-3.9e - 04$	$-2.1e - 05$	$-3.3e - 07$	$-1.4e - 07$
16	$2.4e - 04$	$-2.1e - 05$	$6.1e - 07$	$-4.1e - 08$
17	$6.5e - 05$	$-5.1e - 06$	$1.2e - 06$	$-5.3e - 08$
18	$-3.0e - 04$	$-3.2e - 06$	$1.9e - 07$	$1.3e - 06$
19	$-3.8e - 04$	$1.1e - 06$	$-1.2e - 07$	$-3.4e - 08$
20	$-1.2e - 04$	$1.3e - 06$	$4.1e - 07$	$-5.9e - 08$

noise with standard deviation 1×10^{-6}, the unmodified formula (4.4.2) results in an error of 4.2×10^{-4}. However, when modified by the truncated singular value expansion with $\alpha \approx 10^{-10}$ in (4.7.5), the error is 9.1×10^{-6}. Thus, the result is two orders of magnitude better when modified, with error near the noise level.

To summarize the error in the prediction relative to different noise levels, Table 4.10 is presented. These computational results are for the band-limited signal $f(t) = \text{sinc}^2(t/2)$ under uniform sampling, $TW = 0.20$, and using the truncated singular value expansion, with $\alpha = \delta^{1.9}$. The different noise levels presented in the table are for normally distributed random noise with standard deviations of $1.0e - 04$, $1.0e - 06$, $1.0e - 08$, and $1.0e - 12$.

Note that the results listed in Table 4.10 show that accuracy comparable to the noise level is attained by the prediction at one step ahead for a small number of samples when $TW = 0.20$. For all of the high noise cases, the prediction has comparable accuracy to the sample data.

4.8 Non-uniformly spaced samples

Examples of computational results presented in previous sections have been for prediction based on equally spaced sample data, but there is no necessity for this restriction. In the case of non-uniformly spaced samples, the system (4.3.3) is no longer Toeplitz but is symmetric, and the svd-based prediction formula (4.4.2) applies as before. In this section, we consider two modifications of the equally spaced set of samples. The modifications considered are those which may realistically occur in practice: (i) gaps in data points, and (ii) jittered samples. For both of these cases, application of the methods in the previous sections are shown to achieve levels of accuracy similar to the equally spaced samples case. See also (Wingham 1992).

The first case considered in this section is a modification of uniform sampling that assumes samples are uniformly spaced with sampling interval T, but there is a significant gap in the time data when samples of the signal are unknown. As before, the prediction coefficients a_n^N may be found from the expression (4.4.3) and the error estimated from the error estimate (4.2.3).

As a computational example, consider a prediction of $f(t) = \text{sinc}^2(t/2)$ at $t = 0$, with samples spaced so that $TW = 0.10$. In this case, only $N = 10$ samples are used, with samples taken at times $t - nT$ for $n \in \{1, 2, 3, 4, 8, 9, 10, 11, 12, 13\}$. As for uniform sampling, the methods of Section 4.7 can be applied to suppress the error due to noise. In this example, iterated Tikhonov regularization (with $j = 5$ iterations) is used with regularization parameter $\alpha = \delta^{1.9}$ for normally distributed noise with standard deviation $1.0e - 06$. The emphasis in this chapter is to determine the prediction just ahead of the data, and Table 4.11 gives the prediction for points about twenty steps ahead of the data.

Note that the errors given in Table 4.11 are on the order of the errors in signal samples due to noise for the first three steps ahead of the data. The error gets successively worse as one looks at predictions further ahead of the data, but these errors remain relatively stable. This could be generalized further to extrapolation methods, as discussed by Wingham (1992).

The error estimate for a signal prediction may be done nearly *independently of the particular signal* using (4.2.2). For example, Table 4.12 presents first the error bound (as the numerically integrated value of the integral part of the error bound from (4.2.2)) and the actual prediction error for a prediction for $f(t) = \text{sinc}^2(t/2)$. Each of these values is presented for the three cases of (i) no gap in data, (ii) gap of size 3 starting with fifth data point, and (iii) gap of size 6 starting with fifth data point. For example, for case (iii) and $N = 8$, the sample points are at $t = -0.2, -0.4, -0.6, -0.8, -2.2, -2.4, -2.6$, and -2.8. Each of these is for the case that $TW = 0.20$.

Note that Table 4.12 shows that as the size of the gap increases, both the prediction error estimate and the resulting prediction error are degraded slightly, but are nearly as good as before. The complete error bound is as specified in (4.2.2), and includes the signal energy as well as the error integral value that is reported in Table 4.12.

Second, we consider the case of time-jittered samples. We assume that samples are taken uniformly, but that there is a jitter error in the precise time at which the sample is taken. We suppose that the time at which the samples are taken is recorded, so that the jittered times t_n in the prediction data are known. As a computational example, consider

Table 4.11 *Error in prediction (noise level 10^{-6}) with a gap between data*

t	0.0	0.1	0.2	0.3	0.4	0.5	0.6
error	$2.4e - 06$	$2.5e - 06$	$3.2e - 06$	$2.5e - 05$	$8.0e - 05$	$2.0e - 04$	$4.1e - 04$

t	0.7	0.8	0.9	1.0	1.1	1.2	1.3
error	$7.8e - 04$	$1.3e - 03$	$2.2e - 03$	$3.4e - 03$	$5.0e - 03$	$7.1e - 03$	$9.8e - 03$

t	1.4	1.5	1.6	1.7	1.8	1.9	2.0
error	$1.3e - 02$	$1.7e - 02$	$2.1e - 02$	$2.6e - 02$	$3.1e - 02$	$3.6e - 02$	$4.1e - 02$

Table 4.12 *Error bound and actual prediction error for increasing gap sizes in a prediction of $f(t) = sinc^2(t/2)$*

N	8	9	10	11	12
no gap. error integral	1.0e − 04	3.2e − 05	9.8e − 06	3.2e − 06	6.6e − 06
actual error	5.1e − 06	8.5e − 07	3.1e − 07	3.7e − 07	1.2e − 07
gap size 3. error integral	4.5e − 04	1.8e − 04	7.2e − 05	2.8e − 05	1.1e − 05
actual error	3.0e − 05	7.2e − 06	2.9e − 06	7.4e − 07	2.8e − 07
gap size 6. error integral	1.3e − 03	6.3e − 04	3.0e − 04	1.4e − 04	6.2e − 05
actual error	1.2e − 04	4.0e − 05	1.5e − 05	5.4e − 06	2.2e − 06

Table 4.13 *Prediction error for jittered data points (noise level 10^{-8})*

jitter/N	8	9	10	11	12
0.0	5.3e − 06	1.0e − 06	2.2e − 06	1.3e − 06	9.1e − 07
1.0e − 04	5.7e − 06	3.0e − 06	2.0e − 06	3.8e − 07	8.1e − 07
1.0e − 02	5.6e − 06	2.8e − 06	1.8e − 06	4.3e − 08	5.0e − 07

a prediction one step ahead of the data for the case when $TW = 0.20$ for $f(t) = sinc^2(t/2)$. As before, we assume random noise in the signal samples that is normally-distributed of standard deviation $1.0e - 08$ and apply the Tikhonov regularization method with regularization parameter $\alpha = 1.0e - 13$. In Table 4.13, the error is listed in this prediction for the two cases when the jitter error is applied randomly with standard deviations $1.0e - 04$ and $1.0e - 02$. For example, in the case of random jitter of level $1.0e - 02$, this makes the time data points at -0.19749, -0.40310, -0.60923, -0.80385, -0.98842, -1.19137, -1.41035, and -1.60193 for the case of $N = 8$ samples.

The errors listed in Table 4.13 are on about the same order as for the non-jittered case. In addition, the error bounds obtained from equation (4.2.2) can again be applied to this case. Similar to Table 4.12, the error integral could be numerically integrated to give an error upper bound on any prediction calculation, with that error bound nearly independent of the individual signal, depending only on the signal energy, the prediction point t and the set of sample points t_n.

4.9   Conclusion

Theory and computational methods for linear prediction of the values of a band-limited signal based on past samples have been presented in the previous sections. The advantage of a prediction for this class of signals is that the error (4.2.2) is not dependent on the particular signal but only on its energy and the highest frequency, as well as the set of sample times chosen for data collection.

Different numerical methods for the computation of the optimal set of prediction coefficients were discussed in Sections 4.3, 4.4, and 4.6. As noted by the growth of the values in Table 4.3, even if the appropriate set of linear equations (4.3.4) was able to be numerically solved accurately, the resulting prediction formula would not be optimal with respect to sensitivity to noise or quantization error in the data. The use of the singular value expansion of (4.4.2) and the regularization methods of Section 4.7 to suppress the effects of noise or quantization in the data can gain accuracy of several orders of magnitude. One of the chief difficulties in the regularization methods is the choice of the value of the regularization parameter, and the Morozov discrepancy principle was shown in Section 4.7 to apply to the prediction in a particularly useful form. Examples were presented that indicate that this principle may be quite valid in producing the best value of the regularization parameter based on available information.

Further study of the sensitivity of the regularization parameter α to the estimate of data error δ as computed from the Morozov discrepancy principle in (4,7.2) is needed to determine if the application of this principle can be done in general. With reference to non-uniform sampling, it is not clear that the system (4.3.3) is as badly conditioned as it is for the case of equally spaced samples, as some initial work has shown it to be better conditioned. Whether that case is actually better conditioned than the case of equally spaced samples, and the implications for the singular value expansion method of solution, would also require further study.

5

INTERPOLATION AND SAMPLING THEORIES AND LINEAR ORDINARY BOUNDARY VALUE PROBLEMS

5.1 Introduction

This chapter is devoted to a study of the original Kramer sampling theorem, the link with the interpolation theorem of Shannon–Whittaker, and the connection of these two significant results with boundary value problems associated with linear ordinary differential equations as defined on intervals of the real line. The keystone of this chapter is the connection between these boundary value problems and a generalized form of the original Kramer theorem; this generalization introduces analytic dependence of the kernel on the sampling parameter.

The results given in this chapter are connected with boundary value problems that generate unbounded, self-adjoint differential operators in Hilbert function spaces but confined to those operators that have a simple, discrete spectrum, i.e. the spectrum contains only eigenvalues all of which are of the first order.

5.1.1 *Notation*

$\mathbb{Z}, \mathbb{N}, \mathbb{N}_0$ denote the sets of all integers, the positive integers and the non-negative integers respectively; let $\mathbb{R}, \mathbb{C}$ denote the real and complex number fields respectively. The symbol $\mathbf{H}(\mathbb{C})$ denotes the class of all Cauchy analytic functions on $\mathbb{C}$. The symbols $I = (a, b)$ and $[\alpha, \beta]$ represent open and compact intervals of $\mathbb{R}$. The symbol "$(x \in K)$" is to be read as "for all elements x belonging to the set K".

Let $w : I \to \mathbb{R}$ be Lebesgue measurable and satisfy $w(x) > 0$ for almost all $x \in I$; the Hilbert function space $L^2(I; w)$ is the set of all Lebesgue measurable functions $f : I \to \mathbb{C}$ such that $\int_I w(x)|f(x)|^2 \, dx < \infty$, and then, with due regard to equivalence classes, the norm and inner product are given by

$$\|f\|^2 = \int_I w(x)|f(x)|^2 \, dx \qquad (f, g) = \int_I w(x)f(x)\overline{g}(x) \, dx. \qquad (5.1.1)$$

5.1.2 *The Kramer theorem*

This theorem has played a very significant role in sampling theory, interpolation theory, signal analysis and, generally, in mathematics; see the survey articles by Butzer and Nasri-Roudsari (1997) and Butzer et al. (1988), and the book by Higgins (1996). The original form of this theorem is given by Kramer (1959); we restate the result here with some minor changes.

Theorem 5.1 *Let $I = (a, b)$ be given and let the mapping $K : I \times \mathbb{C} \to \mathbb{C}$ satisfy the following given properties:*

   (i) *K is Lebesgue measurable on $I \times \mathbb{C}$;*

(ii) $K(\cdot, t) \in L^2(I)$ ($t \in \mathbb{R}$);

(iii) *for a given sequence* $\{ t_n \in \mathbb{R} : n \in \mathbb{Z} \}$ *of real numbers, with* $t_r \neq t_s$ ($r \neq s$), *the collection* $\{ K(\cdot, t_n) : n \in \mathbb{Z} \}$ *forms a complete, orthogonal set in* $L^2(I)$.

Define

(a) $\{ S_n : \mathbb{R} \to \mathbb{C} \mid n \in \mathbb{Z} \}$ *by*

$$S_n(t) := \| K(\cdot, t_n) \|^{-2} \int_I K(x, t) \overline{K}(x, t_n) \, dx \quad (t \in \mathbb{R} \text{ and } n \in \mathbb{Z}); \quad (5.1.2)$$

(b) $\{K\}$ *as the collection of functions* $F : \mathbb{R} \times L^2(I) \to \mathbb{C}$ *by*

$$F(t) := \int_I K(x, t) f(x) \, dx \quad (t \in \mathbb{R}; \ f \in L^2(I)). \quad (5.1.3)$$

Then for all $F \in \{K\}$

$$F(t) = \sum_{n \in \mathbb{Z}} F(t_n) S_n(t) \quad (t \in \mathbb{R}) \quad (5.1.4)$$

where the series is absolutely convergent in $\mathbb{C}$ *for each* $t \in \mathbb{R}$.

Additionally if $\{ c_n \in \mathbb{C} : n \in \mathbb{Z} \}$ *is a sequence of complex numbers such that*

$$\sum_{n \in \mathbb{Z}} \frac{|c_n|^2}{\| K(\cdot, t_n) \|^2} < \infty \quad (5.1.5)$$

then there exists a unique $F \in \{K\}$, *equivalently a unique* $f \in L^2(I)$ *with* F *given by* (5.1.3), *such that*

$$F(t_n) = c_n \quad (n \in \mathbb{Z}). \quad (5.1.6)$$

Remarks

(a) For a proof of (5.1.4) see Kramer (1959), and for a proof of (5.1.6) see Everitt, Schöttler and Butzer (1994, Section 10).

(b) The result (5.1.4) is the essential result of the Kramer sampling theorem; given $F \in \{K\}$ it is possible to recover all the values of F from a knowledge of the sequence of values $\{ F(t_n) : n \in \mathbb{Z} \}$.

(c) The result (5.1.6) is an interpolation property; with the restriction (5.1.5) it is possible to interpolate from the given sequence $\{ c_n : n \in \mathbb{Z} \}$ to a unique $F \in \{K\}$.

(d) The form of this Kramer theorem given above uses the index set $\mathbb{Z}$ to label the sequence $\{ t_n : n \in \mathbb{Z} \}$; this has been done to cover the general situations that appear in practice, particularly for boundary value problems when cases arise for which $\lim_{n \to \pm\infty} t_n = \pm\infty$; the results are essentially the same as for the index sets $\mathbb{N}$ and $\mathbb{N}_0$, for which $\lim_{n \to \infty} t_n = \infty$, as considered originally in the examples by Kramer (1959).

5.1.3 *Boundary value problems*

The richest source for the construction of Kramer kernels is the study of symmetric (self-adjoint) boundary value problems generated by ordinary, linear differential equations defined on intervals of the real line $\mathbb{R}$. These boundary value problems are best considered in the context of the general theory of quasi-derivatives; see Everitt (1986) and Everitt and Zettl (1991). These problems are described below in their most general form and then illustrated with two significant and well-known examples.

Let $n \in \mathbb{N}$ be given and let M_A be a Lagrange symmetric, quasi-differential expression of order n on an open interval $I = (a, b)$ of the real line $\mathbb{R}$; in M_A the $n \times n$ matrix A is in the Shin–Zettl class $Z_A(I)$ and satisfies the symmetry requirement $A^+ = A$. For $n \geq 2$ full details of the definition of M_A are given by Everitt (1986, Section 10); for the classical case of $n \geq 1$ see the standard text by Coddington and Levinson (1955, Chapter 7), and Everitt and Race (1987).

The differential equation associated with M_A is

$$M_A[y] = \lambda w y \quad \text{on} \quad (a, b) \tag{5.1.7}$$

with $\lambda = \mu + i\nu \in \mathbb{C}$ as the spectral parameter and w a positive weight as given in Section 5.1.1 above.

The solutions of (5.1.7) are considered in the Hilbert function space

$$L^2((a, b); w),$$

see Section 5.1.1 above. To define symmetric boundary value problems in this space, linear boundary conditions have to be adjoined; these conditions are of the form (note that these conditions involve both the end-points a and b of (a, b))

$$[y, \beta_r]_A \equiv [y, \beta_r]_A(b) - [y, \beta_r]_A(a) = 0 \quad (r = 1, 2, \ldots, d). \tag{5.1.8}$$

Here $[\cdot, \cdot]_A$ is the skew-symmetric bilinear form taken from the Green's formula for M_A; the family $\{\beta_r : r = 1, 2, \ldots, d\}$ is a linearly independent set of maximal domain functions chosen to satisfy the symmetry condition

$$[\beta_r, \beta_s]_A = 0 \quad (r, s = 1, 2, \ldots, d). \tag{5.1.9}$$

The integer $d \in \mathbb{N}_0$ is the common deficiency index of the differential equation (5.1.7), determined in the space $L^2(I; w)$, and specifies the number of boundary conditions required for the boundary value problem (5.1.7 and 5.1.8) to be symmetric, i.e. to generate a self-adjoint operator in $L^2(I; w)$; if $d = 0$ then no boundary conditions are required at either end-point.

Within this framework the boundary value problem generates a uniquely determined, unbounded, self-adjoint operator T in the space $L^2(I; w)$. For details of the results leading to the determination of the operator T see Everitt and Zettl (1991, Section 1).

If the problem is regular on an interval (a, b), in which case this interval has to be bounded, then $d = n$ and the generalized boundary conditions (5.1.8) involve the point-wise values of the solution y and its quasi-derivatives at the end-points a and b. In this regular case when the order $n = 2m$ is even and the Lagrange symmetric matrix is real-valued, the corresponding formulae are given in the book by Naimark (1968, Section 18.2).

In the case $n = 2$, essentially the Sturm–Liouville case, the index d may take the value 0, 1 or 2; this value depends on the regular/limit-point/limit-circle classification, in $L^2(I; w)$, at the end-points a and b of the differential expression M_A; see Titchmarsh (1962, Chapter II), Coddington and Levinson (1955, Chapter 9) and Everitt (1995).

All the classical boundary value problems for symmetric (i.e. formally self-adjoint) differential equations on intervals of $\mathbb{R}$, in both the regular and singular cases, are contained within the general formulation given by (5.1.7) and (5.1.8) of quasi-differential problems; for the link between the classical and quasi-differential systems see the paper by Everitt and Race (1987).

Two examples of this general boundary value problem are now given; both problems are classical, in the sense that classical derivatives are to be observed, but are formulated in the more general quasi-differential sense.

Example 5.2 For $\sigma > 0$ consider the boundary value problem

$$iy'(x) = \lambda y(x) \quad (x \in [-\sigma, \sigma]) \tag{5.1.10}$$

$$y(-\sigma) = y(\sigma). \tag{5.1.11}$$

For the differential equation (5.1.10) the general solution has a basis

$$\{\exp(-ix\lambda) : x \in [-\sigma, \sigma] \quad \text{and} \quad \lambda \in \mathbb{C}\};$$

all solutions are in $L^2(I; w) \equiv L^2(-\sigma, \sigma)$ and thus $d = 1$. The bilinear form for (5.1.10) is given by

$$[f, g](x) = if(x)\overline{g}(x) \quad (x \in [-\sigma, \sigma])$$

and the boundary condition (5.1.11) can be rewritten in the form

$$[y, 1] = [y, 1](\sigma) - [y, 1](-\sigma) = 0$$

where 1 represents the unit function on $\mathbb{R}$. The self-adjoint operator T for this example is determined by

$$D(T) := \{f : [-\sigma, \sigma] \to \mathbb{C} \mid f \in AC[-\sigma, \sigma], \ f' \in L^2(-\sigma, \sigma), \ [f, 1] = 0\}$$

$$Tf := if' \quad (f \in D(T)).$$

From this representation of T, or by a direct calculation from the classical formulation (5.1.10 and 5.1.11) of the problem, it follows that ($\sigma(T)$ is here used for "spectrum of T")

$$\sigma(T) = \{\lambda_n = n\pi/\sigma : n \in \mathbb{Z}\}; \tag{5.1.12}$$

this spectrum is discrete, simple and satisfies

$$\lim_{n \to \pm\infty} \lambda_n = \pm\infty;$$

for the latter property see Remark (d), Section 5.1.2 above.

The corresponding eigenfunctions are given explicitly by (for completeness of this set in $L^2(-\sigma, \sigma)$ see Higgins (1977, p. 36)),

$$\psi_n(x) = \exp(-ixn\pi/\sigma) \quad (x \in [-\sigma, \sigma] \quad \text{and} \quad n \in \mathbb{Z}).$$

This example is considered again below when the Shannon sampling theorem is discussed.

Example 5.3 This is the classical Legendre differential equation

$$-((1 - x^2) y'(x))' + \tfrac{1}{4}y(x) = \lambda y(x) \quad (x \in (-1, 1)) \tag{5.1.13}$$

with singular boundary conditions

$$[y, \beta_1] = [y, \beta_2] = 0. \tag{5.1.14}$$

This problem is considered in the space $L^2(-1, 1)$, since the weight $w \equiv 1$ on $(-1, 1)$. The differential equation (5.1.13) is in the limit-circle classification in $L^2(-1, 1)$ at both end-points ± 1 and so for this boundary value problem $d = 2$.

In (5.1.14) β_1 is in the maximal domain for (5.1.13), and is defined by

$$\beta_1(x) = 1 \quad (x \in [\tfrac{1}{2}, 1]) \quad \text{and} \quad \beta_1(x) = 0 \quad (x \in [-1, 0])$$

and by a real-valued polynomial in $[0, \tfrac{1}{2}]$ so that $\beta_1 \in C^{(2)}[-1, 1]$; β_2 is then defined by

$$\beta_2(x) = \beta_1(-x) \quad (x \in [-1, 1]).$$

Then the pair $\{\beta_1, \beta_2\}$ is a linearly independent set and satisfies the symmetry condition

$$[\beta_r, \beta_s] = 0 \quad (r, s = 1, 2).$$

The symmetric boundary conditions (5.1.14) reduce to

$$\left.\begin{array}{l} 0 = [y, \beta_1] = [y, \beta_1](1) - [y, \beta_1](-1) \quad \text{i.e. } [y, 1](1) = 0 \\ 0 = [y, \beta_2] = [y, \beta_2](1) - [y, \beta_2](-1) \quad \text{i.e. } [y, 1](-1) = 0. \end{array}\right\} \tag{5.1.15}$$

These boundary conditions are equivalent to the conditions requiring that the eigenfunctions be bounded on the interval $(-1, 1)$.

The symmetric boundary value problem (5.1.13) and (5.1.15) is considered in Everitt et al. (1994, Section 11, Example 3).

The corresponding self-adjoint operator T is defined by

$$D(T) := \{ f : (-1, 1) \to \mathbb{C} \mid f, f' \in AC_{loc}(-1, 1) \text{ and}$$
$$f, ((1 - x^2)f')' \in L^2(-1, 1) \text{ and}$$
$$[f, \beta_1] = [f, \beta_2] = 0 \}$$

and

$$Tf := -((1 - x^2)f')' + \tfrac{1}{4}f \quad (f \in D(T)).$$

From known results

$$\sigma(T) = \{ \lambda_n = (n + \tfrac{1}{2})^2 : n \in \mathbb{N}_0 \}; \qquad (5.1.16)$$

this spectrum is simple and discrete. The corresponding eigenfunctions are given by

$$\psi_n(x) = P_n(x) \quad (x \in (-1, 1) \text{ and } n \in \mathbb{N}_0),$$

i.e. the Legendre polynomials; for completeness in $L^2(-1, 1)$ see Higgins (1977, p. 33).

As indicated earlier these examples yield Kramer kernels that satisfy all the requirements of Theorem 5.1, but now with the parameter t replaced by the spectral parameter λ. For the two examples of boundary value problems considered in this section it is possible to give explicit forms of the Kramer kernel as follows:

(a) The kernel for Example 5.2 is determined as follows

$$K(x, t) = K(x, \lambda) = \exp(-ix\lambda) \quad (x \in [-\sigma, \sigma] \text{ and } \lambda \in \mathbb{C})$$

with

$$t_n = \lambda_n = n\pi/\sigma \quad (n \in \mathbb{Z})$$

and

$$K(x, \lambda_n) = \exp(-ixn\pi/\sigma) \quad (x \in [-\sigma, \sigma] \text{ and } \lambda \in \mathbb{C}).$$

(b) The kernel for Example 5.3 is determined as follows (here $_2F_1$ is the Gaussian hypergeometric function),

$$K(x, t) = K(x, \lambda) = {}_2F_1\left(-\sqrt{\lambda} + \tfrac{1}{2}, \sqrt{\lambda} + \tfrac{1}{2}; 1; \tfrac{1}{2}(1 - x)\right)$$

with $x \in (-1, 1)$ and $\lambda \in \mathbb{C}$, and

$$t_n = \lambda_n = (n + \tfrac{1}{2})^2 \quad (n \in \mathbb{N}_0)$$

$$K(x, \lambda_n) = P_n(x).$$

For additional details of this example see Everitt et al. (1994, Example 3, Section 11).

5.1.4   *The analytic form of the Kramer theorem*

The Kramer kernels that arise from the two examples considered in the previous section have an important property not foreseen in the statement of Theorem 5.1, i.e.

$$K(x, \cdot) \in \mathbf{H}(\mathbb{C}) \quad (x \in I).$$

Here $\mathbf{H}(\mathbb{C})$ denotes the class of Cauchy analytic functions that are holomorphic (regular) on the whole complex plane $\mathbb{C}$. This additional analytic property of the kernel is observed to be typical in the Kramer kernels generated by the symmetric boundary value problems given in (5.1.7), (5.1.8) and (5.1.9).

We now state an analytic form of the Kramer theorem to allow for this additional property.

Theorem 5.4 *Let* $I = (a, b)$ *and the space* $L^2(I; w)$ *be given; let the mapping* $K : I \times \mathbb{C} \to \mathbb{C}$ *satisfy the following properties:*

(i) *K is Lebesgue measurable on* $I \times \mathbb{C}$;

(ii) $K(\cdot, \lambda) \in L^2(I; w)$ $(\lambda \in \mathbb{C})$;

(iii) *let a given sequence* $\{ \lambda_n \in \mathbb{R} : n \in \mathbb{Z} \}$ *of real numbers satisfy the conditions*

$$\lambda_n < \lambda_{n+1} \quad (n \in \mathbb{Z}) \qquad \lim_{n \to \pm\infty} \lambda_n = \pm\infty;$$

(iv) *let the collection* $\{ K(\cdot, \lambda_n) : n \in \mathbb{Z} \}$ *form a complete orthogonal set in* $L^2(I; w)$;

(v) $K(x, \cdot) \in \mathbf{H}(\mathbb{C})$ $(x \in I)$;

(vi) *let K satisfy the boundedness condition*

$$\lambda \longmapsto \int_I |K(x, \lambda)|^2 w(x)\, dx \text{ is locally bounded on } \mathbb{C}.$$

Define

(1) $\{ S_n : \mathbb{C} \to \mathbb{C} \mid n \in \mathbb{Z} \}$ *by*

$$S_n(\lambda) := \| K(\cdot, \lambda_n) \|_w^{-2} \int_I K(x, \lambda)\overline{K}(x, \lambda_n) w(x)\, dx \quad (\lambda \in \mathbb{C}; n \in \mathbb{Z});$$

(2) $\{K\}$ *as the collection of functions* $F : \mathbb{C} \to \mathbb{C}$ *with*

$$F(\lambda) := \int_I K(x, \lambda) f(x) w(x)\, dx \quad (\lambda \in \mathbb{C}; f \in L^2(I; w)).$$

Then

(a) $S_n \in \mathbf{H}(\mathbb{C})$ $(n \in \mathbb{Z})$;

(b) $F(\cdot) \in \mathbf{H}(\mathbb{C})$ $(F \in \{K\}$ *i.e.* $f \in L^2(I; w))$;

(c) *for all* $F \in \{K\}$

$$F(\lambda) = \sum_{n \in \mathbb{Z}} F(\lambda_n) S_n(\lambda) \quad (\lambda \in \mathbb{C})$$

where the series is absolutely convergent in $\mathbb{C}$ *for each* λ, *and locally uniformly convergent on* $\mathbb{C}$;

(d) *if* $\{ c_n \in \mathbb{C} : n \in \mathbb{Z} \}$ *is a sequence of complex numbers that satisfies the condition*

$$\sum_{n \in \mathbb{Z}} \frac{|c_n|^2}{\| K(\cdot, \lambda_n) \|_w^2} < \infty$$

then there exists a unique $F \in \{K\}$, *equivalently a unique* $f \in L^2(I; w)$, *such that*

$$F(\lambda_n) = c_n \quad (n \in \mathbb{Z}).$$

Proof For a proof of this result and, in particular, the need for the local boundedness condition (vi) in the statement of Theorem 5.4, see the note of Everitt et al. (to appear). The proof of result (c) follows the same lines as the original Kramer proof of Theorem 5.1; however the local uniform convergence of the sampling series requires the local boundedness condition (vi).

The proof of the holomorphic properties (a) and (b) also requires the local boundedness condition (vi). For example to prove that $F \in \mathbf{H}(\mathbb{C})$ it is possible to prove that F is continuous on $\mathbb{C}$; this requires the condition (vi), the Cauchy integral representation theorem and the Fubini double integral theorem. With this continuity established an application of the theorem of Morera together with another application of Fubini's theorem yields the result that $F \in \mathbf{H}(\mathbb{C})$; see also the results in Titchmarsh (1939, Chapter II, Sections 2.82–2.84).

Full details of the proof of Theorem 5.4 are given in the note of Everitt, Nasri-Roudsari and Rehberg (to appear).

Remarks

(a) The remarks (b), (c) and (d) stated after Theorem 5.1 remain valid for Theorem 5.4.

(b) The holomorphic property (v) carries over to the families of functions $\{\, S_n : n \in \mathbb{Z} \,\}$ and $\{K\}$.

(c) The conditions and consequences of this theorem can be shown to hold for the kernels generated by the linear boundary value problems considered earlier in this section; for some results see the papers by Everitt et al. (1994) and Everitt and Nasri-Roudsari (to appear (a)).

(d) The result (d) interpolates over the set of points $\{\, \lambda_n : n \in \mathbb{Z} \,\}$ to the chosen set of values $\{c_n : n \in \mathbb{Z}\}$, subject to the growth condition, with a function F that is holomorphic on $\mathbb{C}$. This is a special case of a general result due to C. Guichard (1884) concerning the construction of functions that are holomorphic on $\mathbb{C}$ and that take chosen values on a given discrete set of points; see the reference to Guichard in the book of Davis (1965).

(e) The significance of the locally bounded condition (vi) is best seen in the proof as given in Everitt et al. (to appear); however it is noted here that this condition is often satisfied in the case of kernels generated by the linear boundary value problems as given in Section 5.1.3 above; in the second-order case see Everitt et al. (1994, Section 3).

(f) It is shown in Everitt et al. (to appear) that the local boundedness condition (vi) is not only sufficient to prove the holomorphic properties (a) and (b), but that this condition is also necessary.

The explicit form of the functions $\{S_n\}$ and the sampling series for F in (c) of Theorem 5.4, in the case of Examples 5.2 and 5.3, can be given as in the following results.

(a) Example 5.2. By direct computation

$$S_n(\lambda) = \frac{\sin(\sigma\lambda - n\pi)}{\sigma\lambda - n\pi} \quad (\lambda \in \mathbb{C} \text{ and } n \in \mathbb{C})$$

and then the sampling series takes the form, for $F \in \{K\}$,

$$F(\lambda) = \sum_{n \in \mathbb{Z}} F\left(\frac{n\pi}{\sigma}\right) \frac{\sin(\sigma\lambda - n\pi)}{\sigma\lambda - n\pi} \quad (\lambda \in \mathbb{C});$$

this is the famous Shannon sampling theorem; it is referred to again below in Section 5.1.6, and in greater detail in Section 5.3 of this chapter.

(b) Example 5.3. For this example it may be shown that

$$S_n(\lambda) = \frac{(2n+1)\sin\left(\pi(\sqrt{\lambda} - n - \frac{1}{2})\right)}{\pi\left(\lambda - (n + \frac{1}{2})^2\right)} \quad (\lambda \in \mathbb{C} \text{ and } n \in \mathbb{N}_0)$$

and then

$$F(\lambda) = \sum_{n \in \mathbb{N}_0} F\left((n + \tfrac{1}{2})^2\right) \frac{(2n+1)\sin\left(\pi(\sqrt{\lambda} - n - \frac{1}{2})\right)}{\pi\left(\lambda - (n + \frac{1}{2})^2\right)} \quad (\lambda \in \mathbb{C}).$$

For this latter result see Everitt et al. (1994, Section 11, Example 3).

5.1.5   The interpolation function G

The original interpolation theorem of Lagrange involved the class of complex-valued polynomials defined on the comlex plane $\mathbb{C}$. For $n \in \mathbb{N}_0$ let $\{\lambda_r \in \mathbb{C} : r = 0, 1, 2, \ldots, n\}$ be $n + 1$ distinct complex numbers, and let $\{c_r : r = 0, 1, 2, \ldots, n\}$ be any set of $n + 1$ complex numbers. Then the polynomial P of degree n defined by

$$P(\lambda) := \sum_{r=0}^{n} c_r \frac{\omega(\lambda)}{(\lambda - \lambda_r)\omega'(\lambda_r)} \quad (\lambda \in \mathbb{C}), \tag{5.1.17}$$

with the interpolation function ω defined by

$$\omega(\lambda) := \prod_{r=0}^{n} (\lambda - \lambda_r) \quad (\lambda \in \mathbb{C}),$$

satisfies

$$P(\lambda_r) = c_r \quad (r = 0, 1, 2, \ldots, n). \tag{5.1.18}$$

Note that $\omega \in \mathbf{H}(\mathbb{C})$ and that the zeros of ω are all simple, i.e. $\omega'(\lambda_r) \neq 0$ ($r = 0, 1, 2, \ldots, n$).

The analytic Kramer theorem allows of the introduction of an analytic interpolation function G, in certain cases, that plays the same role as ω in (5.1.17) but now for the infinite set $\{\lambda_n : n \in \mathbb{Z}\}$.

Definition 5.5 *Let K be an analytic Kramer kernel in the sense of Theorem 5.4, for the interval I, weight w and sequence $\{\lambda_n : n \in \mathbb{Z}\}$; let $\{S_n(\cdot) : n \in \mathbb{Z}\}$ denote the sequence of interpolation functions; then the function $G : \mathbb{C} \to \mathbb{C}$ is called an* *interpolation function for the kernel K if*

(i) $G \in \mathbf{H}(\mathbb{C})$;

(ii) $G(\lambda) = 0$ *if and only if* $\lambda \in \{ \lambda_n : n \in \mathbb{Z} \}$;

(iii) $G'(\lambda_n) \neq 0$ $(n \in \mathbb{Z})$;

(iv) $S_n(\lambda) = \dfrac{G(\lambda)}{(\lambda - \lambda_n)G'(\lambda_n)}$ $(\lambda \in \mathbb{C} : n \in \mathbb{Z})$.

The existence of interpolation functions G for Kramer kernels generated by linear boundary value problems, of the form considered in Section 5.1.3 above, is shown in the many examples given in the papers by Butzer and Schöttler (1994), Everitt and Nasri-Roudsari (to appear (a)), Everitt et al. (1994), Zayed (1991, 1993a), Zayed et al. (1990) and in the book by Zayed (1993b).

When K has an interpolation function G the sampling series expansion (Theorem 5.4 (d)) takes the form

$$F(\lambda) = \sum_{n \in \mathbb{Z}} F(\lambda_n) \frac{G(\lambda)}{(\lambda - \lambda_n)G'(\lambda_n)} \quad (\lambda \in \mathbb{C}) \tag{5.1.19}$$

with the same convergence properties. The form of the results in (5.1.17) and (5.1.19) give a countably infinite extension of the original Lagrange interpolation result given in (5.1.17).

In a number of cases involving linear boundary value problems the interpolation function G appears naturally from the properties of the differential equation and the boundary conditions; for examples, see Butzer and Schöttler (1994) and Everitt and Nasri-Roudsari (to appear (a); (1994)). In other cases it may be necessary to construct G; this is effected by constructing an entire (integral) function on $\mathbb{C}$, with simple zeros at the eigenvalues of the boundary value problem, from the general theory of entire functions, see Copson (1944, Chapter VII) and Titchmarsh (1939, Chapter VIII); examples of this procedure are given in Zayed (1991, 1993a).

For the two examples:

(a) Example 5.2

$$G(\lambda) = \sin(\sigma\lambda) \quad (\lambda \in \mathbb{C}).$$

(b) Example 5.3

$$G(\lambda) = \sin\left(\pi(\sqrt{\lambda} - \tfrac{1}{2})\right) \quad (\lambda \in \mathbb{C}).$$

5.1.6 *The Shannon interpolation theorem*

This result is mentioned above in Example 5.2; it is given in that case as an example of the Kramer analytic theorem. However the result is earlier; the first appearance is due to E.T. Whittaker (1915), and discovered later and independently by C.E. Shannon (1949). For such information see the references in the paper by Campbell (1964).

These earlier results are stated in the real form but the theorem is best seen in the analytic form; this is an extension of the original result; the notation is that of Example 5.2.

Theorem 5.6 *Let $\sigma > 0$ be given; define the Shannon–Whittaker class of functions $\{S - W\}$ as the set of all $F : \mathbb{C} \to \mathbb{C}$ by $F \in \{S - W\}$ if for some $f \in L^2(-\sigma, \sigma)$*

$$F(\lambda) = \int_{-\sigma}^{\sigma} \exp(-ix\lambda) f(x)\, dx \quad (\lambda \in \mathbb{C}).$$

Then

$$F(\lambda) = \sum_{n \in \mathbb{Z}} F\left(\frac{n\pi}{\sigma}\right) \frac{\sin(\sigma\lambda - n\pi)}{\sigma\lambda - n\pi} \quad (\lambda \in \mathbb{C})$$

where the series converges absolutely for each $\lambda \in \mathbb{C}$, and locally uniformly in $\mathbb{C}$. From the definition of F it follows that

$$F \in \mathbf{H}(\mathbb{C}).$$

Proof The original proof, in the real case, is due to Whittaker and subsequently to Shannon; see the remarks above for references. In the complex case the result is best seen as an example of the Kramer analytic form as given in Theorem 5.4, and as discussed in Example 5.2.

Remarks
   (a) This is the first of the sampling/interpolation theory results; the sampling points $\{n\pi/\sigma : n \in \mathbb{Z}\}$ are evenly spaced over the whole real line $\mathbb{R}$; if F (recall that F has to belong to the $\{S - W\}$ class) can be "measured" at these sampling points then F can be constructed uniquely at every point of the real line $\mathbb{R}$.
   (b) For a certain class of linear boundary value problems, as considered in Section 5.1.3 above, there is a link between the Kramer analytic theorem and Shannon "type" results; this was first observed by Campbell (1964) and is discussed in greater detail in Section 5.3 below.

5.2   Second-order problems

In this section we study in greater detail the connection between the Kramer sampling theorem and linear ordinary differential equations of second order. There will be a discussion on the problem of the existence of kernels K in these cases. Of special interest is the question of whether there exist representations of the sampling series as Lagrange interpolation series involving interpolation functions as given in Definition 5.5.

5.2.1   Brief survey of literature

As mentioned in Section 5.1.3 symmetric boundary value problems may generate suitable kernels for Kramer's sampling theorem. In his paper Kramer (1959) considered self-adjoint eigenvalue problems of nth order and studied as a special case the Shannon sampling theorem (see Example 5.2), and the classical Bessel differential equation and showed the existence of kernels in these cases.

   Campbell (1964) additionally considered the classical Legendre differential equation (compare with Example 5.3). Both authors gave kernels for the examples they considered but did not give a general method to construct such kernels from a given boundary value problem.

This was first done in the paper by Zayed et al. (1990). They also considered the sampling theorem for regular eigenvalue problems of second order with separated boundary conditions, under the aspect of Lagrange interpolation as given in Section 5.1.5.

Zayed continued this work in his papers (1991, 1993a) where he studied singular eigenvalue problems of second order and higher-order problems with coupled boundary conditions (under certain assumptions on the eigenvalues). For his studies in this respect see also Zayed (1993b). Moreover there is a discussion on boundary value problems when the eigenparameter appears in the boundary condition in the work of Annaby (to appear).

In their paper Everitt et al. (1994), under minimal conditions on the coefficients in the differential equation, considered regular or limit-circle eigenvalue problems of second order with separated boundary conditions given in the form (5.1.8). These results were extended to coupled boundary conditions under the assumption of a simple spectrum in Everitt and Nasri-Roudsari (to appear (a)).

5.2.2 *Regular or limit-circle case*

The results in the case of regular or limit-circle eigenvalue problems of second order can be divided into two parts: separated and coupled boundary conditions.

The spectra of all the boundary value problems with these end-point classifications is discrete.

In the case considered in Theorems 5.7, 5.10 and 5.11 below we assume that the spectrum limits at both $\pm\infty$, and we index the eigenvalues as follows

$$\{\lambda_n : n \in \mathbb{Z}\} \quad \text{with} \quad \lim_{n \to \pm\infty} \lambda_n = \pm\infty. \tag{5.2.1}$$

The results given hold for both the other two cases when the spectrum limits either only at $+\infty$ or only at $-\infty$, with minor modifications to the notation.

Separated boundary conditions

In Everitt et al. (1994) the situation is as follows. The Sturm–Liouville differential equation is given by

$$-(p(x)y'(x))' + q(x)y(x) = \lambda w(x)y(x) \quad (x \in (a, b)) \tag{5.2.2}$$

where:

   (i) the spectral parameter $\lambda \in \mathbb{C}$;
   (ii) (a, b) is an open interval of $\mathbb{R}$ with $-\infty \le a < b \le \infty$;
   (iii) the coefficients $p, q, w : (a, b) \to \mathbb{R}$;   (5.2.3)
   (iv) $p^{-1}, q, w \in L^1_{\text{loc}}(a, b)$;
   (v) $w(x) > 0$ for almost all $x \in (a, b)$.

There are two **structural conditions** on the boundary value problem:

  (a) The end-point a of the differential equation is to be regular, or limit-circle in $L^2((a, b); w)$; independently end-point b is to be regular or limit-circle in $L^2((a, b); w)$.

(b) The two linearly independent symmetric boundary conditions are separated, with one condition at end-point a and one condition at end-point b.

The separated boundary conditions are thus given by

$$[y, \kappa_-](a) = [y, \kappa_+](b) = 0, \tag{5.2.4}$$

where for a given pair of functions $\{\kappa_-, \chi_-\}$ the following conditions are satisfied:

$$\left.\begin{array}{l}
\text{(i)}\ \ \kappa_-, \chi_- : (a, b) \to \mathbb{R}; \\
\text{(ii)}\ \ \kappa_-, \chi_-\ \text{are maximal domain functions;} \\
\text{(iii)}\ \ [\kappa_-, \chi_-](a) = 1.
\end{array}\right\} \tag{5.2.5}$$

The pair $\{\kappa_+, \chi_+\}$ satisfies analogous conditions at the end-point b.

This symmetric boundary value problem generates a self-adjoint differential operator T having the following properties (see Naimark 1968 for these results):

(a) T is self-adjoint and unbounded in $L^2((a, b); w)$,
(b) the spectrum of T is real, and discrete with limit-points at $+\infty$ or $-\infty$ or both;
(c) the spectrum of T is simple, i.e. each eigenvalue is of multiplicity one;
(d) the eigenvalues and eigenvectors of T satisfy the boundary value problem.

The results in Everitt et al. (1994) read as follows.

Theorem 5.7 *Let (a, b) be an open interval of the real line; let the coefficients p, q and w satisfy the basic conditions as given in (5.2.3); let the Sturm–Liouville quasi-differential equation (5.2.2) satisfy the end-point classification (a) above; let the separated boundary conditions be given by (5.2.4) where the boundary condition functions $\{\kappa_-, \chi_-\}$ and $\{\kappa_+, \chi_+\}$ satisfy the conditions (5.2.5); let the self-adjoint differential operator T be determined by the separated, symmetric boundary value problem; let the simple, discrete spectrum of T be given by $\{\lambda_n : n \in \mathbb{Z}\}$; let $\{\psi_n : n \in \mathbb{Z}\}$ be the eigenvectors of T; let the pair of basis solutions $\{\varphi_1, \varphi_2\}$ of the differential equation (5.2.2) satisfy the initial conditions, for some point $c \in (a, b)$,*

$$\varphi_1(c, \lambda) = 1, \quad (p\varphi_1')(c, \lambda) = 0, \quad \varphi_2(c, \lambda) = 0, \quad (p\varphi_2')(c, \lambda) = 1.$$

Define the analytic Kramer kernel $K_- : (a, b) \times \mathbb{C} \to \mathbb{C}$ by

$$K_-(x, \lambda) := [\varphi_1(\cdot, \lambda), \kappa_-](a)\varphi_2(x, \lambda) - [\varphi_2(\cdot, \lambda), \kappa_-](a)\varphi_1(x, \lambda). \tag{5.2.6}$$

Then K_- has the following properties:

(i) *$K_-(\cdot, \lambda)$ is a solution of (5.2.2) for all $\lambda \in \mathbb{C}$, and $K_-(\cdot, \lambda) \in \mathbb{R}$ ($\lambda \in \mathbb{R}$);*
(ii) *$K_-(\cdot, \lambda)$ is an element of the maximal domain and in particular of $L^2((a, b); w)$, for all $\lambda \in \mathbb{C}$;*
(iii) *$[K_-(\cdot, \lambda), \kappa_-](a) = 0$ ($\lambda \in \mathbb{C}$);*
(iv) *$[K_-(\cdot, \lambda), \kappa_+](b) = 0$ if and only if $\lambda \in \{\lambda_n : n \in \mathbb{Z}\}$;*
(v) *$K_-(x, \cdot) \in \mathbf{H}(\mathbb{C})$ ($x \in (a, b)$);*
(vi) *$K_-(\cdot, \lambda_n) = k_n\psi_n$ where $k_n \in \mathbb{R}\backslash\{0\}$ ($n \in \mathbb{Z}$);*

(vii) K_- *is unique up to multiplication by a factor* $e(\cdot) \in \mathbf{H}(\mathbb{C})$ *with* $e(\lambda) \neq 0$ $(\lambda \in \mathbb{C})$
and $e(\lambda) \in \mathbb{R}$ $(\lambda \in \mathbb{R})$.

Define the interpolation function $G : \mathbb{C} \to \mathbb{C}$ *by*

$$G(\lambda) := [K_-(\cdot, \lambda), \kappa_+](b) \quad (\lambda \in \mathbb{C}). \tag{5.2.7}$$

Then G has the following properties:

(i) $G \in \mathbf{H}(\mathbb{C})$, $G(\lambda) \in \mathbb{R}$ $(\lambda \in \mathbb{R})$;
(ii) $G(\lambda) = 0$ *if and only if* $\lambda \in \{\lambda_n : n \in \mathbb{Z}\}$;
(iii) $G'(\lambda_n) \neq 0$ $(n \in \mathbb{Z})$.

All the results of Theorem 5.4 and Definition 5.5 now follow.

Remarks

(a) The notation K_- is chosen for technical reasons; there is a kernel K_+, with similar properties, but with a and κ_- interchanged with b and κ_+.

(b) The kernel K_- (also K_+) satisfies the requirements of Theorem 5.4 and also of the original ideas of Kramer in Theorem 5.1.

(c) The function G is an interpolation function for the kernel K_- (and also for K_+) in the sense of Definition 5.5.

(d) For a proof of Theorem 5.7 see Everitt et al. (1994).

(e) These results contain the results of Zayed et al. (1990) on regular and limit-circle eigenvalue problems as a special case; however in these references it is required for the interpolation function G to be constructed by the use of infinite products rather than by invoking the analytic definition given in (5.2.7).

(f) Example 5.3 and the corresponding kernel and interpolation function is an example for Theorem 5.7.

Examples for separated boundary conditions

In Everitt et al. (1994) the following two examples can be found.

Example 5.8 Consider the self-adjoint eigenvalue problem

$$-(py')' = \lambda y \quad \text{on} \quad [-1, 1], \quad y(-1) = y(1) = 0,$$

where

$$p(x) = \begin{cases} -1 & (x \in [-1, 0)) \\ 0 & x = 0 \\ 1 & (x \in (0, 1]). \end{cases}$$

Both endpoints are regular. A basis at $c = 0$ is given by

$$\varphi_1(x, \lambda) = \begin{cases} \cosh(x\sqrt{\lambda}) & (x \in [-1, 0]) \\ \cos(x\sqrt{\lambda}) & (x \in (0, 1]); \end{cases}$$

$$\varphi_2(x, \lambda) = \begin{cases} -\frac{1}{\sqrt{\lambda}} \sinh(x\sqrt{\lambda}) & (x \in [-1, 0]) \\ \frac{1}{\sqrt{\lambda}} \sin(x\sqrt{\lambda}) & (x \in (0, 1]). \end{cases}$$

Choose as boundary condition functions $\kappa_-(x) = x + 1$ and $\kappa_+(x) = x - 1$. The eigenvalues are determined by the zeros of

$$\rho(\lambda) = \frac{1}{\sqrt{\lambda}} \left[\sinh(\sqrt{\lambda}) \cos(\sqrt{\lambda}) - \cosh(\sqrt{\lambda}) \sin(\sqrt{\lambda}) \right];$$

all these zeros are real and simple, and 0 is one of them. The analytic Kramer kernel is given by

$$K_-(x, \lambda) = \frac{1}{\sqrt{\lambda}} \sinh(\sqrt{\lambda}) \, \varphi_1(x, \lambda) - \cosh(\sqrt{\lambda}) \, \varphi_2(x, \lambda)$$

and the interpolation function is given by $G(\lambda) = \rho(\lambda)$.

Example 5.9 Consider the Legendre differential equation

$$-((1 - x^2)y')' = \lambda y \quad \text{on } (-1, 1), \quad [y, \kappa_+](1) = [y, \kappa_-](-1) = 0$$

where

$$\kappa_+(x) = \log(1 - x) \quad \text{and} \quad \kappa_-(x) = \log(1 + x).$$

Both endpoints are limit-circle and non-oscillatory. A fundamental system of the differential equation is given by

$$\varphi_1(x, \lambda) = \frac{1}{2\rho(\lambda)} \left\{ {}_2F_1\left(-\sqrt{\lambda} + \frac{1}{2}, \sqrt{\lambda} + \frac{1}{2}; 1; \frac{1-x}{2}\right) \right.$$

$$\left. + {}_2F_1\left(-\sqrt{\lambda} + \frac{1}{2}, \sqrt{\lambda} + \frac{1}{2}; 1; \frac{1+x}{2}\right) \right\}$$

$$\varphi_2(x, \lambda) = \frac{1}{2\sigma(\lambda)} \left\{ {}_2F_1\left(-\sqrt{\lambda} + \frac{1}{2}, \sqrt{\lambda} + \frac{1}{2}; 1; \frac{1-x}{2}\right) \right.$$

$$\left. - {}_2F_1\left(-\sqrt{\lambda} + \frac{1}{2}, \sqrt{\lambda} + \frac{1}{2}; 1; \frac{1+x}{2}\right) \right\}$$

where

$$\rho(\lambda) = {}_2F_1\left(-\sqrt{\lambda} + \frac{1}{2}, \sqrt{\lambda} + \frac{1}{2}; 1; \frac{1}{2}\right) = \frac{\sqrt{\pi}}{\Gamma(\frac{3}{4} - \frac{\sqrt{\lambda}}{2})\Gamma(\frac{3}{4} + \frac{\sqrt{\lambda}}{2})},$$

$$\sigma(\lambda) = \frac{1}{2}\left(\lambda - \frac{1}{4}\right) {}_2F_1\left(-\sqrt{\lambda} + \frac{3}{2}, \sqrt{\lambda} + \frac{3}{2}; 2; \frac{1}{2}\right) = \frac{2\sqrt{\pi}}{\Gamma(\frac{1}{4} - \frac{\sqrt{\lambda}}{2})\Gamma(\frac{1}{4} + \frac{\sqrt{\lambda}}{2})}.$$

The eigenvalues are determined by the zeros of the interpolation function

$$G(\lambda) = \frac{2}{\pi} \cos \pi \sqrt{\lambda} \left\{ \frac{\pi^2}{\cos^2 \pi \sqrt{\lambda}} - (k_0(\lambda) + \ln 2)^2 \right\}$$

with $k_0(\lambda) = 2\psi(1) - \psi(\frac{1}{2} - \sqrt{\lambda}) - \psi(\frac{1}{2} + \sqrt{\lambda})$, where $\psi(x)$ is the logarithmic derivative of the Γ-function. The analytic Kramer kernel is given by

$$
\begin{aligned}
K_-(x, \lambda) &= (k_0(\lambda) + \ln 2)_2F_1\left(-\sqrt{\lambda} + \frac{1}{2}, \sqrt{\lambda} + \frac{1}{2}; 1; \frac{1 + x}{2}\right) \\
&\quad - \Gamma\left(\frac{1}{2} - \sqrt{\lambda}\right)\Gamma\left(\frac{1}{2} + \sqrt{\lambda}\right){}_2F_1\left(-\sqrt{\lambda} + \frac{1}{2}, \sqrt{\lambda} + \frac{1}{2}; 1; \frac{1 - x}{2}\right).
\end{aligned}
$$

Coupled boundary conditions

In the case of coupled boundary conditions the situation is different. Let the Sturm–Liouville differential equation be given by (5.2.2):

$$
-(p(x)y'(x))' + q(x)y(x) = \lambda w(x)y(x) \quad (x \in (a, b))
$$

satisfying the properties following (5.2.2). Let the boundary conditions be given in the form

$$
\mathbf{y}(b) = e^{i\alpha}T\mathbf{y}(a) \quad \text{for some } \alpha \in [-\pi, \pi] \tag{5.2.8}
$$

with the 2×2 matrix $T = [t_{rs}]$ where $t_{rs} \in \mathbb{R}$ $(r, s = 1, 2)$, $\det(T) = 1$, and the 2×1 vector $\mathbf{y}$ is defined by

$$
\mathbf{y}(t) := \begin{bmatrix} [y, \theta](t) \\ [y, \varphi](t) \end{bmatrix} \quad (t \in (a, b)); \tag{5.2.9}
$$

T is called the boundary condition matrix. The functions θ and φ are chosen such that the following given properties are satisfied:

  (i) θ, φ are real-valued maximal domain functions;
  (ii) $[\theta, \varphi](a) = \lim_{t \to a+}[\theta, \varphi](t) = 1$;
  (iii) $[\theta, \varphi](b) = \lim_{t \to b-}[\theta, \varphi](t) = 1$.

For example θ, φ can be real-valued solutions of (5.2.2) on (a, b); in particular choose $\theta(x) = \varphi_1(x, \mu)$, $\varphi(x) = \varphi_2(x, \mu)$ for any real μ, with φ_1 and φ_2 determined as in Theorem 5.7 above.

The boundary conditions (5.2.8) are coupled and self-adjoint, for each end-point either regular or limit-circle, and are in canonical form; see the results in Bailey et al. (1996).

Let the pair of basis solutions $\{u, v\}$ of the differential equation (5.2.2) be determined by the, possibly singular, initial conditions, see Bailey et al. (1996), for all $\lambda \in \mathbb{C}$,

$$
[u, \theta](a, \lambda) = 0, \quad [u, \varphi](a, \lambda) = 1, \quad [v, \theta](a, \lambda) = 1, \quad [v, \varphi](a, \lambda) = 0. \tag{5.2.10}
$$

To define a differential operator A select any boundary condition matrix T and any $\alpha \in [-\pi, \pi]$; the boundary value problem defines a differential operator having the following properties:

  (a) A is self-adjoint and unbounded in $L^2((a, b); w)$;
  (b) the spectrum of A is real, discrete with limit-points at $+\infty$ or $-\infty$ or both;

(c) the corresponding eigenvalues and eigenvectors satisfy the boundary value problem;

(d) the multiplicity of any eigenvalue does not exceed two.

There is additional information on the multiplicity of the eigenvalues in Bailey et al. (1996): for **complex** boundary conditions, i.e. for α satisfying $0 < \alpha < \pi$ or $-\pi < \alpha < 0$, the spectrum of A is always simple, i.e. each eigenvalue is of multiplicity one. The spectrum may or may not be simple in the case of **real** boundary conditions, i.e. $\alpha = -\pi, 0, \pi$. For this reason, in this section, the following **structural condition** is made (see the second paragraph of the introduction):

$$\left. \begin{array}{l} \text{in the real case } \alpha = -\pi, 0, \pi \\ \text{all the eigenvalues are assumed to be simple} \end{array} \right\} . \tag{5.2.11}$$

This condition is satisfied by many examples; however examples exist with these real boundary conditions for which all eigenvalues are of multiplicity two.

In Bailey et al. (1996) it is also shown that the boundary conditions given by (5.2.8) are equivalent to the general conditions given in Section 5.1.

The results in Everitt and Nasri-Roudsari, (to appear (a)) are given in two parts. The first one is the result in the complex case when $0 < \alpha < \pi$ or $-\pi < \alpha < 0$ and this yields

Theorem 5.10 *Let* (a, b) *be an open interval of the real line; let the coefficients* p, q *and* w *satisfy the basic conditions* (5.2.3); *let the Sturm–Liouville differential equation, see* (5.2.2),

$$-(p(x)y'(x))' + q(x)y(x) = \lambda w(x)y(x) \quad (x \in (a, b))$$

satisfy the structural condition:

the end-point a *is either regular or limit-circle in* $L^2((a, b); w)$; *independently the end-point* b *is either regular or limit-circle in* $L^2((a, b); w)$;

let the symmetric, coupled and complex boundary condition be given by, see (5.2.8),

$$\mathbf{y}(b) = e^{i\alpha} T \mathbf{y}(a) \quad \text{for some } \alpha \in (-\pi, 0) \cup (0, \pi);$$

let A *be the unique self-adjoint, unbounded differential operator in* $L^2((a, b); w)$, *determined by* (5.2.2) *and* (5.2.8); *let the discrete spectrum* σ *of* A *be denoted by* $\{\lambda_n : n \in \mathbb{Z}\}$ *with* $\lim_{n \to \pm\infty} \lambda_n$, *and let* $\{\psi_n : n \in \mathbb{Z}\}$ *denote the corresponding eigenfunctions.*

Let the analytic function $D(T, \cdot) : \mathbb{C} \to \mathbb{C}$ *be defined by, with solutions* u, v *determined by* (5.2.10),

$$\begin{aligned} D(T, \lambda) := \quad & t_{11}[u(\cdot, \lambda), \varphi](b) + t_{22}[v(\cdot, \lambda), \theta](b) \\ & -t_{12}[v(\cdot, \lambda), \varphi](b) - t_{21}[u(\cdot, \lambda), \theta](b) \end{aligned} \tag{5.2.12}$$

Then

(i) $D(T, \cdot) \in \mathbf{H}(\mathbb{C})$;

(ii) λ *is an eigenvalue of* A *if and only if* λ *is a zero of* $D(T, \lambda) - 2\cos(\alpha)$;

(iii) *the zeros of $D(T, \lambda) - 2\cos(\alpha)$ are real and simple*;

(iv) *the eigenvalues of A are simple.*

Let the above stated definitions and conditions hold; then the boundary value problem (5.2.2) and (5.2.8) generates two independent analytic Kramer kernels K_1 and K_2, see Theorem 5.4, defined by, for all $x \in (a, b)$ and all $\lambda \in \mathbb{C}$,

$$K_1(x, \lambda) := \left([u(\cdot, \lambda), \theta](b) - e^{i\alpha} t_{12} \right) v(x, \lambda)$$
$$- \left([v(\cdot, \lambda), \theta](b) - e^{i\alpha} t_{11} \right) u(x, \lambda) \tag{5.2.13}$$

$$K_2(x, \lambda) := \left([u(\cdot, \lambda), \varphi](b) - e^{i\alpha} t_{22} \right) v(x, \lambda)$$
$$- \left([v(\cdot, \lambda), \varphi](b) - e^{i\alpha} t_{21} \right) u(x, \lambda) \tag{5.2.14}$$

For both K_1 and K_2 there is a common interpolation function G, see Definition 5.5 above, defined by

$$G(\lambda) := D(T, \lambda) - 2\cos(\alpha) \quad (\lambda \in \mathbb{C}). \tag{5.2.15}$$

All the results of Theorem 5.4 and Definition 5.5 now follow.

Proof See Everitt and Nasri-Roudsari (to appear (a)).

The second part of Everitt and Nasri-Roudsari, (to appear (a)) is concerned with real boundary value problems, i.e. $\alpha = -\pi, 0, \pi$ for which condition (5.2.11) is satisfied. The results in this case are similar to the results stated in Theorem 5.10 except that a phenomenon of **degeneracy** may occur, now to be explained.

The two kernels K_1, K_2 are defined as in (5.2.13) and (5.2.14). An eigenvalue λ, say, taken from the set $\{\lambda_n : n \in \mathbb{Z}\}$ is said to be degenerate for kernel K_r if $K_r(\cdot, \lambda)$ is identically zero on (a, b). We denote the index set of **non-degenerate** eigenvalues of K_r by $\mathbb{Z}_r \subseteq \mathbb{Z}$, $r = 1, 2$; note degeneracy may or may not occur, i.e. $\mathbb{Z}_r \subset \mathbb{Z}$ or $\mathbb{Z}_r = \mathbb{Z}$. Recall that the eigenfunctions of the boundary value problem are denoted by $\{\psi_n : n \in \mathbb{Z}\}$; in the notation of (vi) of Theorem 5.10 we have, for $r = 1, 2$,

$$K_r(\cdot, \lambda_n) = k_{r,n} \psi_n \quad \text{with } k_{r,n} \in \mathbb{R} \setminus \{0\} \text{ for } n \in \mathbb{Z}_r, \tag{5.2.16}$$

but

$$K_r(\cdot, \lambda_n) = 0 \quad (x \in (a, b)) \quad \text{for } n \in \mathbb{Z} \setminus \mathbb{Z}_r. \tag{5.2.17}$$

Theorem 5.11 *Let all the conditions of Theorem 5.10 above hold with the addition of condition (5.2.11); let the kernels K_1, K_2 and the analytic function G be defined as in (5.2.13), (5.2.14) and (5.2.15).*

Let the phenomenon of degeneracy be defined as above; see in particular the definition of the index sets $\mathbb{Z}_r$ ($r = 1, 2$) and (5.2.16) and (5.2.17), i.e. $\{\lambda_n : n \in \mathbb{Z}_r \subseteq \mathbb{Z}\}$ denotes the set of non-degenerate eigenvalues for K_r.

For $r = 1, 2$, let the subspaces $L_r^2((a, b); w)$ of $L^2((a, b); w)$ be defined by

$$L_r^2((a, b); w) := \text{span}\{\psi_n; n \in \mathbb{Z}_r\} \quad (r = 1, 2). \tag{5.2.18}$$

Then

(i) $\mathbb{Z} = \mathbb{Z}_1 \cup \mathbb{Z}_2$, *i.e. every eigenvalue in* $\{\lambda_n : n \in \mathbb{Z}\}$ *is non-degenerate for at least one* K_r;

(ii) *for* $r = 1, 2$ *the kernel* K_r *is an analytic Kramer kernel for the subspace* $L_r^2((a, b); w)$;

(iii) *for both* K_1 *and* K_2 *acting on their respective subspaces* $L^2((a, b); w)$, *the analytic function G is an interpolation function.*

Finally for an appropriate choice of numbers $\alpha_1, \alpha_2 \in \mathbb{R}$ *the kernel K defined by*

$$K(x, \lambda) = \alpha_1 K_1(x, \lambda) + \alpha_2 K_2(x, \lambda) \quad (x \in (a, b); \lambda \in \mathbb{C}) \qquad (5.2.19)$$

is an analytic Kramer kernel for the whole space $L^2((a, b); w)$, *and G is an interpolation function for K.*

Proof See Everitt and Nasri-Roudsari (to appear (a)).

Remark The case of real boundary conditions but with spectral multiplicity two will be considered in Everitt and Nasri-Roudsari (to appear (b)). However an example to illustrate this case is given in the next subsection.

Examples for coupled boundary conditions

Let us consider some examples to illustrate Theorems 5.10 and 5.11. In all examples we consider the regular differential equation

$$-y''(x) = \lambda y(x) \quad (x \in [-\pi, \pi]).$$

We choose $\theta(x) = \cos x$ and $\varphi(x) = \sin x$ to give the boundary conditions (see (5.2.8))

$$\mathbf{y}(\pi) \equiv \begin{bmatrix} y'(\pi) \\ -y(\pi) \end{bmatrix} = e^{i\alpha} T \begin{bmatrix} y'(-\pi) \\ -y(-\pi) \end{bmatrix} \equiv e^{i\alpha} T \mathbf{y}(-\pi).$$

The system of functions $\{u, v\}$, given by

$$u(x, \lambda) = -\cos\left(\sqrt{\lambda}(x + \pi)\right), \quad v(x, \lambda) = \frac{1}{\sqrt{\lambda}} \sin\left(\sqrt{\lambda}(x + \pi)\right)$$

satisfies the initial conditions (5.2.10).

Example 5.12 For $\alpha = \frac{\pi}{4}$ the boundary conditions are **complex** and Theorem 5.10 can be applied. Choose

$$T = \begin{bmatrix} \frac{1}{\sqrt{2}} & 0 \\ 0 & \sqrt{2} \end{bmatrix}.$$

The analytic Kramer kernels are then given by

$$K_1(x, \lambda) = \sqrt{\lambda} \sin\left(2\pi \sqrt{\lambda}\right) u(x, \lambda) - \cos\left(\left(2\pi \sqrt{\lambda}\right) - \frac{1}{2}(1 + i)\right) v(x, \lambda),$$

$$K_2(x, \lambda) = \left(\cos\left(2\pi\sqrt{\lambda}\right) - (1+i)\right) u(x, \lambda) + \frac{1}{\sqrt{\lambda}}\sin\left(2\pi\sqrt{\lambda}\right) v(x, \lambda).$$

None of them degenerates. The common interpolation function is given by

$$G(\lambda) = (1+i)\left(\frac{3}{2}\cos\left(2\pi\sqrt{\lambda}\right) - 1\right)$$

The eigenvalues are real and simple.

Example 5.13 For $\alpha = 0$ the boundary conditions are **real** and Theorem 5.11 can be applied.

(i) Choose

$$T = \begin{bmatrix} \frac{1}{2} & 0 \\ 0 & 2 \end{bmatrix}.$$

The analytic Kramer kernels are given by

$$K_1(x, \lambda) = \sqrt{\lambda}\sin\left(2\pi\sqrt{\lambda}\right) u(x, \lambda) - \left(\cos\left(2\pi\sqrt{\lambda}\right) - \frac{1}{2}\right) v(x, \lambda),$$

$$K_2(x, \lambda) = \left(\cos\left(2\pi\sqrt{\lambda}\right) - 2\right) u(x, \lambda) + \frac{1}{\sqrt{\lambda}}\sin\left(2\pi\sqrt{\lambda}\right) v(x, \lambda).$$

Neither of the kernels is degenerate. The common interpolation function is given by

$$G(\lambda) = \frac{5}{2}\cos\left(2\pi\sqrt{\lambda}\right) - 2.$$

The eigenvalues are real and simple.

(ii) Choose

$$T = \begin{bmatrix} 1 & 0 \\ -4\pi & 1 \end{bmatrix}.$$

The Kramer kernels are given by

$$K_1(x, \lambda) = \sqrt{\lambda}\sin\left(2\pi\sqrt{\lambda}\right) u(x, \lambda) - \left(\cos\left(2\pi\sqrt{\lambda}\right) - \frac{1}{2}\right) v(x, \lambda),$$

$$K_2(x, \lambda) = \left(\cos\left(2\pi\sqrt{\lambda}\right) - 1\right) u(x, \lambda) + \left(\frac{1}{\sqrt{\lambda}}\sin\left(2\pi\sqrt{\lambda}\right) - 4\pi\right) v(x, \lambda).$$

The eigenvalues λ_k, $k \in \mathbb{Z}$, are determined as the zeros of

$$2\pi\sqrt{\lambda}\sin\left(2\pi\sqrt{\lambda}\right) + \cos\left(2\pi\sqrt{\lambda}\right) = 1.$$

They are all real and simple, 0 is one of them. For 0 the kernel K_1 degenerates but K_2 does not. At all the other eigenvalues neither of the kernels degenerates. The common interpolation function is given by

$$G(\lambda) = 2\cos\left(2\pi\sqrt{\lambda}\right) + 4\pi\sqrt{\lambda}\sin\left(2\pi\sqrt{\lambda}\right) - 2.$$

(iii) Set $T = (t_{rs})_{r,s=1}^2$ with

$$t_{11} = \cos(2\pi s_0), \quad t_{12} = s_0 \sin(2\pi s_0),$$
$$t_{22} = \cos(2\pi t_0), \quad t_{21} = -\frac{1}{t_0} \sin(2\pi t_0),$$

for some fixed $t_0 \in (0, \frac{1}{4})$. Let s_0 be defined as a zero of the function

$$f(s) = 1 - \cos(2\pi s)\cos(2\pi t_0) - \frac{s}{t_0} \sin(2\pi s)\sin(2\pi t_0).$$

(There exists at least one zero in the interval $[n_0 + \frac{1}{4}, n_0 + \frac{3}{4}]$ for n_0 large enough.) Then the two analytic Kramer kernels are given by

$$K_1(x, \lambda) = \sqrt{\lambda}\left(\sin\left(2\pi\sqrt{\lambda}\right) - s_0 \sin(2\pi s_0)\right) u(x, \lambda)$$
$$- \left(\cos\left(2\pi\sqrt{\lambda}\right) - \cos(2\pi s_0)\right) v(x, \lambda),$$
$$K_2(x, \lambda) = \left(\cos\left(2\pi\sqrt{\lambda}\right) - \cos(2\pi t_0)\right) u(x, \lambda)$$
$$+ \left(\left(\sqrt{\lambda}\right)^{-1/2} \sin\left(2\pi\sqrt{\lambda}\right) - \frac{1}{t_0}\sin(2\pi t_0)\right) v(x, \lambda),$$

$K_1(x, \lambda)$ degenerates at $\lambda_1 = s_0^2$ but $K_2(x, \lambda)$ does not, and $K_2(x, \lambda)$ degenerates at $\lambda_2 = t_0^2$ but $K_1(x, \lambda)$ does not. All the eigenvalues are real and simple and degeneracy does not occur at other eigenvalues (apart from λ_1 and λ_2). The common interpolation function is given by

$$G(\lambda) = (\cos(2\pi s_0) + \cos(2\pi t_0))\cos\left(2\pi\sqrt{\lambda}\right)$$
$$+ \left(s_0\left(\sqrt{\lambda}\right)^{-1/2}\sin(2\pi s_0) + \frac{1}{t_0}\sqrt{\lambda}\sin(2\pi t_0)\right)\sin\left(2\pi\sqrt{\lambda}\right).$$

(iv) Choose

$$T = \begin{bmatrix} 1 & 0 \\ 0 & 1 \end{bmatrix}.$$

The eigenvalues are determined as the zeros of $\cos(2\pi\sqrt{\lambda}) = 1$; here 0 is a simple eigenvalue but k^2 ($k \in \mathbb{N}$) are double eigenvalues. Both the functions

$$K_1(x, \lambda) = \sqrt{\lambda}\sin\left(2\pi\sqrt{\lambda}\right) u(x, \lambda) - \left(\cos\left(2\pi\sqrt{\lambda}\right) - 1\right) v(x, \lambda),$$
$$K_2(x, \lambda) = \left(\cos\left(2\pi\sqrt{\lambda}\right) - 1\right) u(x, \lambda) + \left(\sqrt{\lambda}\right)^{-1/2}\sin\left(2\pi\sqrt{\lambda}\right) v(x, \lambda)$$

vanish for k^2 ($k \in \mathbb{N}$).

However setting $K_1(x, \lambda) = \cos x\sqrt{\lambda}$ and $K_2(x, \lambda) = \sqrt{\lambda}\sin(x\sqrt{\lambda})$ the following situation occurs:

$$K_1 \text{ and } K_2 \text{ are entire functions}$$

$$K_1(x, k^2) = \cos(kx) \ (k \in \mathbb{N}_0) \quad \text{and} \quad K_2(x, k^2) = k\sin(kx) \ (k \in \mathbb{N}),$$

taken together the functions form a complete, orthogonal set in $L^2(-\pi, \pi)$.

Lemma 5.14 *Let f_1 and f_2 be represented by*

$$f_1(\lambda) = \int_{-\pi}^{\pi} \cos\left(x\sqrt{\lambda}\right) g_1(x)\, dx \quad (\lambda \in \mathbb{C});$$

$$f_2(\lambda) = \int_{-\pi}^{\pi} \sqrt{\lambda}\sin\left(x\sqrt{\lambda}\right) g_2(x)\, dx \quad (\lambda \in \mathbb{C})$$

for $g_1 \in H_1$, $g_2 \in H_2$ with

$$H_1 = \mathrm{span}\{\cos(kx) : k \in \mathbb{N}_0\} \quad \text{and} \quad H_2 = \mathrm{span}\{k\sin(kx) : k \in \mathbb{N}\}.$$

Then the following sampling expansions are valid

$$f_1(\lambda) = \sum_{k=0}^{\infty} f_1(k^2) \frac{G(\lambda)}{G'(k^2)(\lambda - k^2)} \quad (\lambda \in \mathbb{C});$$

$$f_2(\lambda) = \sum_{k=1}^{\infty} f_2(k^2) \frac{G(\lambda)}{G'(k^2)(\lambda - k^2)} \quad (\lambda \in \mathbb{C})$$

with $G(\lambda) = \sqrt{\lambda}\sin\left(\pi\sqrt{\lambda}\right)$, and $L^2(-\pi, \pi) = H_1 \oplus H_2$.

5.2.3 Some results in the limit-point case

If for the second-order differential equation (5.2.2) one or both end-points a, b are in the limit-point case in $L^2((a, b); w)$, then the spectrum of any self-adjoint operator, generated by (5.2.2) and symmetric boundary conditions, may have a non-empty essential spectrum. The Kramer type theorems require the operator to have a discrete spectrum; thus in these limit-point cases it is necessary to impose as a condition on the boundary value problem, that the spectrum is discrete.

In these limit-point cases not all the solutions of the differential equation (5.2.2) are in the space $L^2((a, b); w)$ and this situation imposes certain difficulties on the construction of the Kramer analytic kernel.

The first results in this area are given in Zayed et al. (1990). The main results are due to Zayed (1991, 1993a and 1993b).

One result is presented here in the form of

Theorem 5.15 *Let* $q : [0, \infty) \to \mathbb{R}$ *and let* $q \in C[0, \infty)$, *then the differential equation*

$$-y'' + qy = \lambda y \quad on \ [0, \infty) \tag{5.2.20}$$

is regular at the end-point 0; *assume that* (5.2.20) *is in the* **limit-point** *classification at* $+\infty$, *for the Hilbert function space* $L^2(0, \infty)$.

Then the boundary value problem, with $\alpha \in [0, \pi)$

$$-y'' + qy = \lambda y \quad on \ [0, \infty) \tag{5.2.21}$$

$$y(0) \cos \alpha + y'(0) \sin \alpha = 0 \tag{5.2.22}$$

is self-adjoint in $L^2[0, \infty)$.

Assume that the spectrum of (5.2.21) *and* (5.2.22) *is discrete and bounded below; denote the spectrum by* $\{\lambda_n : n \in \mathbb{N}_0\}$ *with* $\lim_{n \to \infty} \lambda_n = +\infty$, *then all the eigenvalues are simple and the spectrum* $\sigma(T)$ *of the associated self-adjoint operator* T *has multiplicity one; assume without loss of generality, that* $\lambda_0 = 0$; *let the eigenfunctions of* T *be denoted by* $\{\psi_n : n \in \mathbb{N}_0\}$.

Assume that there exists a positive number $\sigma > 0$ *such that*

$$\sum_{n=1}^{\infty} \frac{1}{\lambda_n^{\sigma}} < \infty. \tag{5.2.23}$$

Then

(a) *there exists an analytic Kramer kernel* K, *see Theorem 5.4, with the properties*
  (i) $K(\cdot, \lambda)$ *is a solution of* (5.2.2) *for all* $\lambda \in \mathbb{C}$, *and* $K(\cdot, \lambda) \in \mathbb{R}$ $(\lambda \in \mathbb{R})$;
  (ii) $K(\cdot, \lambda)$ *is an element of the maximal domain and in particular of* $L^2((a, b); w)$ *for all* $\lambda \in \mathbb{C}$;
  (iii) $K(x, \cdot) \in \mathbf{H}(\mathbb{C})$ $(x \in (a, b))$;
  (iv) $K(\cdot, \lambda_n) = k_n \psi_n$ *where* $k_n \in \mathbb{R} \setminus \{0\}$ $(n \in \mathbb{Z})$;
(b) *there exists an interpolation function* G, *see Definition 5.5, for* K.

Proof See the papers by Zayed et al. (1990), Zayed (1991) and the book by Zayed (1993b).

Remarks
(a) There is an extension of this result to the case when the differential equation (5.2.21) is defined on $(-\infty, \infty)$ and is in the limit-point classification at both end-points $\pm\infty$ in $L^2(-\infty, \infty)$. For details see Zayed (1991) and the book by Zayed (1993b).
(b) Examples of these results are considered in the next subsection.
(c) There exist potentials q, with q satisfying all the required conditions for Theorem 5.15, such that the infinite series (5.2.23) is divergent for all $\sigma > 0$; see the remarks in Everitt et al. (1994).

5.2.4 *Examples in the limit-point case*

A detailed discussion of the following two examples can be found in Zayed (1991).

Example 5.16 Consider the self-adjoint Bessel differential equation

$$-y''(x) + \left(v^2 - \tfrac{1}{4}\right) x^{-2} y(x) = \lambda y(x) \quad (x \in (0, b]) \tag{5.2.24}$$

for $v \geq 1$ with the boundary condition $y(b) = 0$ at b. The end-point 0 is in the limit-point case in $L^2(0, 1)$. The problem is equivalent to a singular Sturm–Liouville problem on $(0, \infty)$ and Theorem 5.15 can be applied.

The simple eigenvalues are given by the zeros of $\lambda^{-v/2} J_v(b\sqrt{\lambda})$, where $J_v(z)$ is the classical Bessel function of order v. An analytic Kramer kernel is given by

$$K(x, \lambda) = \frac{2^v \Gamma(v+1)}{b^{v+\frac{1}{2}}} \sqrt{x} \lambda^{-v/2} J_v(x\sqrt{\lambda})$$

and an interpolation function is given by the canonical product

$$G(\lambda) = \prod_{n=1}^{\infty} \left(1 - \frac{\lambda b^2}{\alpha_{v,n}^2}\right),$$

where $\alpha_{v,n}$ is the nth positive zero of $J_v(z)$, $n \in \mathbb{N}$.

Example 5.17 Consider the self-adjoint Laguerre differential equation

$$-y''(x) + \left(x^4 + \alpha - \frac{1}{4}\right) x^{-2} y(x) = \lambda y(x) \quad (x \in (0, \infty)) \tag{5.2.25}$$

for $\alpha \geq -1$. Both end-points are in the limit-point case and results similar to Theorem 5.15 can be applied since the spectrum is discrete.

The simple eigenvalues are given by $\lambda_n = 4n + 2\alpha + 2$ ($n \in \mathbb{N}_0$) with the corresponding eigenfunctions $\psi_n(x) = x^{\alpha+1/2} e^{-x^2/2} L_n^\alpha(x^2)$, where $L_n^\alpha(z)$ is the generalized Laguerre polynomial of degree n.

An analytic Kramer kernel is given by

$$K(x, \lambda) = x^{-\frac{1}{2}} W_{\frac{\lambda}{4}, \frac{\alpha}{2}}(x^2),$$

where $W_{k,\mu}(x) = \exp(-\frac{x}{2}) x^{\frac{c}{2}} \Psi(a, c; x)$, with $k = -a + \frac{c}{2}$, $\mu = \frac{c}{2} - \frac{1}{2}$, is the Whittaker function related to the confluent hypergeometric function $\Psi(a, c; x)$.

An interpolation function is given by

$$G(\lambda) = \frac{\exp\left(-\frac{\lambda}{4} \psi\left(\frac{\alpha}{2} + \frac{1}{2}\right)\right) \Gamma\left(\frac{\alpha}{2} + \frac{1}{2}\right)}{\Gamma\left(\frac{\alpha}{2} + \frac{1}{2} - \frac{\lambda}{4}\right)},$$

where $\psi(z)$ is the logarithmic derivative of the gamma function $\Gamma(z)$.

5.3   Shannon-type interpolation formulae

5.3.1   *The first-order regular case*

The Shannon–Whittaker interpolation theorem is discussed in Section 5.1.6 above. In an important contribution to the general area covered by the contents of this chapter, Campbell (1964) discussed a comparison of the Shannon–Whittaker type formulae to the Kramer type formulae, the latter as given in Sections 5.1.2 and 5.1.4 above.

In (Campbell, 1964, Section 2), in the notation of Theorem 5.4, a Kramer analytic kernel K_1 is obtained from the first-order, regular, self-adjoint boundary value problem

$$iv'(\xi) + [\lambda + Q(\xi)]v(\xi) = 0 \quad (\xi \in [-\Omega, \Omega]) \tag{5.3.1}$$

$$v(-\Omega) = \exp(i\varphi)v(\Omega) \tag{5.3.2}$$

where (i) $\Omega > 0$, (ii) $Q : [-\Omega, \Omega] \to \mathbb{R}$ and $Q \in C[-\Omega, \Omega]$, and (iii) the parameter $\varphi \in [0, 2\pi]$.

The problem (5.3.1) and (5.3.2) has a discrete spectrum, say, $\{\lambda_n : n \in \mathbb{Z}\}$ given by the formula

$$\lambda_n = \frac{n\pi}{\Omega} - \frac{M(\Omega) + \varphi}{2\Omega} \quad (n \in \mathbb{Z}) \tag{5.3.3}$$

where $M : [-\Omega, \Omega] \to \mathbb{R}$ is defined by

$$M(\xi) := \int_{-\Omega}^{\xi} Q(t)\, dt \quad (\xi \in [-\Omega, \Omega]). \tag{5.3.4}$$

This boundary value problem leads to the definition of the Kramer analytic kernel $K_1 : [-\Omega, \Omega] \times \mathbb{R}$, as a solution of (5.3.1), satisfying all the required properties for Theorem 5.4 of Section 5.1.4 above. If the function class $\{K_1\}$ is then defined as the set of all $F : \mathbb{C} \to \mathbb{C}$ where for some $G \in L^2[-\Omega, \Omega]$

$$F(\lambda) = \int_{-\Omega}^{\Omega} K_1(\xi, \lambda)G(\xi)\, d\xi \quad (\lambda \in \mathbb{C}); \tag{5.3.5}$$

here K_1 has the form

$$K_1(\xi, \lambda) = c(\lambda)\exp(i[\lambda\xi + M(\xi)]) \quad (\xi \in [-\Omega, \Omega],\ \lambda \in \mathbb{C})$$

for some $c(\cdot) \in \mathbf{H}(\mathbb{C})$. This leads to a Kramer sampling result with the functions $\{S_n : n \in \mathbb{Z}\}$ given in the form

$$S_n(\lambda) = \frac{c(\lambda)}{c(\lambda_n)} \frac{\sin((\lambda - \lambda_n)\Omega)}{(\lambda - \lambda_n)\Omega} \quad (\lambda \in \mathbb{C}\ \text{and}\ n \in \mathbb{Z}); \tag{5.3.6}$$

here $c(\lambda)/c(\lambda_n)$ is interpreted as unity if $c(\lambda_n) = 0$. In turn this leads to a sampling result, for all $F \in \{K_1\}$,

$$F(\lambda) = \sum_{n \in \mathbb{Z}} F\left(\frac{n\pi}{\Omega}\right) \frac{c(\lambda)}{c(\lambda_n)} \frac{\sin((\lambda - \lambda_n)\Omega)}{(\lambda - \lambda_n)\Omega} \quad (\lambda \in \mathbb{C}) \tag{5.3.7}$$

with the same convergence properties as given in Theorem 5.6.

Clearly the expansion (5.3.7) is of a similar type to the original Shannon–Whittaker result; note in particular that the sampling points $\{\lambda_n : n \in \mathbb{Z}\}$ are equally spaced on the line $\mathbb{R}$, i.e. we have $\lambda_{n+1} - \lambda_n = \pi/\Omega$ $(n \in \mathbb{Z})$. However it seems, in general, impossible to avoid the introduction of the holomorphic function $c(\cdot)$ into the formula; essentially this is due to the introduction of the coefficient Q into the differential equation (5.3.1).

It is not clear, in the account given by Campbell (1964), how the determination of the holomorphic function $c(\cdot)$ is made; in particular there is no discussion of the consequences for the expansion (5.3.7) of zeros of $c(\cdot)$ additional to possible zeros at the points $\{\lambda_n : n \in \mathbb{Z}\}$.

5.3.2 σ-Shannon-type kernels

Before passing to consider Shannon-type formulae for the regular and singular second-order cases it is convenient to introduce a definition that is consistent with the results of Campbell (1964), Zayed et al. (1990) and Genuit and Schöttler (1995). This is an appropriate point to state that an extensive historical perspective of the equivalence of the Kramer and Shannon sampling theorems, in the second-order case, can be found in Genuit and Schöttler (1995); there also may be found additional references.

To give the definition below it is necessary to make definite the square-root function $\sqrt{\cdot} : \mathbb{C} \to \mathbb{C}$; let the polar form of $\lambda \in \mathbb{C}$ be written as $\lambda = r \exp(i\theta)$ with $r \geq 0$ and $\theta \in [0, 2\pi)$; then

$$\sqrt{\lambda} := +\sqrt{r} \exp(\tfrac{1}{2}i\theta) \quad (\lambda \in \mathbb{C}). \tag{5.3.8}$$

We note that $\sqrt{\cdot} \notin \mathbf{H}(\mathbb{C})$ but $\sqrt{\cdot} \in \mathbf{H}(\mathbb{C}\setminus [0, \infty))$.

We now make a definition that is designed to conform with ideas introduced by Campbell (1964).

Definition 5.18 *Let* $K : I \times \mathbb{C} \to \mathbb{C}$ *be an analytic Kramer kernel, i.e. let K satisfy conditions (i)–(vi) of Theorem 5.4 above; let the set $\{K\}$ be defined as in (2) of Theorem 5.4; let $\sigma \in \mathbb{R}$ with $\sigma > 0$. Then K is a σ-Shannon-type kernel if*

(i) *there exists* $c(\cdot) \in \mathbf{H}(\mathbb{C})$*with* $c(\lambda) \neq 0$ $(\lambda \in \mathbb{C}\setminus \{\lambda_n : n \in \mathbb{Z}\})$;
(ii) *for each* $F \in \{K\}$ *there exists* $\varphi_F \in L^2[-\sigma, \sigma]$
   such that

$$F(\lambda) = c(\lambda) \int_{-\sigma}^{\sigma} \varphi_F(x) \exp\left(ix\sqrt{\lambda}\right) dx \quad (\lambda \in \mathbb{C}). \tag{5.3.9}$$

·Remarks

(a) Note that the notation φ_F indicates that this function depends on the element $F \in \{K\}$. On the other hand the entire function $c(\cdot)$ depends only on the kernel K.
(b) In the requirement (5.3.9) the term $\exp\left(ix\sqrt{\lambda}\right)$ can be replaced by the function $\exp\left(-ix\sqrt{\lambda}\right)$ and the resulting new definition is equivalent to the given definition; this follows from the change of variable $x \to -x$ in the integral.
(c) This definition imposes an additional condition on the representation (5.3.9) in that the entire multiplier $c(\cdot)$ is required to have no zeros on the set $\mathbb{C} \setminus \{\lambda_n : n \in \mathbb{Z}\}$; this is to avoid the difficulties raised at the end of Section 5.3.1 above. Indeed it

is not difficult to show that if there is a representation of the form (5.3.9) for all $F \in \{K\}$, then to suppose that for some $n \in \mathbb{Z}$ it is the case that $c(\lambda_n) = 0$ leads to a contradiction on condition (iv) of Theorem 5.4 above. This condition on the multiplier $c(\cdot)$ is satisfied in the examples considered below.

(d) Since $F \in \{K\}$ this definition implies that the integral term on the right-hand side of (5.3.9) is holomorphic on $\mathbb{C}$; see (b) of Theorem 5.4. This places a hidden restriction on the function φ_F since, in general, this integral term is not holomorphic on $\mathbb{C}$, due to the fact that $\sqrt{\cdot}$ is only holomorphic on $\mathbb{C} \setminus [0, \infty)$. In the results given below this requirement is met by the additional property of evenness in the function φ_F, i.e.

$$\varphi_F(-x) = \varphi_F(x) \quad (x \in [-\sigma, \sigma]). \tag{5.3.10}$$

If this property is satisfied then

$$\int_{-\sigma}^{\sigma} \varphi_F(x) \exp\left(ix\sqrt{\lambda}\right) dx = 2 \int_0^{\sigma} \varphi_F(x) \cos\left(ix\sqrt{\lambda}\right) dx \quad (\lambda \in \mathbb{C})$$

and the right-hand side is seen to be in $\mathbf{H}(\mathbb{C})$ since

(i) $\cos\left(x\sqrt{\cdot}\right) \in \mathbf{H}(\mathbb{C}) \quad (x \in [-\sigma, \sigma])$;

(ii) $\lambda \longmapsto \int_0^{\sigma} \left|\cos\left(x\sqrt{\lambda}\right)\right|^2 dx$ is locally bounded on $\mathbb{C}$.

A combination of this definition and Theorem 5.6 of Section 5.1.6 leads to

Theorem 5.19 *Let $K : I \times \mathbb{C} \to \mathbb{C}$ satisfy*

(i) *K is a Kramer analytic kernel*;
(ii) *K is a σ-Shannon-type kernel with multiplier $c(\cdot)$.*

Then for all $F \in \{K\}$

1. $F(\lambda) = \displaystyle\sum_{n \in \mathbb{Z}} F(\lambda_n) S_n(\lambda)$;

2. $F(\lambda) = \displaystyle\sum_{n \in \mathbb{Z}} F\left((n\pi/\sigma)^2\right) \dfrac{c(\lambda)}{c\left((n\pi/\sigma)^2\right)} \dfrac{\sin\left(\sigma\sqrt{\lambda} - n\pi\right)}{\sigma\sqrt{\lambda} - n\pi}$, $\tag{5.3.11}$

where the multiplier ratio is interpreted with the value 1 if $c\left((n\pi/\sigma)^2\right) = 0$ for some $n \in \mathbb{Z}$.

There is absolute and local uniform convergence in both infinite series as given in Theorem 5.4 of Section 5.1.

Proof This follows from the results of Theorems 5.4 and 5.6 of Section 5.1 and Definition 5.18 above.

Remark For some examples the two expansions in (5.3.11) have an asymptotic equivalence; some indication of results of this kind may be seen in (Campbell, 1964, Sections 2 and 3).

5.3.3 The second-order regular case

This case is considered in some detail by Campbell (1964); see in particular Section 3 and Appendices 1 and 2 of this paper by Campbell.

The boundary value problem considered by Campbell (1964) involves the Sturm–Liouville differential equation

$$-(p(x)u'(x))' + q(x)u(x) = \lambda\rho(x)u(x) \quad (x \in [a, b]) \tag{5.3.12}$$

and two separated boundary conditions, one each at the end-points a and b of the compact interval. The coefficients p, q and ρ are real-valued on the compact interval $[a, b]$; additionally p', q and ρ' are continuous on $[a, b]$.

In the notation of this chapter this boundary value problem yields an analytic Kramer kernel on $L^2([a, b] : \rho) \times \mathbb{C}$.

By consideration of the Riemann–Green function of the partial differential equation

$$\frac{\partial}{\partial x}\left[p(x)\frac{\partial w(x, y)}{\partial x}\right] - \rho(x)\frac{\partial^2 w(x, y)}{\partial y^2} - q(x)w(x, y) = 0$$
$$((x, y) \in [a, b] \times [a, b]) \tag{5.3.13}$$

it is shown by Campbell (1964), again in the notation of this chapter, that for some $\sigma > 0$, any $F \in \{K\}$ can also be represented in the form

$$F(\lambda) = c(\lambda) \int_{-\sigma}^{\sigma} h(x) \exp\left(ix\sqrt{\lambda}\right) dx \quad (\lambda \in \mathbb{C}); \tag{5.3.14}$$

here the multiplier $c(\cdot)$ satisfies the conditions given in Definition 5.18 above. However it is not clear from the analysis by Campbell (1964), in the representation (5.3.14), that $h \in L^2[-\sigma, \sigma]$, but it does seem that this result holds; in this respect see the comment made by Genuit and Schöttler (1995, p. 34).

From results of this form it appears that second-order boundary value problems generated by regular Sturm–Liouville differential equations with "smooth" coefficients, yield analytic Kramer kernels that are also σ-Shannon-type kernels, for some $\sigma > 0$.

It is an open question as to whether or not these results hold for regular Sturm–Liouville boundary value problems when the coefficients satisfy only the minimal conditions on the interval $[a, b]$ and the boundary conditions are separated; for a statement of the boundary value problem under these conditions see Everitt et al. (1994).

Likewise there is an open question on the extension of these results to Sturm–Liouville problems with coupled boundary conditions; see Everitt et al. (to appear) and Everitt and Nasri-Roudsari (to appear (a)).

5.3.4 The second-order singular case

There are as yet no general results to the effect that all, or a well-defined subclass, of analytic Kramer kernels, generated by singular second-order Sturm–Liouville problems, are also σ-Shannon-type kernels, according to Definition 5.18 above. However there is an interesting set of examples of these singular problems for which this result holds.

The first examples are considered by Campbell (1964) and concern the Bessel and Legendre differential equations. These results are followed by the work of Jerri (1969) and of Zayed et al. (1990).

We report here on the results of Genuit and Schöttler (1995) since this work contains essentially all the earlier examples. Moreover, their methods are more complete than those used in earlier work, in that:

(a) The regular/limit-point/limit-circle classification of the differential equation is included for the differential equations;

(b) Consequent upon (a) above the boundary value problems are correctly stated in terms of the Glazman–Krein–Naimark form of the regular and singular boundary conditions, as given in the book by Naimark (1968, Chapter V).

The Bessel differential equation

The boundary value problem is the differential equation

$$-y''(x) + \left(v^2 - \tfrac{1}{4}\right)x^{-2}y(x) = \lambda y(x) \quad (x \in (0, 1]) \tag{5.3.15}$$

with the boundary conditions

$$\lim_{x \to 0+} [y(x), x^{v+\frac{1}{2}}] = 0 \quad y(1) = 0; \tag{5.3.16}$$

here $[\cdot, \cdot]$ is the bilinear form associated with (5.3.15). This problem is self-adjoint for all $v \in \mathbb{R}$, but note that the first boundary condition in (5.3.16) is superfluous when $v \geq 1$ as the point 0 is then in the limit-point case in $L^2(-1, 1)$. The end-point classification of (5.3.15) and other details of this equation are given by Genuit and Schöttler (1995, Section 2).

The spectrum of the problem (5.3.15) and (5.3.16) is bounded below on $\mathbb{R}$ and is discrete, say $\{\lambda_n(v) : n \in \mathbb{N}_0\}$; these eigenvalues have the asymptotic form (Niessen and Zettl 1992),

$$\lambda_n(v) \sim k(v)n^2 \quad (n \to \infty) \tag{5.3.17}$$

where $k(v) \in (0, \infty)$ depends only on the parameter v.

This problem generates an analytic Kramer kernel on $L^2(0, 1) \times \mathbb{C}$ given explicitly by (here J_v is the classical Bessel function of order v)

$$K(x, \lambda) = \lambda^{-v/2}\sqrt{x}J_v\left(x\sqrt{\lambda}\right) \quad (x \in (0, 1]; \ \lambda \in \mathbb{C})$$

where $\lambda \longmapsto \lambda^{v/2}$ is defined by $\lambda^{v/2} := \left(\sqrt{\lambda}\right)^v$ $(\lambda \in \mathbb{C})$.

The associated class $\{K\}$ is then the set of all F determined by (here $f \in L^2(0, 1)$),

$$F(\lambda) = \int_0^1 \lambda^{-v/2}\sqrt{x}J_v\left(x\sqrt{\lambda}\right)f(x)\,dx \quad (\lambda \in \mathbb{C}). \tag{5.3.18}$$

The result of Genuit and Schöttler (1995, Section 2) shows that if

$$v > -\tfrac{1}{2} \tag{5.3.19}$$

then this Bessel kernel K is a 1-Shannon-type kernel, i.e. for each $F \in \{K\}$ there exists $H_F \in L^2(-1, 1)$ such that

$$F(\lambda) = \int_{-1}^{1} H_F(x) \exp\left(ix\sqrt{\lambda}\right) dx \quad (\lambda \in \mathbb{C}) \tag{5.3.20}$$

where, with f determined by the representation (5.3.18),

$$H_F(x) = \int_{|x|}^{1} (x^2 - u^2)^{\nu - \frac{1}{2}} x^{\frac{1}{2} - \nu} f(x)\, dx \quad (x \in [-1, 1]). \tag{5.3.21}$$

Note that H_F is an even function on $[-1, 1]$; see Remark (c) in Section 5.3.2 above.

In this example note that the multiplier $c(\cdot)$ is determined by $c(\lambda) = 1$ $(\lambda \in \mathbb{C})$ and thereby satisfies all the requirements given in Definition 5.18 of Section 5.3.2.

There is a detailed analytical proof of Genuit and Schöttler (1995, Section 2) of the critical result that in (5.3.20) we have $H_F \in L^2(-1, 1)$.

The result (5.3.20) leads to an expansion for all $F \in \{K\}$ of the form given in Theorem 5.19 above; see (5.3.11, 2)

The Jacobi differential equation

In some respects this case is similar to the Bessel differential equation so that results are given in outline; full details may be found in Genuit and Schöttler (1995, Section 3).

The differential equation is

$$-\left((1-x)^{\alpha+1}(1+x)^{\beta+1} y'(x)\right)'$$
$$= \left(\lambda - \tau(\alpha, \beta)^2\right)(1-x)^{\alpha}(1+x)^{\beta} y(x) \quad (x \in (-1, 1)) \tag{5.3.22}$$

where the parameters $\alpha, \beta \in \mathbb{R}$ with $\alpha, \beta > -1$, and $\tau(\alpha, \beta) := \frac{1}{2}(\alpha + \beta + 1)$.

For this problem there is a weight function

$$w_{\alpha,\beta}(x) = (1-x)^{\alpha}(1+x)^{\beta} \quad (x \in (-1, 1)).$$

In the space $L^2((-1, 1) : w_{\alpha,\beta})$ the equation (5.3.22) is regular or limit-circle at $+1^-$ if $-1 < \alpha < 0$ and limit-point if $\alpha \geq 1$; there is a similar result at -1^+ in terms of β.

A self-adjoint boundary value problem is determined if to the equation (5.3.22) two separated boundary conditions are added of the form

$$\lim_{x \to -1^+} [y, 1] = 0 \qquad \lim_{x \to +1^-} [y, 1] = 0 \tag{5.3.23}$$

where $[\cdot, \cdot]$ is the bilinear form associated with the differential equation. These boundary conditions are superfluous in the limit-point cases.

The spectrum of the problem (5.3.22) and (5.3.23) is bounded below on $\mathbb{R}$ and discrete, say $\{\lambda_n(\alpha, \beta) : n \in \mathbb{N}_0\}$; as in the Bessel case these eigenvalues have an asymptotic form (Niessen and Zettl 1992),

$$\lambda_n(\alpha, \beta) \sim k(\alpha, \beta)n^2 \quad (n \to \infty).$$

This problem generates an analytic Kramer kernel K on $L^2((-1, 1) : w_{\alpha,\beta}) \times \mathbb{C}$ given by (here $_2F_1$ is the Gaussian hypergeometric function)

$$K(x, \lambda) = {_2F_1}\left(\tau(\alpha, \beta) - \sqrt{\lambda}, \ \tau(\alpha, \beta) + \sqrt{\lambda}; \ \alpha + 1; \ \tfrac{1}{2}(1 - x)\right).$$

The result by Genuit and Schöttler (1995, Section 3) shows that if

$$\alpha > -\tfrac{1}{2} \quad \text{and} \quad -1 < \beta < +1$$

then for each $F \in \{K\}$ there exists $H_F \in L^2(-1, 1)$ such that

$$F(\lambda) = \int_{-1}^{1} H_F(x) \exp\left(ix\sqrt{\lambda}\right) dx \quad (\lambda \in \mathbb{C}). \tag{5.3.24}$$

There is an explicit formula for H_F in terms of f, see Genuit and Schöttler (1995, p. 441); this formula is similar to (5.3.21) above and shows that H_F is an even function on $[-1, 1]$.

Again the multiplier $c(\cdot)$ in this case is the unit constant, as in the Bessel example.

There is a detailed analytical proof by Genuit and Schöttler (1995, Section 3) that $H_F \in L^2(-1, 1)$.

The result (5.3.24) allows of an expansion for all $F \in \{K\}$ of the form given in (5.3.11, 2.).

5.3.5 Remarks

1. The σ-Shannon-type kernels have the advantage that in the expansion for $F \in \{K\}$

$$F(\lambda) = \sum_{n \in \mathbb{Z}} F((n\pi/\sigma)^2) \frac{c(\lambda)}{c((n\pi/\sigma)^2)} \frac{\sin\left(\sigma\sqrt{\lambda} - n\pi\right)}{\sigma\sqrt{\lambda} - n\pi}$$

the sampling points $\{(n\pi/\sigma)^2 : n \in \mathbb{Z}\}$ are given explicitly, provided that the end-point σ is known; so also are the interpolation functions involving the $\sin(\cdot)$ function. There may be difficulties in obtaining explicit information about the multiplier function $c(\cdot)$ but in many of the known examples this turns out to be the unit function on $\mathbb{R}$.

2. The question has been asked as to whether or not all Kramer analytic kernels generated by second-order boundary value problems, either regular or singular, are also σ-Shannon-type kernels?

The work of Campbell (1964) indicates that there is a positive answer to this question in the case of regular problems; however the Campbell results need to be reworked under the minimal conditions on the coefficients of the differential equation, i.e. under the L^1 conditions for p, q, w on $[a, b]$.

This question in the singular case may have a positive answer when only non-oscillatory limit-circle end-points are involved; all the known examples that have been considered have yielded a σ-Shannon-type kernel, at least for some range of the parameters involved in the problem; see the results of Genuit and Schöttler (1995). The asymptotic form of the eigenvalues in the general limit-circle case suggests that these problems have all the required characteristics for the existence of σ-Shannon-type kernels, but as yet there is no method to establish this result in general.

This question in the singular case may have a negative answer when limit-point or oscillatory limit-circle end-points are involved; however no example has yet been produced to confirm this conjecture. There is no general form of the asymptotic behaviour of the eigenvalues for these problems but examples show that there is a marked departure from the regular/limit-circle case; this is an indication that there are Kramer kernels from such boundary value problems for which there are no σ-Shannon-type kernels.

On this interesting question see the remarks by Genuit and Schottler (1995, Section 1).

5.4 Higher-order problems

In this section we study results in the higher-order case, i.e. the case where the symmetric differential equation is of order greater than two. The results in this case are not as complete as in the second-order case discussed in Section 5.2.

There is some discussion on the higher-order case in the papers of Butzer and Schöttler (1994), Zayed (1993 a) and Zayed et al. (1991), and the book by Zayed (1993b). In Butzer and Schöttler (1994) the situation is as follows.

The symmetric differential equation, of order $2n$ with $n \in \mathbb{N}$, is given by

$$M[y] = \lambda w y \quad \text{on} \quad (a, b), \tag{5.4.1}$$

where

$$M[y] := (p_0 y^{(n)})^{(n)} + \cdots + (p_{n-1} y')' + p_n y \tag{5.4.2}$$

and

(i) the spectral parameter $\lambda \in \mathbb{C}$;
(ii) $[a, b]$ is a compact interval of $\mathbb{R}$;
(iii) the coefficients $p_j, w : [a, b] \to \mathbb{R}$ ($j = 0, 1, 2, \ldots, n$); $\qquad$ (5.4.3)
(iv) $p_j \in C^{(n-j)}[a, b]$ ($j = 0, 1, 2, \ldots, n$);
(v) $w \in C[a, b]$ and $w(x) > 0$ for almost all $x \in [a, b]$.

The boundary conditions are given by $2n$ linearly independent linear forms

$$U_j(y) = \sum_{k=1}^{2n} \{\alpha_{jk} y^{(k-1)}(a) + \beta_{jk} y^{(k-1)}(b)\} = 0 \quad (j = 1, 2, \ldots, 2n) \tag{5.4.4}$$

with $\alpha_{jk}, \beta_{jk} \in \mathbb{C}$ ($j, k = 1, 2, \ldots, 2n$). Let the boundary value problem (5.4.1), (5.4.4) be self-adjoint. (See Coddington and Levinson (1955, Chapter 7, Section 2) for the definition of self-adjointness of boundary conditions.) It generates a self-adjoint differential operator T having the following properties:

(a) T is self-adjoint and unbounded in $L^2((a, b); w)$;
(b) the spectrum of T is real, and discrete with limit-points at $+\infty$ or $-\infty$ or both;
(c) the eigenvalues and eigenvectors of T satisfy the boundary value problem.

For the results in Theorem 5.20 to hold it is necessary to impose the following given structural condition:

The eigenvalues of the operator T generated by the boundary value problem (5.4.1), (5.4.4) are simple, i.e. the spectrum $\sigma(T)$ of T is of multiplicity one.

The results of Butzer and Schöttler (1994) read as follows:

Theorem 5.20 *Let $[a, b]$ be a compact interval of the real line; let the coefficients $\{p_j : (j = 0, 1, 2, \dots n)\}$, and w satisfy the basic conditions as given in (5.4.3); let the differential equation (5.4.1) satisfy the end-point classification (a) above; let the boundary conditions be given by (5.4.4); let the self-adjoint differential operator T be determined by the symmetric boundary value problem (5.4.1), (5.4.4); let the simple, discrete spectrum of T be given by $\{\lambda_n : n \in \mathbb{Z}\}$; let $\{\psi_n : n \in \mathbb{Z}\}$ be the eigenvectors of T; let the system of basis solutions $\{\varphi_k : k = 1, 2, \dots, 2n\}$ of the differential equation (5.4.1) satisfy the initial conditions*

$$\varphi_k^{(j-1)}(a, \lambda) = \begin{cases} 0, & k \neq j \\ 1, & k = j. \end{cases}$$

Assume that for each eigenvalue λ at least one of the first minors of the determinant

$$\det\left[U_j \left(\varphi_k(x, \lambda) : j, k = 1, 2, \dots, 2n \right]\right.$$

does not vanish.
Define the analytic Kramer kernel $K : (a, b) \times \mathbb{C} \to \mathbb{C}$ by

$$K(x, \lambda) := \det \begin{bmatrix} \varphi_1(x, \lambda) & \dots & \varphi_{2n}(x, \lambda) \\ U_1(\varphi_1) & \dots & U_1(\varphi_{2n}) \\ \vdots & \ddots & \vdots \\ U_{2n-1}(\varphi_1) & \dots & U_{2n-1}(\varphi_{2n}) \end{bmatrix}. \tag{5.4.5}$$

Then K has the following properties:

(i) $K(\cdot, \lambda)$ *is a solution of (5.4.1) for all $\lambda \in \mathbb{C}$, and $K(\cdot, \lambda) \in \mathbb{R}$ $(\lambda \in \mathbb{R})$;*
(ii) $K(\cdot, \lambda)$ *is an element of the maximal domain and in particular of $L^2((a, b); w)$ for all $\lambda \in \mathbb{C}$;*
(iii) $U_j(K(\cdot, \lambda)) = 0$ *($j = 1, 2, \dots, 2n - 1$ and $\lambda \in \mathbb{C}$);*
(iv) $U_n(K(\cdot, \lambda)) = 0$ *if and only if $\lambda \in \{\lambda_n : n \in \mathbb{Z}\}$;*
(v) $K(x, \cdot) \in \mathbf{H}(\mathbb{C})$ *($x \in (a, b)$);*
(vi) $K(\cdot, \lambda_n) = k_n \psi_n$ *where $k_n \in \mathbb{R} \setminus \{0\}$ $(n \in \mathbb{Z})$.*

Define the interpolation function $G : \mathbb{C} \to \mathbb{C}$ by

$$G(\lambda) := U_{2n}(K(\cdot, \lambda)) \quad (\lambda \in \mathbb{C}); \qquad (5.4.6)$$

then G is an interpolation function for the Kramer kernel K in the sense of Definition 5.5. All the results of Theorem 5.4 and Definition 5.5 now follow.

Remark There exist kernels K_j $(j = 1, 2, \ldots, 2n - 1)$, defined in a similar way as K but leaving out the linear form U_j instead of U_{2n}. The interpolation function $G_j := U_j(K_j(\cdot, \lambda))$ $(j = 1, 2, \ldots, 2n - 1)$ may differ from G by a factor -1.

In Zayed (1993a) and the book by Zayed (1993b) the following results are derived by means of the Green's function.

The differential equation is given by

$$M[y] = \lambda y \quad \text{on } [a, b], \qquad (5.4.7)$$

where

$$M[y] := p_0 y^{(2n)} + \cdots + p_{2n-1} y' + p_{2n} y \qquad (5.4.8)$$

and

$$\left.\begin{array}{l}
\text{(i) the spectral parameter } \lambda \in \mathbb{C}; \\
\text{(ii) } [a, b] \text{ is a compact interval of } \mathbb{R}; \\
\text{(iii) the coefficients } p_j : [a, b] \to \mathbb{C} \ (j = 0, 1, 2, \ldots, 2n); \\
\text{(iv) } p_j \in C^{(2n-j)}[a, b] \ (j = 0, 1, 2, \ldots, 2n); \\
\text{(v) } p_0(x) \neq 0 \ (x \in [a, b]).
\end{array}\right\} \qquad (5.4.9)$$

Thus the differential expression is not necessarily symmetric, and the coefficients $\{p_j\}$ are allowed to be complex-valued. The boundary conditions are given by (5.4.4). The following result can be found in Zayed (1993a).

Theorem 5.21 *Let $[a, b]$ be a compact interval of the real line; let the coefficients $\{p_j : j = 0, 1, 2, \ldots, 2n\}$ satisfy the basic conditions as given in (5.4.9); let the boundary value problem be given by (5.4.7) and (5.4.4); let the differential operator T be determined by the boundary value problem; let the discrete spectrum of T be given by $\{\lambda_n : n \in \mathbb{Z}\}$; let $\{\psi_n : n \in \mathbb{Z}\}$ be the eigenvectors of T; let the system of basis solutions $\{\varphi_k : k = 1, 2, \ldots, 2n\}$ of the differential equation (5.4.7) satisfy the initial conditions*

$$\varphi_k^{(j-1)}(a, \lambda) = \begin{cases} 0, & k \neq j \\ 1, & k = j. \end{cases}$$

Assume that the Green's function of the boundary value problem has only simple poles, and assume that the asymptotic behaviour of the eigenvalues is given by

$$\lambda_k = \mathcal{O}(k\pi/(b-a)^n) \quad \text{as } |k| \to \infty.$$

Then there exists an analytic Kramer kernel K in the sense of Theorem 5.4 and a corresponding interpolation function in the sense of Definition 5.5.

Remark For details see Zayed (1993a, 1993b).

6

SAMPLING BY GENERALIZED KERNELS

6.1 Introduction

One version of the Shannon sampling theorem in Euclidean n-space $\mathbb{R}^n$ states that if a signal $f \in L^2(\mathbb{R}^n) \cap C(\mathbb{R}^n)$ is band-limited to some n-dimensional rectangle $[-\pi W, \pi W]$ for some $W \in \mathbb{R}^n$ with positive components W_j, $j = 1, 2, \ldots, n$, then it can be completely reconstructed from its samples at the points k/W, $k \in \mathbb{Z}^n$, in terms of

$$f(t) = \sum_{k \in \mathbb{Z}^n} f\left(\frac{k}{W}\right) \prod_{j=1}^{n} \operatorname{sinc}(W_j t_j - k_j) \quad (t \in \mathbb{R}^n) \tag{6.1.1}$$

the series being absolutely and uniformly convergent (cf. Parzen 1956; Peterson and Middleton 1962; Splettstößer 1982; Higgins 1996, Chapter 14; and for the one-dimensional case see Whittaker 1915; Kotel'nikov 1933; Shannon 1949; Higgins 1985; Butzer et al. 1988; Higgins 1996, Chapter 6). Here

$$\operatorname{sinc}(x) := \begin{cases} \dfrac{\sin(\pi x)}{\pi x}, & x \in \mathbb{R} \setminus \{0\}, \\ 1, & x = 0. \end{cases}$$

There are several possible extensions of this theorem. For example, one may use other than rectangular lattices, which in general are more efficient (see e.g. Peterson and Middleton 1962; Mersereau 1979; Mersereau and Speake 1983; Stark 1979; Higgins 1985; Butzer and Hinsen 1989; Higgins 1996, Chapter 14). However, in any case, band-limitation is a rather restrictive condition, since a band-limited function is in particular an entire function which cannot be simultaneously duration-limited, in view of the uniqueness theorem for analytic functions. So the question arises whether it is possible to have representations similar to (6.1.1) for $W \to \infty$ and to investigate whether

$$f(t) = \lim_{W \to \infty} \sum_{k \in \mathbb{Z}^n} f\left(\frac{k}{W}\right) \prod_{j=1}^{n} \operatorname{sinc}(W_j t_j - k_j) \quad (t \in \mathbb{R}^n). \tag{6.1.2}$$

It can be readily shown that continuity of f alone does not suffice for (6.1.2) to hold; more restrictive conditions upon f are needed (cf. Theis 1919; Prosser 1966; Brown 1967/68; Butzer and Splettstößer 1977; Splettstößer 1979; Stens 1980a,b, 1984; Splettstößer et al. 1981; Ries and Stens 1987; Butzer and Stens 1992). Hence it seems to be convenient to replace the product of sinc functions by some other function $\varphi : \mathbb{R}^n \to \mathbb{C}$ and ask for conditions upon φ such that

$$f(t) = \lim_{W \to \infty} \frac{1}{(\sqrt{2\pi})^n} \sum_{k \in \mathbb{Z}^n} f\left(\frac{k}{W}\right) \varphi(Wt - k) \quad (t \in \mathbb{R}^n) \tag{6.1.3}$$

is valid for every continuous signal function f. The first result in this respect seems to be due to Theis (1919) who considered $\varphi(t) = (\operatorname{sinc} t)^2$ and showed that in this case (6.1.3) holds for every continuous and bounded function f.

It will turn out that the condition $\sum_{k \in \mathbb{Z}^n} \varphi(t-k) = (2\pi)^{n/2}, t \in \mathbb{R}^n$, is necessary and sufficient provided the series converges absolutely and uniformly on compact subsets of $\mathbb{R}^n$. In this case φ is called a kernel(-function). The series in (6.1.3) is a discrete convolution or convolution sum of f and the kernel φ. Continuous versions of those sums, namely convolution integrals, are well known in approximation theory (cf. (6.3.2) below).

The most elementary example of such a kernel φ in the one dimensional instance is the hat function $\varphi(t) := \sqrt{2\pi}(1-|t|), t \in (-1, 1)$, and $:= 0$ elsewhere. The series in (6.1.3) then represents the linear spline function interpolating f at the knots $k/W, k \in \mathbb{Z}, W > 0$. Particular examples of multivariate kernels φ are product kernels $\varphi(t) = \prod_{j=1}^{n} \psi_j(t_j)$, $t = (t_1, \ldots, t_n) \in \mathbb{R}^n$, where $\psi_j(t_j)$ are one-dimensional kernels. Other examples are radial kernels, i.e., those φ for which $\varphi(t) = \psi(|t|), t \in \mathbb{R}^n$, for some univariate ψ. However, the theory to be presented will be built up in such a fashion that it will cover not only those particular cases but arbitrary multivariate functions φ.

Whereas Section 6.3 deals with the reconstruction of *not necessarily band-limited* functions in the sense (6.1.3), Section 6.4 is devoted to the particular case of reconstructing *band-limited* functions by means of convolution sums. Indeed, due to several drawbacks of the Shannon series (6.1.1) in practice, one uses other techniques in order to recover a signal from its samples. In the case of univariate signal functions f, a common technique is to use smoothed step functions built up from the samples $f(k/W)$ as an approximation to f.

A suitable step function $a(t)$ is given by the convolution sum

$$a(t) := \sum_{k=-\infty}^{\infty} f\left(\frac{k}{V}\right) \chi_{[-1/2V, 1/2V)}\left(t - \frac{k}{V}\right) \quad (t \in \mathbb{R}), \qquad (6.1.4)$$

where χ_B denotes the characteristic function of the set B. The parameter $V > 0$ is the width of the steps and has to be chosen less than or equal to the bandwidth parameter W of the function f. This step function coincides with f at the midpoint k/V of each interval $[(k-1/2)/V, (k+1/2)/V], k \in \mathbb{Z}$.

In order to smooth the function $a(t)$, its Fourier transform is multiplied by $\chi_{(-\pi W, \pi W)}$ and one ends up with $(a^\wedge \chi_{(-\pi W, \pi W)})^\vee(t)$ as an approximation to f. Here $^\wedge$ and $^\vee$ denote the Fourier transform and inverse Fourier transform, respectively. It is shown in Section 6.4 that the characteristic function $\chi_{[-1/2V, 1/2V)}$ in (6.1.4) can be replaced by more general functions $h \in L^2(\mathbb{R})$ yielding smaller errors.

The approximation of functions by series like that in (6.1.3) have been studied extensively by many authors in different settings. Let us mention here in particular Chapter 7 of the present volume by A. Fischer and Chapter 8 by N. Dyn. Fischer obtains results similar to ours using an approach via wavelets, multiresolution analysis and shift-invariant spaces. Dyn's chapter, based upon work done by herself, C. de Boor, M.D. Buhmann, D. Levin, I.R.H. Jackson and A. Ron, deals with the approximation by *radial* kernels on regular and quasi-uniform grids.

These two chapters as well as ours consider pointwise or uniform convergence of the sampling series to f (with the exception of our Corollary 6.22). For convergence with respect to L^p-norms see, e.g. Dinh-Dung (1992), Ron (1992), and Buhmann and Ron (1994).

6.2 Notation and preliminary results

In the following, let $\mathbb{N}$, $\mathbb{N}_0$, $\mathbb{Z}$ denote the sets of all naturals, all non-negative integers, and all integers, respectively, $\mathbb{R}$, $\mathbb{R}_+$, $\mathbb{C}$ being the sets of real, positive real and complex numbers, respectively. Let $\mathbb{N}^n$ denote the set of all n-tuples $k = (k_1, \ldots, k_n)$ of elements from $\mathbb{N}$; $\mathbb{N}_0^n$, $\mathbb{Z}^n$, etc. are defined analogously. In particular, $\mathbb{R}^n$ is the Euclidean n-space endowed with the norm $\|u\|_2 := (u_1^2 + \cdots + u_n^2)^{1/2}$, where $u = (u_1, \ldots, u_n)$, $u_j \in \mathbb{R}$, $j \in \{1, 2, \ldots, n\}$. Further, $\alpha u := (\alpha u_1, \ldots, \alpha u_n)$ is the product of u with $\alpha \in \mathbb{R}$, $u \cdot v := \sum_{j=1}^n u_j v_j$ is the scalar product of $u, v \in \mathbb{R}^n$, but $uv := (u_1 v_1, \ldots, u_n v_n)$; u/v denotes the vector of fractions $(u_1/v_1, \ldots, u_n/v_2)$, and u^{-1} will be used for $(1/u_1, \ldots, 1/u_n)$. Also $\lfloor u \rfloor$ is the vector $(\lfloor u_1 \rfloor, \lfloor u_2 \rfloor, \ldots, \lfloor u_j \rfloor)$, where $\lfloor u_j \rfloor$ is the largest integer not bigger than u_j. For $u, v \in \mathbb{R}^n$, $u > v$ iff $u_j > v_j$, and $u > x$ for $x \in \mathbb{R}$ iff $u_j > x$ for $1 \le j \le n$; likewise for $\ge$, $<$ and $\le$. By $[a, b]$, $a, b \in \mathbb{R}^n$, we understand the n-dimensional rectangle given by all vectors $u \in \mathbb{R}^n$ with $a \le u \le b$, and $[c, d]^n$, $c, d \in \mathbb{R}$, is the n-dimensional square $\{u \in \mathbb{R}^n; c \le u \le d\}$. Further, standard multi-index notation is used, i.e., for $k \in \mathbb{N}_0^n$, $u \in \mathbb{R}^n$, let $|k| := k_1 + \cdots + k_n$, $u^k := u_1^{k_1} \cdot \ldots \cdot u_n^{k_n}$ and $k! := k_1! \cdot \ldots \cdot k_n!$.

For a function $f : \mathbb{R}^n \to \mathbb{C}$,

$$D^k f := \frac{\partial^{|k|}}{\partial u^k} f := \frac{\partial^{|k|}}{\partial u_1^{k_1} \ldots \partial u_n^{k_n}} f \quad (|k| = r)$$

is called an rth-order derivative of f. Let $C(\mathbb{R}^n)$ be the space of all uniformly continuous and bounded functions $f : \mathbb{R}^n \to \mathbb{C}$, endowed with the usual supremum norm $\|f\|_C$, and let $C^r(\mathbb{R}^n) := \{f \in C(\mathbb{R}^n); D^k f \in C(\mathbb{R}^n)$ for all $|k| \le r\}$, $r \in \mathbb{N}_0$, be the subspace of all r-fold continuously differentiable functions.

Below, Taylor's formula with integral remainder term will be used in the form: For $g \in C^r(\mathbb{R}^n)$, $u, t \in \mathbb{R}^n$, there holds (see e.g. Hoffmann 1975, p. 195)

$$g(u) - g(t) = \sum_{0 < |m| < r} \frac{(u - t)^m}{m!} D^m g(t) + r \sum_{|m| = r} \frac{(u - t)^m}{m!}$$

$$\times \int_0^1 (1 - s)^{r-1} D^m g\big(t(1 - s) + us\big) ds. \qquad (6.2.1)$$

$L^p(\mathbb{R}^n)$ denotes the space of all measurable functions $f : \mathbb{R}^n \to \mathbb{C}$ for which the norm

$$\|f\|_{L^p} := \left\{ \frac{1}{(\sqrt{2\pi})^n} \int_{\mathbb{R}^n} |f(u)|^p \, du \right\}^{1/p} \quad (1 \le p < \infty),$$

$$\|f\|_{L^\infty} := \operatorname*{ess\,sup}_{u \in \mathbb{R}^n} |f(u)|$$

is finite. L_λ^p, $\lambda \in \mathbb{R}_+^n$, $1 \leq p < \infty$, is the space of all λ-periodic measurable functions $f : \mathbb{R}^n \to \mathbb{C}$ that are integrable to the pth power over the rectangle $(0, \lambda)$ with the norm

$$\|f\|_{L_\lambda^p} := \left\{ \left(\prod_{j=1}^n \lambda_j \right)^{-1} \int_{(0,\lambda)} |f(u)|^p du \right\}^{1/p},$$

and L_λ^∞ is defined appropriately.

The Fourier transform $f^\wedge$ of a function $f \in L^1(\mathbb{R}^n)$ is defined by

$$f^\wedge(v) := \frac{1}{\left(\sqrt{2\pi}\right)^n} \int_{\mathbb{R}^n} f(u)e^{-iv\cdot u} du \quad (v \in \mathbb{R}^n), \qquad (6.2.2)$$

and the same notation is used for the Fourier–Plancherel transform of $f \in L^p(\mathbb{R}^n)$, $1 < p \leq 2$, i.e.,

$$\lim_{\rho \to \infty} \left\| \frac{1}{\left(\sqrt{2\pi}\right)^n} \int_{K_\rho(0)} f(u)e^{-iv\cdot u} du - f^\wedge(v) \right\|_{L^q} = 0,$$

$K_\rho(0) \subset \mathbb{R}^n$ being the ball of radius $\rho > 0$ centred at the origin, and $1/p + 1/q = 1$. The finite Fourier transform (or k-th Fourier coefficient) of $f \in L_\lambda^1$ is given by

$$[f]_\lambda^\wedge(k) := \left(\prod_{j=1}^n \lambda_j \right)^{-1} \int_{(0,\lambda)} f(u)e^{-i2\pi k\cdot(u/\lambda)} du \quad (k \in \mathbb{Z}^n), \qquad (6.2.3)$$

and $f(t) \sim \sum_{k \in \mathbb{Z}^n} [f]_\lambda^\wedge(k)e^{i2\pi k\cdot(t/\lambda)}$, $t \in \mathbb{R}^n$, is the associated Fourier series. In general, $\sum_{k \in \mathbb{Z}^n}$ means the limit $\lim_{N \to \infty} \sum_{k_1=-N}^N \cdots \sum_{k_n=-N}^N$.

The connection between the $L^1(\mathbb{R}^n)$-Fourier transform and the L_λ^1-transform is revealed by the Poisson summation formula. Indeed, if $f \in L^1(\mathbb{R}^n)$, then the series

$$f_\lambda^*(t) := \frac{\prod_{j=1}^n \lambda_j}{\left(\sqrt{2\pi}\right)^n} \sum_{k \in \mathbb{Z}^n} f(t + \lambda k) \quad (t \in \mathbb{R}^n)$$

converges absolutely a.e., and thus defines a function $f_\lambda^* \in L_\lambda^1$ with

$$\left(\prod_{j=1}^n \lambda_j \right)^{-1} \int_{(0,\lambda)} g(u)f_\lambda^*(u)du = \frac{1}{\left(\sqrt{2\pi}\right)^n} \int_{\mathbb{R}^n} g(u)f(u)du, \qquad (6.2.4)$$

for each $g \in L_\lambda^\infty$. Choosing in particular $g(u) = \exp\{-i2\pi k \cdot (u/\lambda)\}$ in (6.2.4) yields

$$(f_\lambda^*)^\wedge(k) = f^\wedge\left(\frac{2\pi k}{\lambda} \right) \quad (k \in \mathbb{Z}^n). \qquad (6.2.5)$$

Here the hat on the left-hand side denotes the Fourier coefficient of the λ-periodic function f_λ^* defined by (6.2.3), and the hat on the right-hand side is the Fourier transform of

$f \in L^1(\mathbb{R})$ defined by (6.2.2). Thus f_λ^* has the Fourier series expansion

$$f_\lambda^*(t) = \frac{\prod_{j=1}^n \lambda_j}{(\sqrt{2\pi})^n} \sum_{k \in \mathbb{Z}^n} f(t + \lambda k) \sim \sum_{k \in \mathbb{Z}^n} f^\wedge\left(\frac{2k\pi}{\lambda}\right) e^{i2\pi k \cdot (t/\lambda)} \quad (t \in \mathbb{R}^n). \quad (6.2.6)$$

If the series on both sides of (6.2.6) are uniformly convergent on compact subsets of $\mathbb{R}^n$, then "$\sim$" can be replaced by "$=$", i.e., f_λ^* is represented by its Fourier series.

For this material see e.g. Butzer and Nessel (1971, Sections 3.1.2, 5.1.5), Stein and Weiss (1971, Section VII.2), and Higgins (1996, Section 2.3). Let us mention that Poisson's formula can already be found in a note written by Gauss on the cover of a book between 1799 and 1813 (cf. Gauss 1900, p. 88; Higgins 1996, p. 19).

The convolution $f_1 * f_2$ of $f_1, f_2 : \mathbb{R}^n \to \mathbb{C}$ is defined by $(f_1 * f_2)(t) := (2\pi)^{-n/2} \int_{\mathbb{R}^n} f_1(u) f_2(t - u) du$ whenever the integral exists. If $f_1 \in L^1(\mathbb{R}^n)$, $f_2 \in L^p(\mathbb{R}^n)$, $1 \le p \le \infty$, or $f_2 \in C(\mathbb{R}^n)$ then $f_1 * f_2$ belongs to $L^p(\mathbb{R}^n)$ or $C(\mathbb{R}^n)$, respectively, and one has $\|f_1 * f_2\|_{L^p} \le \|f_1\|_{L^1} \|f_2\|_{L^p}$ or $\|f_1 * f_2\|_C \le \|f_1\|_{L^1} \|f_2\|_C$, respectively. For $1 \le p \le 2$ one additionally has the convolution theorem $(f_1 * f_2)^\wedge(v) = f_1^\wedge(v) f_2^\wedge(v)$ a.e. (see Butzer-Nessel 1971, Section 0.2, pp. 189, 212; Stein and Weiss, 1971, pp. 3, 18).

We also need the following result on the convolution of two $L^2(\mathbb{R}^n)$-functions. For a proof see Hettich and Stens (to appear).

Lemma 6.1 *If $f, \phi \in L^2(\mathbb{R})$ with $\phi^\wedge \in L^\infty(\mathbb{R})$, then $f * \phi \in L^2(\mathbb{R})$ and $\|f * \phi\|_2 \le \|f\|_2 \|\phi^\wedge\|_\infty$. Furthermore, $f * \phi$ is continuous and there holds the convolution theorem $(f * \phi)^\wedge(v) = f^\wedge(v) \phi^\wedge(v)$ a.e.*

For $\sigma \in \mathbb{R}_+^n$ and $1 \le p \le \infty$ let B_σ^p be the class of entire functions on $\mathbb{C}^n$ of exponential type σ, i.e.,

$$|f(z)| \le \|f\|_{C(\mathbb{R}^n)} \exp\left\{ \sum_{j=1}^n \sigma_j |y_j| \right\} \quad (z \in \mathbb{C}^n),$$

where $z = (z_1, \ldots, z_n)$, $z_j = x_j + iy_j$, which belong to $L^p(\mathbb{R}^n)$ when restricted to $\mathbb{R}^n$. One has

$$B_\sigma^1 \subset B_\sigma^p \subset B_\sigma^{p'} \subset B_\sigma^\infty \quad (1 \le p \le p' \le \infty),$$

and for any $h \in \mathbb{R}_+^n$ there holds

$$\|f\|_{L^p} \le \sup_{u \in \mathbb{R}^n} \left\{ \frac{\prod_{j=1}^n h_j}{(\sqrt{2\pi})^n} \sum_{k \in \mathbb{Z}^n} |f(u + hk)|^p \right\}^{1/p}$$

$$\le \prod_{j=1}^n (1 + \sigma_j h_j) \|f\|_{L^p} \quad (f \in B_\sigma^p). \quad (6.2.7)$$

The classes B_σ^p, $1 \le p \le 2$, can be characterized in terms of Fourier transforms by the Paley–Wiener theorem. It states that a function $f \in L^p(\mathbb{R}^n)$, $1 \le p \le 2$, has an extension to the whole of $\mathbb{C}^n$ as an element of B_σ^p iff $f^\wedge$ vanishes almost everywhere outside the rectangle $[-\sigma, \sigma]$, i.e., iff

$$f(t) = \frac{1}{(\sqrt{2\pi})^n} \int_{(-\sigma,\sigma)} f^{\wedge}(v) e^{iv\cdot t} dv \quad (t \in \mathbb{R}^n).$$

As usual we do not distinguish such a function f from its unique extension to $\mathbb{C}^n$. For these facts on entire functions see Boas (1954, Chapter 6) and Nikol'skiĭ (1975, Chapter 3).

The following lemma is an easy consequence of Poisson's summation formula (6.2.6) applied to functions in B_σ^p (see Stens 1980a; Butzer and Stens 1983; Butzer et al. 1988 for the one-dimensional case).

Lemma 6.2 *Let* $f_1 \in B_{\pi W}^p$, $f_2 \in B_{\pi W}^q$ *for some* $W \in \mathbb{R}_+^n$, $1 \leq p \leq \infty$, $1/p + 1/q = 1$, *then*

$$(f_1 * f_2)(t) = \frac{1}{(\sqrt{2\pi})^n \prod_{j=1}^n W_j} \sum_{k \in \mathbb{Z}^n} f_1\left(\frac{k}{W}\right) f_2\left(t - \frac{k}{W}\right) \quad (t \in \mathbb{R}^n), \quad (6.2.8)$$

the series converging absolutely and uniformly on $\mathbb{R}^n$.

Proof Applying Poisson's summation formula (6.2.6) to some function $g \in B_{2\pi W}^1$ with $\lambda = 1/W$ and $t = 0$ yields

$$\frac{1}{(\sqrt{2\pi})^n \prod_{j=1}^n W_j} \sum_{k \in \mathbb{Z}} g\left(\frac{k}{W}\right) = \frac{1}{(\sqrt{2\pi})^n} \int_{\mathbb{R}^n} g(u) du. \quad (6.2.9)$$

Here "$\sim$" converts to "$=$" in view of (6.2.7), the remark following (6.2.6), and the fact that $g^{\wedge}(2k\pi W) = 0$ for all $k \neq 0$. Now apply (6.2.9) to $g(u) := f_1(u) f_2(t-u) \in B_{2\pi W}^1$ to deduce (6.2.8). $\blacksquare$

This lemma shows that a convolution integral of two band-limited functions can be rewritten as a convolution sum. Noting that a convolution integral and hence a convolution sum of those functions is commutative one obtains a very simple proof of Shannon's theorem for $f \in B_{\pi W}^p$, $1 \leq p < \infty$. Indeed, since $\prod_{j=1}^n \text{sinc}(W_j t_j) \in B_{\pi W}^q$ for $1 < q \leq \infty$ it follows immediately from Lemma 6.2 that

$$\sum_{k \in \mathbb{Z}^n} f\left(\frac{k}{W}\right) \prod_{j=1}^n \text{sinc}\left(W_j\left(t_j - \frac{k_j}{W_j}\right)\right) = \sum_{k \in \mathbb{Z}^n} f\left(t - \frac{k}{W}\right) \prod_{j=1}^n \text{sinc}(k_j) = f(t)$$

noting that $\text{sinc}(k_j) = \delta_{k_j 0}$ (= Kronecker's symbol).

For $f \in C(\mathbb{R}^n)$, $u, h \in \mathbb{R}^n$, $r \in \mathbb{N}$, consider the differences $(\Delta_h f)(t) := f(t + h) - f(t)$ with $\Delta_h^r := \Delta_h \circ \Delta_h^{r-1}$ (Δ_h^0 being the identity). This allows one to define a modulus of continuity for $f \in C(\mathbb{R}^n)$ and $\delta \in \mathbb{R}_+^n$ by

$$\omega_r(f; \delta) := \sup\{|(\Delta_h^r f)(t)|; t \in \mathbb{R}^n, -\delta \leq h \leq \delta\} \quad (r \in \mathbb{N}).$$

Another measure of smoothness needed below is the K-functional

$$K_r(f; \delta) := \inf_{g \in C^r(\mathbb{R}^n)} \left\{ \|f - g\|_C + \sum_{|m|=r} \delta^m \|D^m g\|_C \right\} \quad (r \in \mathbb{N}), \quad (6.2.10)$$

which is equivalent to the modulus of continuity in the sense that

$$c_1\,\omega_r(f;\delta) \le K_r(f;\delta) \le c_2\,\omega_r(f;\delta) \tag{6.2.11}$$

for two constants $c_1, c_2 > 0$ independent of $f \in C(\mathbb{R}^n)$ and $\delta \in \mathbb{R}^n_+$ (cf. Butzer et al. 1993).

6.3 Approximation of non-band-limited functions

We may turn now to the question of what happens if the sinc function in the sampling series (6.1.1) is replaced by some other function φ and under which conditions upon φ does the generalized sampling series

$$(S^\varphi_W f)(t) := \frac{1}{\left(\sqrt{2\pi}\right)^n} \sum_{k\in\mathbb{Z}^n} f\!\left(\frac{k}{W}\right)\varphi(Wt - k) \tag{6.3.1}$$

converge to $f(t)$?

6.3.1 *Convergence of generalized sampling series*

To answer this question let us first consider the continuous version of the series in (6.3.1), namely the convolution integral,

$$(I^\varphi_W f)(t) := \frac{\prod_{j=1}^n W_j}{\left(\sqrt{2\pi}\right)^n} \int_{\mathbb{R}^n} f(u)\varphi\big(W(t - u)\big)du \quad (t \in \mathbb{R}^n; W \in \mathbb{R}^n_+). \tag{6.3.2}$$

These integrals are known to generate an approximation process on $C(\mathbb{R}^n)$ provided $\varphi \in L^1(\mathbb{R}^n)$ and

$$\varphi^\wedge(0) \equiv \frac{1}{\left(\sqrt{2\pi}\right)^n} \int_{\mathbb{R}^n} \varphi(u)du = 1. \tag{6.3.3}$$

More precisely, the I^φ_W are bounded linear operators mapping $C(\mathbb{R}^n)$ into itself and satisfy (cf. Butzer and Nessel 1971, p. 121)

$$\|I^\varphi_W\|_{[C,C]} = \|\varphi\|_{L^1} \quad (W > 0), \tag{6.3.4}$$

$$\lim_{W\to\infty} \|I^\varphi_W f - f\|_C = 0 \quad (f \in C(\mathbb{R}^n)).$$

Here and below $W \to \infty$ always means that each component of $W \in \mathbb{R}^n_+$ tends to infinity.

Returning to the sampling series $S^\varphi_W f$ of (6.3.1), the conditions $\varphi \in L^1(\mathbb{R}^n)$ and (6.3.3) are not sufficient for the convergence of $S^\varphi_W f$ towards f, but they have to be modified in the following way.

Definition 6.3 *If $\varphi : \mathbb{R}^n \to \mathbb{C}$ is a bounded function such that*

$$\frac{1}{\left(\sqrt{2\pi}\right)^n} \sum_{k \in \mathbb{Z}^n} |\varphi(t - k)| < \infty \quad (t \in \mathbb{R}^n) \tag{6.3.5}$$

the absolute convergence being uniform on compact subsets on $\mathbb{R}^n$, and

$$\frac{1}{\left(\sqrt{2\pi}\right)^n} \sum_{k \in \mathbb{Z}^n} \varphi(t - k) = 1 \quad (t \in \mathbb{R}^n), \tag{6.3.6}$$

then φ is said to be a kernel (for a generalized sampling series). The absolute (sum-)moment of order $r \in \mathbb{N}_0$ of a kernel φ is defined by

$$m_r(\varphi) := \max_{|j|=r} \sup_{t \in \mathbb{R}^n} \frac{1}{\left(\sqrt{2\pi}\right)^n} \sum_{k \in \mathbb{Z}^n} |(t - k)^j \varphi(t - k)|.$$

In the following theorem we just need the moment $m_0(\varphi)$ of order zero.

Theorem 6.4 *Let $\varphi \in C(\mathbb{R}^n)$ be a kernel.*
(a) If $f : \mathbb{R}^n \to \mathbb{C}$ is bounded on $\mathbb{R}^n$, then

$$\lim_{W \to \infty} (S_W^\varphi f)(t_0) = f(t_0)$$

at each point $t_0 \in \mathbb{R}^n$ where f is continuous.
(b) $\{S_W^\varphi\}_{W > 0}$ defines a family of bounded linear operators mapping $C(\mathbb{R}^n)$ into itself, having operator norm

$$\|S_W^\varphi\|_{[C,C]} = m_0(\varphi) \quad (W > 0), \tag{6.3.7}$$

and satisfying

$$\lim_{W \to \infty} \|S_W^\varphi f - f\|_C = 0 \quad (f \in C(\mathbb{R}^n)). \tag{6.3.8}$$

Proof First note that the uniform convergence of the series (6.3.5) on compact sets implies $m_0(\varphi) < \infty$, since the function $(\sqrt{2\pi})^{-n} \sum_{k \in \mathbb{Z}^n} |\varphi(t - k)|$ is 1-periodic in each component t_j of t. Regarding part (a), if f is continuous at t_0, given $\varepsilon > 0$ there exists $\delta > 0$ such that $|f(t) - f(t_0)| < \varepsilon$ for all $t \in \mathbb{R}^n$ satisfying $-\delta < t - t_0 < \delta$. Hence

$$\left(\sqrt{2\pi}\right)^n |f(t_0) - (S_W^\varphi f)(t_0)| \leq \left\{ \sum_{k \in S_1} + \sum_{k \in S_2} \right\} \left| f(t_0) - f\left(\frac{k}{W}\right) \right| |\varphi(W t_0 - k)|$$

$$\leq \varepsilon \left(\sqrt{2\pi}\right)^n m_0(\varphi) + 2 \sup_{t \in \mathbb{R}^n} |f(t)| \sum_{k \in S_2} |\varphi(W t_0 - k)|,$$

where $S_1 := \{k \in \mathbb{Z}^n; -\delta W < W t_0 - k < \delta W\}$, and $S_2 := \mathbb{Z}^n \setminus S_1 = \{k \in \mathbb{Z}^n; \exists$ at least one $j \in \{1, \ldots, n\}$ with $|W_j t_{0j} - k_j| \geq \delta W_j\}$. We have to show that the latter sum tends to zero for $W \to \infty$. In view of (6.3.5) one has for all $1 \leq j \leq n$,

$$\sum_{k \in \mathbb{Z}^n} |\varphi(t - k)| = \sum_{k_j = -\infty}^{\infty} \left\{ \sum_{k_{[j]} \in \mathbb{Z}^{n-1}} |\varphi(t - k)| \right\} < \infty,$$

where $k_{[j]} := (k_1, \ldots, k_{j-1}, k_{j+1}, \ldots, k_n)$, and since the convergence is uniform on compact sets there exists a $K_j \in \mathbb{N}$ with

$$\sum_{|k_j| \geq K_j} \left\{ \sum_{k_{[j]} \in \mathbb{Z}^{n-1}} |\varphi(t - k)| \right\} < \varepsilon \quad (t \in [0, 1]^n). \tag{6.3.9}$$

Setting $K := \max_{1 \leq j \leq n} \{K_j\}$, it follows that

$$\sum_{k \in S_2} |\varphi(Wt_0 - k)| \leq \sum_{j=1}^{n} \left\{ \sum_{|W_j t_{j0} - k_j| \geq \delta W_j} \sum_{k_{[j]} \in \mathbb{Z}^{n-1}} |\varphi(Wt_0 - k)| \right\}$$

$$= \sum_{j=1}^{n} \left\{ \sum_{|W_j t_{j0} - k_j| \geq \delta W_j} \sum_{k_{[j]} \in \mathbb{Z}^{n-1}} \left| \varphi(Wt_0 - \lfloor Wt_0 \rfloor - (k - \lfloor Wt_0 \rfloor)) \right| \right\}.$$

Denoting $m := k - \lfloor Wt_0 \rfloor \in \mathbb{Z}^n$, we have, since $Wt_0 - \lfloor Wt_0 \rfloor \in [0, 1]^n$,

$$\sum_{k \in S_2} |\varphi(Wt_0 - k)| \leq \sum_{j=1}^{n} \left\{ \sum_{|m_j| \geq \delta W_j - 1} \sum_{m_{[j]} \in \mathbb{Z}^{n-1}} |\varphi(Wt_0 - \lfloor Wt_0 \rfloor - m)| \right\} \leq n\varepsilon$$

for all $W \geq (K + 1)/\delta$, observing (6.3.9). This yields a). Regarding b), first show that $S_W^\varphi f$ is uniformly continuous on $\mathbb{R}^n$. In view of the identity $(S_W^\varphi f)(t) = (S_1^\varphi g)(Wt)$ with $g(t) = f(t/W)$ it is enough to consider the case $W = (1, \ldots, 1)$. Since the series (6.3.5) converges uniformly on $[-1, 2]^n$, given $\varepsilon > 0$ there is a $K_0 \in \mathbb{N}$ such that

$$\sum_{k \in \mathbb{Z}^n} |\varphi(t - k)| - \sum_{-K_0 \leq k \leq K_0} |\varphi(t - k)| < \frac{\varepsilon(\sqrt{2\pi})^n}{3\|f\|_C} \tag{6.3.10}$$

for all $t \in [-1, 2]^n$. As $\varphi \in C(\mathbb{R}^n)$, there is a $0 < \delta < 1$ such that

$$|\varphi(t) - \varphi(t')| \leq \frac{\varepsilon(\sqrt{2\pi})^n}{3(2K_0 + 1)^n \|f\|_C} \tag{6.3.11}$$

for all $t, t' \in \mathbb{R}^n$ with $-\delta \leq t - t' \leq \delta$. Now let $t, t' \in \mathbb{R}^n$ be arbitrary with $-\delta < t - t' < \delta$. Since $t - \lfloor t \rfloor$ and $t' - \lfloor t \rfloor = (t' - t) + t - \lfloor t \rfloor$ both belong to $[-1, 2]^n$, and since $\lfloor t \rfloor - k \in \mathbb{Z}^n$,

$$(\sqrt{2\pi})^n |(S_1^\varphi f)(t) - (S_1^\varphi f)(t')|$$
$$\leq \sum_{k \in \mathbb{Z}^n} |f(k)| |\varphi(t - \lfloor t \rfloor - (k - \lfloor t \rfloor)) - \varphi(t' - \lfloor t \rfloor - (k - \lfloor t \rfloor))|$$

$$\leq \|f\|_C \sum_{m \in \mathbb{Z}^n} |\varphi(t - \lfloor t \rfloor - m) - \varphi(t' - \lfloor t \rfloor - m)|$$

$$\leq \|f\|_C \Bigg\{ \sum_{-K_0 \leq k \leq K_0} |\varphi(t - \lfloor t \rfloor - k) - \varphi(t' - \lfloor t \rfloor - k)|$$

$$+ \Bigg\{ \sum_{k \in \mathbb{Z}^n} - \sum_{-K_0 \leq k \leq K_0} \Bigg\} \Big(|\varphi(t - \lfloor t \rfloor - k)| + |\varphi(t' - \lfloor t \rfloor - k)| \Big) \Bigg\}$$

$$\leq \|f\|_C \Bigg\{ \sum_{-K_0 \leq k \leq K_0} \frac{\varepsilon(\sqrt{2\pi})^n}{3(2K_0 + 1)^n \|f\|_C} + \frac{2\varepsilon(\sqrt{2\pi})^n}{3\|f\|_C} \Bigg\}$$

$$\leq (\sqrt{2\pi})^n \varepsilon$$

noting (6.3.11) and (6.3.10), as well as the fact that the latter sum consists of $(2K_0 + 1)^n$ terms.

Secondly, $\|S_W^\varphi f\|_C \leq m_0(\varphi)\|f\|_C$, so that S_W^φ is a bounded linear operator on $C(\mathbb{R}^n)$ into itself with $\|S_W^\varphi\|_{[C,C]} \leq m_0(\varphi)$. Regarding the opposite inequality, one may take the function $f(t) := \prod_{j=1}^n (1 - 2|k_j - W_j t_j|) a_k(t')$. Here the $k_j \in \mathbb{Z}$ are such that $k_j - 1/2 \leq W_j t_j < k_j + 1/2$, the a_k are defined as $a_k(t) := \overline{\varphi(Wt - k)}/|\varphi(Wt - k)|$ whenever the denominator is non-zero, and $a_k(t) := 0$ otherwise, the bar denoting the complex conjugate, and $t' \in \mathbb{R}^n$ being chosen such that $m_0(\varphi) = (2\pi)^{-n/2} \sum_{k \in \mathbb{Z}^n} |\varphi(Wt' - k)|$. Thus (6.3.7) follows. The proof of (6.3.8) is analogous to that of part (a).

In the theorem above it is necessary that each component of $W \in \mathbb{R}_+^n$ tends to infinity; it is not sufficient that, e.g. $\|W\|_2 \to \infty$, see Butzer et al. (1993). The case $j = 0, c = 1$ of the following lemma will be useful in order to verify condition (6.3.6).

Lemma 6.5 *Let $\varphi \in C(\mathbb{R}^n)$ be such that $m_r(\varphi) < \infty$ for some $r \in \mathbb{N}_0$, and let $j \in \mathbb{N}_0^n$ be fixed with $|j| \leq r$. The following assertions are equivalent for $c \in \mathbb{R}$:*

(i) $\dfrac{1}{(\sqrt{2\pi})^n} \displaystyle\sum_{k \in \mathbb{Z}^n} (t - k)^j \varphi(t - k) = c$ *a.e. in $\mathbb{R}^n$,*

(ii) $D^j \varphi^\wedge(2k\pi) = \begin{cases} (-i)^{|j|}c, & k = 0, \\ 0, & k \in \mathbb{Z}^n \setminus \{0\}. \end{cases}$

Under the additional assumption that

$$\sum_{k \in \mathbb{Z}^n} |(t - k)^s \varphi(t - k)| \tag{6.3.12}$$

converges uniformly on each compact subset of $\mathbb{R}^n$ for each $s \in \mathbb{N}_0^n$ with $|s| = r$ the restriction a.e. in (i) can be dropped.

Proof First it has to be shown that $m_r(\varphi) < \infty$ implies $m_{r'}(\varphi) < \infty$ for each $0 \leq r' \leq r$, and $t^j \varphi(t) \in L^1(\mathbb{R}^n)$ for all $j \in \mathbb{N}_0^n$ with $|j| \leq r$. For the rather technical proofs of these facts we refer to Butzer et al. (1993). Concerning the equivalence of

(i) and (ii), apply the Poisson summation formula (6.2.6) to $u^j \varphi(u) \in L^1(\mathbb{R}^n)$ and note that $[u^j \varphi(u)]^\wedge(v) = (-i)^{-|j|} D^j \varphi^\wedge(v)$ (cf. Stein and Weiss 1971, p. 5]). This yields

$$\sum_{k \in \mathbb{Z}^n} (t-k)^j \varphi(t-k) \sim \left(\sqrt{2\pi}\right)^n \sum_{k \in \mathbb{Z}^n} [u^j \varphi(u)]^\wedge(2k\pi) e^{i2\pi k \cdot t}$$

$$= \left(\sqrt{2\pi}\right)^n \sum_{k \in \mathbb{Z}^n} i^{|j|} D^j \varphi^\wedge(2k\pi) e^{i2\pi k \cdot t} \qquad (6.3.13)$$

Now if (i) holds, then the right of (6.3.13) reduces to the term for $k = 0$, and one has $D^j \varphi^\wedge(0) = (-i)^{|j|} c$, i.e., (ii) follows. Conversely, if (ii) holds, then the right side is trivially uniformly convergent, and so represents the left side a.e., i.e., (i) results.

If, in addition, the series (6.3.12) is uniformly convergent on compact subsets of $\mathbb{R}^n$ for each $s \in \mathbb{N}_0^n$, $|s| = r$, then for all $|j| \leq r$ it can readily be shown that the series $\sum_{k \in \mathbb{Z}^n} (t-k)^j \varphi(t-k)$ is also uniformly convergent on such subsets. So both sides of (i) are continuous on $\mathbb{R}^n$ and the restriction a.e. can be dropped.

Examples of kernels are the Fejér kernel F

$$F(t) := \frac{1}{\left(\sqrt{2\pi}\right)^n} \prod_{j=1}^{n} \left(\frac{\sin(t_j/2)}{t_j/2}\right)^2 \qquad (t \in \mathbb{R}^n), \qquad (6.3.14)$$

and the kernel of de la Vallée Poussin ϑ

$$\vartheta(t) := \left(\frac{4}{\sqrt{2\pi}}\right)^n \prod_{j=1}^{n} \left(\frac{\sin(t_j/2)\sin(3t_j/2)}{t_j^2}\right) \qquad (t \in \mathbb{R}^n). \qquad (6.3.15)$$

They are products of the well-known one-dimensional versions (cf. Butzer and Nessel 1971, p. 122; Splettstößer et al. 1981), band-limited to $[-1, 1]^n$ and $[-2, 2]^n$, respectively, and they satisfy (6.3.5) and (6.3.6) in view of (6.2.7) and Lemma 6.5, respectively. As an example of a radial kernel one may take the kernel of Bochner and Riesz

$$b^\gamma(t) := 2^\gamma \Gamma(\gamma+1) \|t\|_2^{-((n/2)+\gamma)} J_{(n/2)+\gamma}(\|t\|_2) \qquad (t \in \mathbb{R}^n) \qquad (6.3.16)$$

for $\gamma > (n-1)/2$, where J_λ is the Bessel function of order λ. The assumptions of Theorem 6.4 now hold since $J_\lambda(x) = \mathcal{O}(x^{-1/2})$, $x \to \infty$, for $\lambda > -1/2$, and

$$(b^\gamma)^\wedge(v) = \begin{cases} \left(1 - \|v\|_2^2\right)^\gamma, & \|v\|_2 \leq 1, \\ 0, & \|v\|_2 \geq 1 \end{cases}$$

(cf. Nessel 1967; Nessel and Pawelke 1968; Stein and Weiss 1971, p. 255). Further examples of band-limited kernels can be found in Splettstößer (1978), Gervais et al. (1984), and Butzer and Stens (1985).

Our last examples here are non-band-limited kernels. Let us consider the one-dimensional B-spline of order $r \geq 2$, defined for $x \in \mathbb{R}$ by

$$M_r(x) := \begin{cases} \sqrt{2\pi} \displaystyle\sum_{v=0}^{\lfloor r/2 - |x| \rfloor} \dfrac{(-1)^v r (r/2 - |x| - v)^{r-1}}{v!(r-v)!}, & |x| \leq \dfrac{r}{2}, \\ 0, & |x| > \dfrac{r}{2}. \end{cases} \qquad (6.3.17)$$

Its Fourier transform has the representation

$$M_r^\wedge(v) = \left(\frac{\sin v/2}{v/2}\right)^r = \left(\text{sinc}\frac{v}{2\pi}\right)^r \quad (v \in \mathbb{R})$$

showing by Lemma 6.5 that M_r is a kernel for $n = 1$ in the sense of Definition 6.3. Taking the product

$$M_{r,n}(t) := \prod_{j=1}^{n} M_r(t_j) \quad (t \in \mathbb{R}^n)$$

yields an n-dimensional kernel (see Butzer et al. 1988, 1993).

In order to study kernels which are not built up from one-dimensional ones, let A be an $n \times m$-matrix with column vectors $A_j \in \mathbb{Z}^n \setminus \{0\}, j = 1, 2, \ldots, m$, and rank$(A) = n$. The box spline M_A is then defined via

$$\int_{\mathbb{R}^n} M_A(t)g(t)dt = \int_{Q^m} g(Ax)dx \quad (g \in C(\mathbb{R}^n)), \tag{6.3.18}$$

$Q^m := [-1/2, 1/2]^m$ being the m-dimensional unit cube. Since M_A is defined only a.e. by (6.3.18), it is assumed that M_A is continuous whenever possible. It follows that

$$M_A(t) \geq 0 \quad (t \in \mathbb{R}^n), \qquad \text{supp}(M_A) = AQ^m,$$

in particular, M_A has compact support. If $\rho = \rho(A)$ is the largest integer for which all submatrices generated from A by deleting ρ columns have rank n, then $M_A \in C^{\rho-1}(\mathbb{R}^n)$. Further, the M_A are piecewise polynomials, i.e., polynomial splines of total degree $m - n$. Basic here is that the Fourier transform of M_A is given by

$$M_A^\wedge(v) = \frac{1}{(\sqrt{2\pi})^n} \prod_{j=1}^{m} \text{sinc}\left(\frac{v \cdot A_j}{2\pi}\right) \quad (v \in \mathbb{R}^n).$$

For further properties of the M_A see Micchelli (1979), de Boor and Höllig (1982, 1983), de Boor and DeVore (1983), Höllig (1986), Jetter (1987), Chui (1988, Chapter 2), and de Boor (1990, §13).

Thus if $\rho(A) \geq 1$ and $\varphi(t) := (\sqrt{2\pi})^n M_A(t)$, then $\varphi \in C(\mathbb{R}^n)$ has compact support, and $\varphi^\wedge(0) = 1, \varphi^\wedge(2\pi k) = 0$ for $k \in \mathbb{Z}^n \setminus \{0\}$; the latter holds since the entries of A are integers and rank$(A) = n$. By Lemma 6.5 it follows that φ is a kernel, i.e., there holds (6.3.8); see also Butzer et al. (1990), Fischer (1990), Fischer and Stens (1990).

6.3.2 Order of convergence for band-limited kernels

In this section we investigate the rate of convergence in (6.3.8). It turns out that it is convenient to distinguish between band-limited and non-band-limited kernels φ. In the first case one can transfer the desired results easily from corresponding results on convolution integrals (6.3.2), whereas in the latter case one uses a Taylor expansion approach which is well known in approximation theory.

Theorem 6.6 *Let $\varphi \in B_\pi^1$ with $\varphi^\wedge(0) = 1$. There exist constants $c_1, c_2 > 0$ depending on φ only, such that*

$$c_1 \| I_W^\varphi f - f \|_C \leq \| S_W^\varphi f - f \|_C \leq c_2 \| I_W^\varphi f - f \|_C \quad (f \in C(\mathbb{R}^n); W \in \mathbb{R}_+^n).$$

Proof First we note that φ is a kernel in view of (6.2.7) and Lemma 6.5. Moreover, $I_W^\varphi f$ as well as $S_W^\varphi f$ both belong to $B_{\pi W}^\infty$ (see Nikol'skiĭ 1975, pp. 127, 136). Hence one obtains by Lemma 6.2 that

$$S_W^\varphi I_W^\varphi f = I_W^\varphi I_W^\varphi f, \quad I_W^\varphi S_W^\varphi f = S_W^\varphi S_W^\varphi f \quad (f \in C(\mathbb{R}^n); W \in \mathbb{R}_+^n)$$

and further

$$\| S_W^\varphi f - f \|_C \leq \| S_W^\varphi f - S_W^\varphi I_W^\varphi f \|_C + \| I_W^\varphi I_W^\varphi f - I_W^\varphi f \|_C + \| I_W^\varphi f - f \|_C$$
$$\leq \{ \| S_W^\varphi \|_{[C,C]} + \| I_W^\varphi \|_{[C,C]} + 1 \} \| I_W^\varphi f - f \|_C.$$

Since the operators S_W^φ and I_W^φ are uniformly bounded (cf. Theorem 6.4 and (6.3.4)) the right-hand inequality is established, and the other one follows along the same lines.

This theorem enables one to transfer all results regarding the approximation by singular convolution integrals to that by generalized sampling series. If the kernel φ is band-limited to some $\sigma \in \mathbb{R}_+^n$ with $\pi < \sigma < 2\pi$ rather than to $0 < \sigma \leq \pi$, then one can obtain somewhat weaker results, see Butzer and Stens (1983).

Let us consider the examples of the foregoing section, namely the sampling series based upon Fejér's kernel F (6.3.14), de la Vallée Poussin's kernel ϑ (6.3.15) and the kernel of Bochner and Riesz b^γ (6.3.16) with respect to the order of convergence.

The approximation behaviour of the singular convolution integrals I_W^F and I_W^ϑ can be obtained from the well-known qne-dimensional case (Butzer and Nessel 1971, p. 149; Splettstößer et al. 1981; Splettstößer 1982; Butzer et al. 1990), and then one transfers these results via Theorem 6.6 to the S_W^F and S_W^ϑ, respectively.

Corollary 6.7 *Let $f \in C(\mathbb{R}^n)$ and $W = (W_1, \ldots, W_1)$, $\delta = (\delta_1, \ldots, \delta_1)$, i.e., all components of these vectors are equal.*
(a) For $0 < \alpha < 1$ the following assertions are equivalent:

(i) $\omega_2(f; \delta) = \mathcal{O}(\delta_1^\alpha)$ $(\delta_1 \to 0+)$,

(ii) $\| S_W^F f - f \|_C = \mathcal{O}(W_1^{-\alpha})$ $(W_1 \to \infty)$.

(b) For $r \in \mathbb{N}_0$ and $0 < \alpha \leq 1$ there are equivalent:

(i) $f \in C^r(\mathbb{R}^n)$ and $\omega_2(D^j f; \delta) = \mathcal{O}(\delta_1^\alpha)$, $\delta_1 \to 0+$, for all $j \in \mathbb{N}_0^n$ with $|j| = r$,

(ii) $\| S_W^\vartheta f - f \|_C = \mathcal{O}(W_1^{-r-\alpha})$, $W_1 \to \infty$.

As to the kernel of Bochner and Riesz, one can show by standard arguments that for $f \in C(\mathbb{R}^n)$, $\gamma > (n+3)/2$, $0 < \alpha \leq 2$ and $W, \delta \in \mathbb{R}_+^n$

$$\omega_2(f; \delta) = \mathcal{O}(\|\delta\|_2^\alpha) \ (\delta \to 0+) \iff \| I_W^{b^\gamma} f - f \|_C = \mathcal{O}(\|W^{-1}\|_2^\alpha) \ (W \to \infty),$$

and an application of Theorem 6.6 gives

Corollary 6.8 Let $f \in C(\mathbb{R}^n)$, $\gamma > (n + 3)/2$, $0 < \alpha \leq 2$ and $W, \delta \in \mathbb{R}^n_+$. The following assertions are equivalent:

(i) $\omega_2(f; \delta) = \mathcal{O}(\|\delta\|_2^\alpha)$ $(\delta \to 0+)$,

(ii) $\|S_W^{b\gamma} f - f\|_C = \mathcal{O}(\|W^{-1}\|_2^\alpha)$ $(W \to \infty)$.

For a more detailed treatment of these examples see Stens (1980a), Butzer and Stens (1985), Butzer et al. (1990), and Fischer (1990).

6.3.3 Order of convergence for non-band-limited kernels

Our next problem will be to obtain some information concerning the rate of convergence in (6.3.8) if the kernels are not band-limited. In this case there does not hold a comparison theorem like Theorem 6.6 but one has

Theorem 6.9 Let $\varphi \in C(\mathbb{R}^n)$ be a kernel with $m_r(\varphi) < \infty$ for some $r \in \mathbb{N}$. If, additionally, the moments satisfy

$$\frac{1}{\left(\sqrt{2\pi}\right)^n} \sum_{k \in \mathbb{Z}^n} (t - k)^j \varphi(t - k) = 0 \tag{6.3.19}$$

for all $t \in \mathbb{R}^n$, $j \in \mathbb{N}_0^n$ with $0 < |j| < r$, then

$$\|S_W^\varphi g - g\|_C \leq m_r(\varphi) \sum_{|m|=r} \frac{\|D^m g\|_C}{m! W^m} \quad (g \in C^r(\mathbb{R}^n); \ W > 0), \tag{6.3.20}$$

$$\|S_W^\varphi f - f\|_C \leq M \omega_r(f; W^{-1}) \quad (f \in C(\mathbb{R}^n); \ W > 0). \tag{6.3.21}$$

Proof For $g \in C^r(\mathbb{R}^n)$ take Taylor's formula with integral remainder (6.2.1) and apply the operator S_W^φ to this expansion, considered as a function of u. Then one has by (6.3.6),

$$\left(\sqrt{2\pi}\right)^n \{(S_W^\varphi g)(t) - g(t)\}$$
$$= \left(\sqrt{2\pi}\right)^n (S_W^\varphi [g(\cdot) - g(t)])(t)$$
$$= \sum_{0 < |m| < r} \frac{D^m g(t)}{m!} \sum_{k \in \mathbb{Z}^n} \left(\frac{k}{W} - t\right)^m \varphi(Wt - k)$$
$$+ r \sum_{|m|=r} \sum_{k \in \mathbb{Z}^n} \frac{(k/W - t)^m}{m!}$$
$$\times \left\{ \int_0^1 (1 - s)^{r-1} D^m g\left(t(1 - s) + \frac{k}{W} s\right) ds \right\} \varphi(Wt - k).$$

The first term on the right-hand side vanishes in view of (6.3.19). Thus as $\int_0^1 (1-s)^{r-1}\,ds = 1/r$ there holds the estimate

$$|(S_W^\varphi g)(t) - g(t)|$$

$$\leq r \int_0^1 (1-s)^{r-1}\,ds \sum_{|m|=r} \frac{\|D^m g\|_C}{m!\,W^m} \frac{1}{(\sqrt{2\pi})^n} \sum_{k\in\mathbb{Z}^n} |(Wt-k)^m \varphi(Wt-k)|$$

$$\leq m_r(\varphi) \sum_{|m|=r} \frac{\|D^m g\|_C}{m!\,W^m}.$$

This yields (6.3.20). Concerning (6.3.21), given any $f \in C(\mathbb{R}^n)$, let $g \in C^r(\mathbb{R}^n)$ be arbitrary. Then, by (6.3.20),

$$\|S_W^\varphi f - f\|_C \leq (1 + \|S_W^\varphi\|_{[C,C]}) \left\{ \|f - g\|_C + \|S_W^\varphi g - g\|_C \right\}$$

$$\leq M \left\{ \|f - g\|_C + \sum_{|m|=r} W^{-m} \|D^m g\|_C \right\}.$$

Taking the supremum over all $g \in C^r(\mathbb{R}^n)$ gives by the definition of the K-functional (6.2.10) and by (6.2.11),

$$\|S_W^\varphi f - f\|_C \leq M K_r(f; W^{-1}) \leq M c_2 \omega_r(f; W^{-1}).$$

This completes the proof.

Note that the foregoing proof of (6.3.20) goes through provided that only all r-th order derivatives of g are bounded. Neither g nor the lower order derivatives need be so. This implies that the basic conditions (6.3.6), (6.3.19) upon the moments are equivalent to the condition of polynomial reproduction, namely

$$\frac{1}{(\sqrt{2\pi})^n} \sum_{k\in\mathbb{Z}^n} p(k)\varphi(t-k) = p(t) \quad (t \in \mathbb{R}^n) \tag{6.3.22}$$

for all polynomials p of degree $r - 1$. In fact, the case $p(u) \equiv 1$ in (6.3.22) gives (6.3.6), and $p(u) = (t-u)^j$, $0 < |j| < r$ gives (6.3.19). Conversely, if (6.3.6) and (6.3.19) hold, then (6.3.22) follows by (6.3.20).

The reproduction of polynomials (6.3.22) up to a certain degree is one of the basic conditions in Chapter 8 of this volume and the papers mentioned there. Instead of the finite moment condition $m_r(\varphi) < \infty$ they require a certain decay of the kernels in question.

Let us also mention that condition (6.3.19) is in a certain sense necessary for (6.3.20) to hold; see Ries and Stens (1984).

In order to find examples satisfying the hypotheses of Theorem 6.9, we might first proceed as in some of the examples of Section 6.3.1 and use univariate theory to construct product kernels. In this respect we consider certain linear combinations of univariate B-splines (6.3.17), see Butzer et al. (1988), and for further examples Butzer et al. (1986), Engels et al. (1987).

For $r \in \mathbb{N}$, $r \geq 2$, let $a_{\mu r}$, $\mu = 0, 1, \ldots, \lfloor (r - 1)/2 \rfloor$ be unique solutions of the Vandermonde-type linear system

$$(-1)^{\nu} \sum_{\mu=0}^{\lfloor (r-1)/2 \rfloor} a_{\mu r} \mu^{2\nu} = \left(\frac{1}{M_r^{\wedge}} \right)^{(2\nu)}(0) \quad (\nu = 0, 1, \ldots, \lfloor (r - 1)/2 \rfloor),$$

and define

$$\psi_r(x) := a_{0r} M_r(x) + \frac{1}{2} \sum_{\mu=1}^{\lfloor (r-1)/2 \rfloor} a_{\mu r} \{ M_r(x + \mu) + M_r(x - \mu) \} \quad (x \in \mathbb{R}).$$

Then ψ_r is a polynomial spline of order r having support in $[-r + 1/2, r - 1/2]$, and satisfying $\sum_{k=-\infty}^{\infty} \psi_r(x - k) = \sqrt{2\pi}$ as well as $\sum_{k=-\infty}^{\infty} (x - k)^{\nu} \psi_r(x - k) = 0$ for $x \in \mathbb{R}$ and $0 < \nu \leq r - 1$.

Now set $\psi_r(t) := \prod_{j=1}^{n} \psi_j(t_j)$, $t \in \mathbb{R}^n$. Then it follows that ψ_r is a kernel that satisfies $m_r(\psi_r) < \infty$ and (6.3.19), and Theorem 6.9 can be applied. The first three kernels thus constructed read for $r = 2, 3, 4$

$$\varphi_2(t) := \prod_{j=1}^{n} M_2(t_j),$$

$$\varphi_3(t) := \prod_{j=1}^{n} \left\{ \frac{5}{4} M_3(t_j) - \frac{1}{8} \{ M_3(t_j + 1) + M_3(t_j - 1) \} \right\},$$

$$\varphi_4(t) := \prod_{j=1}^{n} \left\{ \frac{4}{3} M_4(t_j) - \frac{1}{6} \{ M_4(t_j + 1) + M_4(t_j - 1) \} \right\}.$$

As a typical example we state for φ_4 in the two-dimensional instance $n = 2$:

Corollary 6.10 For $\varphi = \varphi_4$ there exist constants $c_1, c_2 > 0$, such that for all $f \in C(\mathbb{R}^2)$, $g \in C^4(\mathbb{R}^2)$, $W = (W_1, W_2) \in \mathbb{R}_+^2$,

(i) $\| S_W^{\varphi} f - f \|_C \leq c_1 \omega_4(f; W^{-1})$,

(ii) $| S_W^{\varphi} g - g \|_C \leq c_2 \{ W_1^{-4} \| D^{(4,0)} g \|_C + W_1^{-3} W_2^{-1} \| D^{(3,1)} g \|_C + W_1^{-2} W_2^{-2} \| D^{(2,2)} g \|_C$

$$+ W_1^{-1} W_2^{-3} \| D^{(1,3)} g \|_C + W_2^{-4} \| D^{(0,4)} g \|_C \}.$$

In particular, one has

(i) $\| S_W^{\varphi} g - g \|_C \leq c_2 (\min\{W_1, W_2\})^{-4} \max_{|m|=4} \{ \| D^m g \|_C \}.$

For a non-product kernel satisfying the assumptions of Theorem 6.9 one starts with the box splines (6.3.18) and constructs linear combinations in a similar fashion as above.

Let $r, m \in \mathbb{N}$, $r \geq 2$, $m \geq n + r - 1$, and A be an $n \times m$-matrix with column vectors $A_j \in \mathbb{Z}^n \setminus \{0\}$, $j = 1, 2, \ldots, m$, and $\rho(A) \geq r - 1$, and let $b_{\mu r}, \mu \in \mathbb{N}_0^n$ with

$|\mu| \leq r_0 := 2\lfloor (r-1)/2 \rfloor$ be the unique solutions of the linear system

$$(-1)^{\lfloor |\nu|/2 \rfloor} \sum_{0 \leq |\mu| \leq r_0} b_{\mu r} \mu^\nu = D^\nu \left(\frac{1}{M_A^\wedge} \right)(0) \quad (\nu \in \mathbb{N}_0^n; |\nu| \leq r_0).$$

Then

$$\varphi_{A,r}(t) := b_{0r} M_A(t) + \frac{1}{2} \sum_{0 < |\mu| \leq r_0} b_{\mu r} \{ M_A(t+\mu) + M_A(t-\mu) \} \quad (t \in \mathbb{R}^n)$$

is a polynomial spline of degree $m - n$ with compact support, $(r-2)$ times continuously differentiable, and satisfying the assumptions of Theorem 6.9. In particular the estimates (6.3.20) and (6.3.21) hold for $\varphi = \varphi_{A,r}$. For details see Fischer (1990), Butzer et al. (1990), and for the existence and uniqueness of the $b_{\mu r}$ see Chui and Lai (1987).

There also exist inverse approximation theorems for non-band-limited kernels. In this respect see Fischer (1990), Fischer and Stens (1990).

6.3.4 *Behaviour of sampling series at jump discontinuities*

Here we discuss how the various sampling series behave for $W \to \infty$ at points t_0 where the underlying function has a jump discontinuity. To this end we have to restrict the matter to the one-dimensional setting. For the Dirichlet integral $(I_W f)(t) := W \int_{-\infty}^{\infty} f(u) \mathrm{sinc}(W(t-u)) du$, $t \in \mathbb{R}$, it is known that

$$\lim_{W \to \infty} (I_W f)(t_0) = \frac{1}{2} \{ f(t_0 + 0) + f(t_0 - 0) \},$$

provided f is of bounded variation in a neighbourhood of $t_0 \in \mathbb{R}$. In contrast to this result de la Vallée Poussin (1908) showed that the Shannon series $(S_W f)(t_0)$ diverges for $W \to \infty$ if f has a jump at t_0. Theis (1919) extended this result to the generalized sampling series with kernel $(\mathrm{sinc}\, t)^2$, and it was shown by Butzer et al. (1987) that a sampling series with a band-limited kernel always diverges for $W \to \infty$ at jump discontinuities of the function f in question. We are going to show now that there are indeed kernels φ such that the associated series $S_W^\varphi f$ converges for $W \to \infty$ at discontinuity points.

For a one-dimensional kernel φ the functions

$$\psi_\varphi^-(t) := \frac{1}{\sqrt{2\pi}} \sum_{k > t} \varphi(t-k), \quad \psi_\varphi^+(t) := \frac{1}{\sqrt{2\pi}} \sum_{k < t} \varphi(t-k), \quad (t \in \mathbb{R})$$

will be needed. These two series converge for all $t \in \mathbb{R}$ and represent functions with period one. If the kernel φ is assumed to be continuous, then it follows from the representation

$$\psi_\varphi^-(t) = \frac{1}{\sqrt{2\pi}} \sum_{k=j+1}^{\infty} \varphi(t-k) \quad (t \in [j, j+1))$$

and the uniform convergence of the series on $[j, j+1)$, $j \in \mathbb{Z}$, that ψ_φ^- is continuous from the right at the integers. Similarly, ψ_φ^+ is continuous from the left there. Further properties of these functions are contained in the following lemma.

Lemma 6.11 *Let $\varphi \in C(\mathbb{R})$ be a kernel, and let $\alpha \in \mathbb{C}$.*

(a) *There holds*

$$\psi_\varphi^-(t) + \psi_\varphi^+(t) = 1 \quad (t \notin \mathbb{Z}). \tag{6.3.23}$$

(b) *If $\psi_\varphi^-(t) = \alpha$ for $t \in (0, 1)$, then $\psi_\varphi^-(t) = \alpha$ for all $t \in \mathbb{R}$, and $\varphi(0) = 0$.*

Proof Part (a) is obvious by (6.3.6). Concerning (b), since $\psi_\varphi^-(t)$ is 1-periodic and continuous from the right, one obtains that $\psi_\varphi^-(t) = \alpha$ for all $t \in \mathbb{R}$. Similarly, one has by (6.3.23) and the continuity from the left that $\psi_\varphi^+(t) = 1 - \alpha$ for all $t \in \mathbb{R}$. The fact that $\varphi(0) = 0$ now follows immediately from (6.3.6) and the identity

$$1 = \frac{1}{\sqrt{2\pi}} \sum_{k=-\infty}^{\infty} \varphi(-k) = \psi_\varphi^-(0) + \frac{\varphi(0)}{\sqrt{2\pi}} + \psi_\varphi^+(0) = \alpha + \frac{\varphi(0)}{\sqrt{2\pi}} + 1 - \alpha.$$

As a counterpart of Theorem 6.4 one now has

Theorem 6.12 *Let $f : \mathbb{R} \to \mathbb{C}$ be a bounded function having a jump at $t_0 \neq 0$, and let $\alpha \in \mathbb{C}$. If $\varphi \in C(\mathbb{R})$ is a kernel, then the following assertions are equivalent:*

(i) $\displaystyle \lim_{W \to \infty} (S_W^\varphi f)(t_0) = \alpha f(t_0 + 0) + (1 - \alpha) f(t_0 - 0),$

(ii) $\psi_\varphi^-(t) = \alpha \ (t \in (0, 1)),$

(iii) $\psi_\varphi^+(t) = 1 - \alpha \ (t \in (0, 1)),$

(iv) $\displaystyle \frac{1}{\sqrt{2\pi}} \int_{-\infty}^{0} \varphi(u) e^{-i2k\pi u} \, du = \alpha \delta_{0,k},$

(v) $\displaystyle \frac{1}{\sqrt{2\pi}} \int_{0}^{\infty} \varphi(u) e^{-i2k\pi u} \, du = (1 - \alpha) \delta_{0,k},$

$\delta_{j,k}$ *being Kronecker's symbol.*

Proof Let us first set

$$g(u) = \begin{cases} f(u) - f(t_0 - 0), & u < t_0, \\ 0, & u = t_0 \\ f(u) - f(t_0 + 0), & u > t_0. \end{cases}$$

Then g is bounded and continuous at t_0 with $g(t_0) = 0$, and hence by Theorem 6.4(a)

$$\lim_{W \to \infty} (S_W^\varphi g)(t_0) = 0. \tag{6.3.24}$$

Moreover, one has the representation

$$(S_W^\varphi f)(t_0) = (S_W^\varphi g)(t_0) + f(t_0 - 0)$$
$$+ \{f(t_0 + 0) - f(t_0 - 0)\} \psi_\varphi^-(Wt_0) \quad (W > 0). \tag{6.3.25}$$

This identity is obvious for $Wt_0 \notin \mathbb{Z}$, and for $Wt_0 \in \mathbb{Z}$ it holds since $\varphi(0) = 0$ by Lemma 6.11 (b).

Now, if (ii) is satisfied, then one has by Lemma 6.11(b) that $\psi_\varphi^-(t) = \alpha$ for all $t \in \mathbb{R}$, and (i) follows from (6.3.24) and (6.3.25) by letting $W \to \infty$. Conversely, if (i) holds, then one has again by (6.3.25) that

$$\alpha f(t_0 + 0) + (1 - \alpha)f(t_0 - 0)$$
$$= f(t_0 - 0) + \{f(t_0 + 0) - f(t_0 - 0)\} \lim_{W \to \infty} \psi_\varphi^-(Wt_0).$$

Since $f(t_0 + 0) \neq f(t_0 - 0)$, this implies $\lim_{W \to \infty} \psi_\varphi^-(Wt_0) = \alpha$. Noting once more that ψ_φ^- has period one, assertion (ii) is obvious.

So far we have proved (i) $\Leftrightarrow$ (ii), and the equivalence of (i) and (iii) holds in view of (6.3.23). Setting now $\varphi_0(t) := \varphi(t)$ for $t < 0$ and $:= 0$ for $t \geq 0$, then $\psi_\varphi^-(t) = (2\pi)^{-1/2} \sum_{k=-\infty}^{\infty} \varphi_0(t - k)$ is a continuous function on $[0, 1)$ with period one. By Poisson's summation formula (6.2.6) its Fourier expansion is given by

$$\psi_\varphi^-(t) \sim \sum_{k=-\infty}^{\infty} \varphi_0^\wedge(2k\pi)e^{i2k\pi t} = \sum_{k=-\infty}^{\infty} \left\{ \frac{1}{\sqrt{2\pi}} \int_{-\infty}^{0} \varphi(u)e^{-i2k\pi u} \, du \right\} e^{i2k\pi t}.$$

Therefore $\psi_\varphi^-(t) = \alpha, t \in [0, 1)$, if and only if the Fourier series reduces to the term for $k = 0$, and this term is equal to α. This is exactly (ii) $\Leftrightarrow$ (iv). Finally, the equivalence (iv) $\Leftrightarrow$ (v) follows from Lemma 6.5 for $c = 1$ and $j = r = 0$.

As a consequence of Theorem 6.12 and Lemma 6.11(b) one has

Corollary 6.13 *If $\varphi \in C(\mathbb{R})$ is a kernel satisfying (i) of Theorem 6.12, then $\varphi(0) = 0$.*

This result is somewhat surprising. It means in particular that a continuous kernel with property (i) cannot have the familiar bell-shaped graph around the origin.

As an example of a kernel generating a sampling series which converges at jump discontinuities one may take linear combinations of translates of B-splines B_r (cf. (6.3.17)), namely,

$$C_r(t) := \alpha M_r\left(t - \frac{r}{2}\right) + (1 - \alpha)M_r\left(t + \frac{r}{2}\right) \quad (t \in \mathbb{R})$$

with $r \geq 2$. One easily verifies that C_r is a continuous kernel which satisfies e.g. (iv) of Theorem 6.12 (see Butzer et al. 1987).

Corollary 6.13 has another interesting consequence. If the sampling series $S_W^\varphi f$ with continuous kernel φ converges at jump discontinuities, then the series $S_W^\varphi f$ cannot interpolate f at the knots $k/W, k \in \mathbb{Z}$. This is due to the fact that the interpolation property is equivalent to $\varphi(k) = \delta_{k,0}, k \in \mathbb{Z}$. Here it is essential to require φ to be continuous. In fact, the situation becomes entirely different if this assumption is dropped. In this case there are kernels φ such that $(S_W^\varphi f)(t)$ interpolates f at $k/W, k \in \mathbb{Z}$, and converges for $W \to \infty$ in the sense of Theorem 6.12 (i). To this end let

$$C_2^*(t) = \frac{1}{2}\{M_2(t + 2) + M_2(t - 2)\}$$

$$= \begin{cases} (|t| - 1)/2, & 1 \leq |t| < 2, \\ (3 - |t|)/2, & 2 \leq |t| < 3, \\ 0, & \text{elsewhere} \end{cases} \quad (t \in \mathbb{R});$$

then C_2^* is a continuous kernel satisfying (i) of Theorem 6.12 with $\alpha = 1/2$. Next consider the step function

$$S(t) = \begin{cases} 1, & |t| < 1, \\ -1, & 1 < |t| < 2, \\ -1/2, & |t| = 2, \\ 0, & \text{elsewhere} \end{cases} \quad (t \in \mathbb{R}).$$

By a more refined analysis than that given above it follows that

$$D(t) := C_2^*(t) + S(t) \quad (t \in \mathbb{R})$$

is again a kernel having the interpolation property $(S_W^D f)(j/W) = f(j/W)$ for $j \in \mathbb{Z}$ and $W > 0$, and there holds

$$\lim_{\substack{W \to \infty \\ Wt \notin \mathbb{Z}}} \sum_{k=-\infty}^{\infty} f\left(\frac{k}{W}\right) D(Wt - k) = \frac{1}{2}\{f(t+0) + f(t-0)\}$$

whenever the right-hand side is meaningful (cf. Butzer et al. 1987).

6.4 Approximation of band-limited functions

Above we have approximated arbitrary continuous and even discontinuous functions by generalized sampling series. In this section we turn to the question of how to approximate, i.e. to reconstruct, band-limited functions in terms of generalized sampling series. Of course, one can apply the techniques presented above, but it seems to be convenient to use a different approach here. Again we restrict the matter to the one-dimensional setting, but we use a slightly extended concept of band-limitation.

6.4.1 The space L_A^2

As a generalization of the space B_σ^2 let us introduce the space $L_A^2 := \{f \in L^2(\mathbb{R}); f^\wedge(v) = 0 \text{ for } v \notin A\}$, A being a measurable subset of $\mathbb{R}$. The functions belonging to this space are called band-limited to A. Furthermore, we use the notation $L^2(A) := \{f \in L^2(\mathbb{R}); f(u) = 0 \text{ for } u \notin A\}$. Throughout A is assumed to satisfy the "disjoint translates" condition

$$A \cap (A + 2\pi kV) = \emptyset \quad (k \in \mathbb{Z} \setminus \{0\}) \tag{6.4.1}$$

for some $V > 0$. This condition implies that A has finite measure, more precisely, $\text{meas}(A) \leq 2\pi V$, but A might be unbounded. If A is bounded, with say $|x| < M$ for all $x \in A$, then (6.4.1) holds for all $V \geq \pi^{-1}M$ (cf. Dodson and Silva 1985; Beaty et al. 1994). For an algorithm for finding the least value of V see Higgins (1996, p. 141). Since a function $f \in L_A^2$ with A satisfying (6.4.1) has the representation $f(t) = (2\pi)^{-1/2} \int_A f^\wedge(v) e^{-ivt} dv$ a.e., we can assume in the sequel that such a function f is continuous.

When dealing with functions in $L^2(A)$ where A and $V > 0$ satisfy (6.4.1), one has $f(x + 2\pi kV)\overline{f(x + 2\pi jV)} = 0$ for $k \neq j$, the bar denoting the complex conjugate. So there holds for $\lambda = 2\pi V$

$$|f_{2\pi V}^*(x)|^2 = 2\pi V^2 \sum_{j,k=-\infty}^{\infty} f(x + 2\pi kV)\overline{f(x + 2\pi jV)}$$

$$= 2\pi V^2 \sum_{k=-\infty}^{\infty} |f(x + 2\pi kV)|^2 = \sqrt{2\pi} V(|f|^2)_{2\pi V}^*(x),$$

and it follows from (6.2.4) with f replaced by $|f|^2$ and $g = 1$ that

$$\frac{1}{2\pi V^2} \int_0^{2\pi V} |f_{2\pi V}^*(u)|^2 du = \int_{-\infty}^{\infty} |f(u)|^2 du. \tag{6.4.2}$$

In particular, one has that $f_{2\pi V}^* \in L_{2\pi V}^2$.

Lemma 6.14

(a) *For $f \in L^2(A)$ and $V > 0$ satisfying (6.4.1) there holds*

$$\|f\|_{L^2} = \left\{ \frac{1}{\sqrt{2\pi} V} \sum_{k=-\infty}^{\infty} \left| f^\wedge\left(\frac{k}{V}\right) \right|^2 \right\}^{1/2}.$$

(b) *For $f \in L_A^2$ and $V > 0$ satisfying (6.4.1) one has*

$$\|f\|_{L^2} = \left\{ \frac{1}{\sqrt{2\pi} V} \sum_{k=-\infty}^{\infty} \left| f\left(\frac{k}{V}\right) \right|^2 \right\}^{1/2}.$$

Proof Concerning the proof of part (a), Parseval's equation (cf. Butzer and Nessel 1971, p. 175) and (6.2.5) yield

$$\int_0^{2\pi V} |f_{2\pi V}^*(u)|^2 du = 2\pi V \sum_{k=-\infty}^{\infty} |(f_{2\pi V}^*)^\wedge(k)|^2$$

$$= 2\pi V \sum_{k=-\infty}^{\infty} \left| f^\wedge\left(\frac{k}{V}\right) \right|^2 \quad (k \in \mathbb{Z}).$$

The assertion now follows from (6.4.2). Note that since meas$(A) < \infty$, the function f belongs to $L^1(\mathbb{R})$ as well, so that one can apply (6.2.5). As to part b), one replaces f by $f^\wedge$ and uses the fact that $f^{\wedge\wedge}(v) = f(-v)$ and that the Fourier transform is an isometry on $L^2(\mathbb{R})$ (cf. Butzer and Nessel 1971, p. 217).

Lemma 6.14 shows in particular that under the given assumptions the sequences $\{f^\wedge(k/V)\}_{k=-\infty}^{\infty}$ and $\{f(k/V)\}_{k=-\infty}^{\infty}$, respectively, belong to the sequence space $l^2(\mathbb{Z}) := \{\{c_k\}; \sum_{k=-\infty}^{\infty} |c_k|^2 < \infty\}$. This can be generalized as follows.

Corollary 6.15

(a) *Let $f \in L^2(A)$ for some bounded set A and $\lambda > 0$. Then there exists a constant $C > 0$ such that*

$$\left\{ \sum_{k=-\infty}^{\infty} |f^{\wedge}(v + \lambda k)|^2 \right\}^{1/2} \leq C \|f\|_{L^2} \quad (v \in \mathbb{R}).$$

(b) *Let $f \in L_A^2$ for some bounded set A and $\lambda > 0$. Then there exists a constant $C > 0$ such that*

$$\left\{ \sum_{k=-\infty}^{\infty} |f(t + \lambda k)|^2 \right\}^{1/2} \leq C \|f\|_{L^2} \quad (t \in \mathbb{R}).$$

Proof Choose $n \in \mathbb{N}$ such that $A \subset (-n\pi/\lambda, n\pi/\lambda)$. Then (6.4.1) holds for $V = n/\lambda$ and Lemma 6.14(a) applied to $f(x)e^{-ivx}$ yields

$$\sum_{k=-\infty}^{\infty} |f^{\wedge}(v + \lambda k)|^2 \leq \sum_{k=-\infty}^{\infty} \left| f^{\wedge}\left(v + \frac{\lambda k}{n}\right) \right|^2 = \frac{\sqrt{2\pi} n}{\lambda} \|f\|_{L^2}^2 \quad (v \in \mathbb{R}).$$

This proves part (a), and (b) follows similarly.

Lemma 6.14 can be easily extended from $|f|^2 = f\bar{f}$ to the product $f\bar{g}$ of two functions using the representation

$$4f\bar{g} = |f + g|^2 - |f - g|^2 + i|f + ig|^2 - i|f - ig|^2.$$

Lemma 6.16 *For $f, g \in L_A^2$ and $V > 0$ for which (6.4.1) holds one has*

$$\int_{-\infty}^{\infty} f(u)\overline{g(u)}du = \frac{1}{V} \sum_{k=-\infty}^{\infty} f\left(\frac{k}{V}\right)\overline{g\left(\frac{k}{V}\right)},$$

where the sum is absolutely convergent.

The absolute convergence of the series follows from the Cauchy–Schwarz inequality in view of $\{f(k/V)\}_{k\in\mathbb{Z}} \in l^2(\mathbb{Z})$ and $\{\overline{g(k/V)}\}_{k\in\mathbb{Z}} \in l^2(\mathbb{Z})$. Note that one cannot replace $\bar{g}$ by g in general, since $g \in L_A^2 \iff \bar{g} \in L_{(-A)}^2$. However, one can replace $g(\cdot)$ by $g(t - \cdot)$; indeed

Lemma 6.17 *If $f, g \in L_A^2$ and $s > 0$ satisfies (6.4.1), then*

$$(f * g)(t) = \frac{1}{\sqrt{2\pi} V} \sum_{k=-\infty}^{\infty} f\left(\frac{k}{V}\right)g\left(t - \frac{k}{V}\right) \quad (t \in \mathbb{R}),$$

where the series is absolutely and uniformly convergent.

Proof For $t \in \mathbb{R}$ fixed replace $g(\cdot)$ in Lemma 6.16 by $\overline{g(t - \cdot)}$. This is possible, since $[g(t - \cdot)]^{\wedge}(v) = e^{-ivt}g^{\wedge}(v)$ and thus $\overline{g(t - \cdot)} \in L_A^2$. The absolute and uniform convergence follow from

$$\sum_{|k|>N} \left| f\left(\frac{k}{V}\right) \right| \left| g\left(t - \frac{k}{V}\right) \right| \le \left\{ \sum_{|k|>N} \left| f\left(\frac{k}{V}\right) \right|^2 \right\}^{1/2} \left\{ \sum_{k=-\infty}^{\infty} \left| g\left(t - \frac{k}{V}\right) \right|^2 \right\}^{1/2}$$

$$= (2\pi V^2)^{1/4} \left\{ \sum_{|k|>N} \left| f\left(\frac{k}{V}\right) \right|^2 \right\}^{1/2} \|g\|_{L^2},$$

where we have again used the Cauchy–Schwarz inequality and Lemma 6.14(b).

This result extends Lemma 6.2 to functions $f, g \in L_A^2$. As a particular case we obtain a generalization of the Whittaker–Kotel'nikov–Shannon sampling theorem due to Lloyd (1959), and Dodson and Silva (1985).

Corollary 6.18 *Let $f \in L_A^2$ and $V > 0$ be such that* (6.4.1) *holds; then*

$$f(t) = \frac{1}{\sqrt{2\pi} V} \sum_{k=-\infty}^{\infty} f\left(\frac{k}{V}\right) \chi_A^{\vee}\left(t - \frac{k}{V}\right) \quad (t \in \mathbb{R}), \tag{6.4.3}$$

$^{\vee}$ *denoting the inverse Fourier transform. The series* (6.4.3) *is absolutely and uniformly convergent.*

Proof Since A has finite measure one has that $f^{\wedge} \in L^1(\mathbb{R})$ and

$$f(t) = \frac{1}{\sqrt{2\pi}} \int_{-\infty}^{\infty} f^{\wedge}(v)\chi_A(v)e^{ivt}dv = [f^{\wedge}\chi_A]^{\vee}(t) = (f * \chi_A^{\vee})(t) \quad (t \in \mathbb{R}),$$

the last equality being valid in view of the convolution theorem of Lemma 6.1. Now apply Lemma 6.17 to $f \in L_A^2$ and $\chi_A^{\vee} \in L_A^2$ and the assertion follows.

For $A = (-\pi W, \pi W)$ one can choose $V = W$ which yields the classical sampling theorem (6.1.1) since $\chi_{(-\pi W, \pi W)}^{\vee}(t) = \sqrt{2\pi} W \mathrm{sinc}(Wt)$.

6.4.2 *More general convolution sums*

The sampling theorem (6.1.1) or its generalization (6.4.3) enable one to recover a signal from its samples. In practice, however, one does not make use of (6.1.1) or (6.4.3) but a standard technique is first to construct a step function in terms of the samples $f(k/W)$ and then to smooth this step function by restricting its Fourier transform to $(-\pi W, \pi W)$ or A, respectively. The resulting function is then used as an approximation to f. Beaty et al. (1994) studied the error involved in this type of approximation by showing that the smoothed step function can be rewritten in terms of a convolution integral.

Our aim here is to replace the step function by more general functions built up from the samples of f. It will be shown that the connection to convolution integrals remains valid in this frame and that one can deduce error estimates in a similar fashion.

Furthermore, we will give examples of approximation processes providing smaller errors than the approximation by step functions.

Our main tool will be Lemma 6.17, where we have shown that the convolution integral of two functions in L_A^2 equals a convolution sum. Of course this cannot hold for arbitrary functions in $L^2(\mathbb{R})$, because the series need not be convergent in this case even if one assumes that the functions involved are continuous. Nevertheless one can consider convolution sums of more general functions.

Lemma 6.19 *Let $f \in L_A^2$ and $V > 0$ satisfy (6.4.1). If $h \in L^2(\mathbb{R})$ is such that*

$$\sum_{k=-\infty}^{\infty} |h^{\wedge}(\cdot + 2\pi k V)|^2 \in L^{\infty}(\mathbb{R}), \tag{6.4.4}$$

then

$$a_h(t) := \frac{1}{\sqrt{2\pi}V} \sum_{k=-\infty}^{\infty} f\left(\frac{k}{V}\right) h\left(t - \frac{k}{V}\right) \quad (t \in \mathbb{R}) \tag{6.4.5}$$

converges in $L^2(\mathbb{R})$.

Proof Since the Fourier transform is isometric on $L^2(\mathbb{R})$ it suffices to prove that

$$\left\| \sum_{n \leq |k| \leq m} f\left(\frac{k}{V}\right) \left[h\left(\cdot - \frac{k}{V}\right) \right]^{\wedge}(v) \right\|_{L^2} = \left\| h^{\wedge}(v) \sum_{n \leq |k| \leq m} f\left(\frac{k}{V}\right) e^{-ikv/V} \right\|_{L^2} < \varepsilon$$

for n, m large enough. Let $s_{n,m}(v) := \sum_{n \leq |k| \leq m} f(k/V) e^{-ikv/V}$; then since $\{f(k/V)\} \in l^2(\mathbb{Z})$ the Riesz–Fischer theorem implies

$$\int_0^{2\pi V} |s_{n,m}(v)|^2 dv < \varepsilon \quad (n, m > N_0),$$

and it follows from (6.2.4) with $f = |h^{\wedge}|^2$, $g = |s_{n,m}|^2$ and $\lambda = 2\pi V$ that

$$\int_{-\infty}^{\infty} |h^{\wedge}(v)s_{n,m}(v)|^2 dv = \int_0^{2\pi V} |s_{n,m}(v)|^2 \sum_{k=-\infty}^{\infty} |h^{\wedge}(v + 2\pi k V)|^2 dv$$

$$< \varepsilon \operatorname{ess\,sup}_{v \in \mathbb{R}} \sum_{k=-\infty}^{\infty} |h^{\wedge}(v + 2\pi k V)|^2 \quad (n, m > N_0).$$

This proves the assertion.

If $h \in L_A^2$ for some bounded set A and $h^{\wedge} \in L^{\infty}(\mathbb{R})$, then (6.4.4) is obviously satisfied. Similarly, if $h \in L^2(A)$ with A bounded, then (6.4.4) holds by Corollary 6.15. This yields in particular that for bounded sets A the series (6.4.3) in Corollary 6.18 converges in $L^2(\mathbb{R})$, too. This is well-known in the case of the Whittaker–Kotel'nikov–Shannon series (6.1.1).

In view of the $L^2(\mathbb{R})$-convergence of the series (6.4.5) the Fourier transform of a_h is given by

$$a_h^\wedge(v) = \frac{1}{\sqrt{2\pi V}} \sum_{k=-\infty}^{\infty} f\left(\frac{k}{V}\right) h^\wedge(v) e^{-ikv/V} \quad (v \in \mathbb{R}),$$

where the series converges in $L^2(\mathbb{R})$ again. This representation shows that in general $a_h^\wedge$ does not vanish outside the set A as $(f * h)^\wedge$ does. Hence the convolution integral $f * h$ is different from the convolution sum (6.4.5). If one, however, restricts the Fourier transform of a_h to the set A, then the resulting function equals $f * h$ as will be seen below. Now we are at the stage to prove the main result of this section.

Theorem 6.20 *Let* $f \in L_A^2$, $V > 0$ *satisfy* (6.4.1), *and* $h \in L^2(\mathbb{R})$ *such that* (6.4.4) *holds; then*

$$(a_h * \chi_A^\vee)(t) = \frac{1}{\sqrt{2\pi V}} \sum_{k=-\infty}^{\infty} f\left(\frac{k}{V}\right)(h * \chi_A^\vee)\left(t - \frac{k}{V}\right)$$

$$= (f * h)(t) \quad (t \in \mathbb{R}),$$

where the series converges uniformly and in $L^2(\mathbb{R})$.

Proof Since the series (6.4.5) converges in $L^2(\mathbb{R})$ and since the convolution with $\chi_A^\vee$ is a bounded operator from $L^2(\mathbb{R})$ into itself (cf. Lemma 6.1), one has

$$(a_h * \chi_A^\vee)(t) = \frac{1}{\sqrt{2\pi V}} \sum_{k=-\infty}^{\infty} f\left(\frac{k}{V}\right)\left[h\left(\cdot - \frac{k}{V}\right) * \chi_A^\vee\right](t)$$

$$= \frac{1}{\sqrt{2\pi V}} \sum_{k=-\infty}^{\infty} f\left(\frac{k}{V}\right)(h * \chi_A^\vee)\left(t - \frac{k}{V}\right) \quad \text{a.e.}$$

On the other hand, $h * \chi_A^\vee$ belongs to L_A^2 by Lemma 6.1 and one can apply Lemma 6.17 to obtain

$$\left(f * (h * \chi_A^\vee)\right)(t) = \frac{1}{\sqrt{2\pi V}} \sum_{k=-\infty}^{\infty} f\left(\frac{k}{V}\right)(h * \chi_A^\vee)\left(t - \frac{k}{V}\right) \quad (t \in \mathbb{R}),$$

where the series is uniformly convergent. So far we have shown that

$$(a_h * \chi_A^\vee)(t) = \frac{1}{\sqrt{2\pi V}} \sum_{k=-\infty}^{\infty} f\left(\frac{k}{V}\right)(h * \chi_A^\vee)\left(t - \frac{k}{V}\right)$$

$$= \left(f * (h * \chi_A^\vee)\right)(t) \quad (t \in \mathbb{R}).$$

Here the first equality now holds for all $t \in \mathbb{R}$, since $a_h * \chi_A^\vee$ as well as the infinite series are continuous.

It remains to show that $f * (h * \chi_A^\vee) = f * h$. Since the assumption (6.4.4) upon h implies $h^\wedge \in L^\infty(\mathbb{R})$, one has again by Lemma 6.1 that $(f * h)^\wedge = f^\wedge h^\wedge$. Similarly there holds $\left(f * (h * \chi_A^\vee)\right)^\wedge = f^\wedge h^\wedge \chi_A = f^\wedge h^\wedge$, the latter equality being valid since $f^\wedge(v) = 0$ outside A. The assertion now follows from the uniqueness theorem of the Fourier transform, noting that $f * (h * \chi_A^\vee)$ and $f * h$ are continuous functions.

Theorem 6.20 generalizes a result of Beaty et al. (1994) who considered $h = (\sqrt{2\pi}V)\chi_{[-1/2V,1/2V]}$. In this case $f * h$ are the integral or Steklov means

$$\sqrt{2\pi}V(f * \chi_{[-1/2V,1/2V]})(t) = V \int_{-1/2V}^{1/2V} f(t-u)du \quad (t \in \mathbb{R}),$$

and a_h is the step function coinciding with f at the midpoints of the intervals $[V^{-1}(k - 1/2), V^{-1}(k+1/2)]$, $k \in \mathbb{Z}$, i.e.,

$$a_h(t) = \sum_{k=-\infty}^{\infty} f\left(\frac{k}{V}\right)\chi_{[-1/2V,1/2V]}\left(t - \frac{k}{V}\right) \quad (t \in \mathbb{R}).$$

As shown in Lemma 6.1 the convolution $f * \chi_A^{\vee}$ can be rewritten in terms of Fourier transform as

$$\left(f * \chi_A^{\vee}\right)(t) = \left(f^{\wedge}\chi_A\right)^{\vee}(t) \qquad \text{a.e.,}$$

i.e., the convolution with $\chi_A^{\vee}$ is the operator that restricts the Fourier transform of f to the set A. Restricting the Fourier transform of a function f to a set A of finite measure means smoothing f since in this case $f^{\wedge}\chi_A \in L^1(\mathbb{R})$ and hence $(f^{\wedge}\chi_A)^{\vee}$ equals a.e. a continuous function. If A is even assumed to be bounded, then $f * \chi_A^{\vee} = (f^{\wedge}\chi_A)^{\vee}$ is a so-called Paley–Wiener function, i.e., it can be extended to the whole complex plane as an entire function of exponential type.

6.4.3 Error estimates

The representation of $a_h * \chi_A^{\vee}$ as $f * h$ enables one to deduce estimates for the error

$$\|f - a_h * \chi_A^{\vee}\|$$

measured in the $L^2(\mathbb{R})$-norm or in the uniform norm.

Theorem 6.21 Let $f \in L_A^2$, $V > 0$ satisfy (6.4.1), and $h \in L^2(\mathbb{R})$ such that (6.4.4) holds. Then

$$\|f - a_h * \chi_A^{\vee}\|_{L^2} = \left\{\frac{1}{\sqrt{2\pi}}\int_A |1 - h^{\wedge}(v)|^2|f^{\wedge}(v)|^2 dv\right\}^{1/2}, \qquad (6.4.6)$$

$$\|f - a_h * \chi_A^{\vee}\|_C \leq \|f\|_{L^2}\left\{\frac{1}{\sqrt{2\pi}}\int_A |1 - h^{\wedge}(v)|^2 dv\right\}^{1/2}. \qquad (6.4.7)$$

Proof Equality (6.4.6) follows immediately from Theorem 6.20, since

$$\|f - a_h * \chi_A^{\vee}\|_{L^2}^2 = \|f - f * h\|_{L^2}^2 = \|f^{\wedge}(1 - h^{\wedge})\|_{L^2}^2$$

$$= \frac{1}{\sqrt{2\pi}}\int_A |1 - h^{\wedge}(v)|^2|f^{\wedge}(v)|^2 dv,$$

where $h^{\wedge} \in L^{\infty}(\mathbb{R})$ and $(f * h)^{\wedge} = f^{\wedge}h^{\wedge}$ by Lemma 6.1.

Concerning (6.4.7), we make use of the Fourier inversion formula to deduce

$$|f(t) - a_h * \chi_A^\vee(t)| = \left| \frac{1}{\sqrt{2\pi}} \int_{-\infty}^{\infty} [f^\wedge(v) - f^\wedge(v)h^\wedge(v)]e^{ivt} dv \right|$$

$$\leq \frac{1}{\sqrt{2\pi}} \int_{-\infty}^{\infty} |f^\wedge(v)||1 - h^\wedge(v)| dv \quad (t \in \mathbb{R}).$$

Since $f^\wedge$ vanishes outside A this can be estimated further by the Cauchy–Schwarz inequality as

$$|f(t) - (a_h * \chi_A^\vee)(t)| \leq \frac{1}{\sqrt{2\pi}} \int_A |f^\wedge(v)||1 - h^\wedge(v)| dv$$

$$\leq \|f^\wedge\|_{L^2} \left\{ \frac{1}{\sqrt{2\pi}} \int_A |1 - h^\wedge(v)|^2 dv \right\}^{1/2} \quad (t \in \mathbb{R}).$$

This establishes inequality (6.4.7).

When dealing with particular functions h and sets A the above error estimates can be made more concrete. In the following we will consider low-pass signals, i.e., $A = (-\pi W, \pi W)$ for some $W > 0$, and for h we will take linear combinations of convolutions of characteristic functions. Let

$$h_j(t) := (\sqrt{2\pi} V)^j \big(\underbrace{\chi_{[-1/2V, 1/2V]} * \cdots * \chi_{[-1/2V, 1/2V]}}_{j\text{-times}} \big)(t) \quad (t \in \mathbb{R}),$$

$$h(t) = \sum_{j=1}^r \binom{r}{j}(-1)^{j+1} h_j(t) \quad (t \in \mathbb{R}); \tag{6.4.8}$$

then $h \in L^2(\mathbb{R})$ and h has compact support, and thus satisfies the assumptions of Theorem 6.21. Noting that $(\sqrt{2\pi} V)\chi_{[-1/2V, 1/2V]}^\wedge(v) = \text{sinc}(v/2\pi V)$, the Fourier transform of h turns out to be

$$h^\wedge(v) = -\sum_{j=1}^r \binom{r}{j}(-1)^j \left(\text{sinc} \frac{v}{2\pi V} \right)^j = 1 - \left(1 - \text{sinc} \frac{v}{2\pi V} \right)^r \quad (v \in \mathbb{R}).$$

For the error estimates we need an upper bound for $\left(1 - \text{sinc}(v/2\pi V)\right)^r$ on $A = (-\pi W, \pi W)$. From the Taylor expansion of the sine function we obtain

$$\left| 1 - \text{sinc} \frac{v}{2\pi V} \right| \leq \frac{1}{6}\left(\frac{v}{2V} \right)^2 = \frac{v^2}{24V^2}.$$

Inserting this inequality into (6.4.6) and (6.4.7), respectively, yields the following estimates.

Corollary 6.22 *Let* $f \in L_A^2$ *with* $A = (-\pi W, \pi W)$ *for some* $W > 0$, *and let* $0 < W \leq V$. *For h given by* (6.4.8) *one has the error estimates*

$$\|f - a_h * \chi_A^\vee\|_{L^2} \leq V^{-2r} W^{2r} \frac{\pi^{2r}}{24^r} \|f\|_{L^2},$$

$$\|f - a_h * \chi_A^\vee\|_C \leq V^{-2r} W^{2r+1/2} \frac{2^{1/4} \pi^{2r+1/4}}{24^r (4r + 1)^{1/2}} \|f\|_{L^2}.$$

In particular, there holds

$$\lim_{V \to \infty} \|f - a_h * \chi_A^\vee\|_{L^2} = \lim_{V \to \infty} \|f - a_h * \chi_A^\vee\|_C = 0.$$

In the low-pass case considered above the estimates are valid provided the sampling rate V satisfies $V \geq W$, since this condition guarantees (6.4.1). On the other hand, this corollary also shows that the error can be reduced by "oversampling", i.e., by sampling at rates which considerably exceed W. Indeed, oversampling is a common technique in electrical engineering.

7

SAMPLING THEORY AND WAVELETS

7.1 Introduction

Over the past decade wavelets and wavelet techniques have become increasingly important in the more applied areas of mathematics. There seems to be hardly any field
untouched by the recent preoccupation with these and related concepts. Their effectiveness in data compression makes them invaluable for the compression of auditory
signals and images (Mallat 1989b; Wickerhauser 1991; Benedetto and Teolis 1993),
their usefulness being further enhanced by their suitability for edge detection (Froment
and Mallat 1992; Antoine et al. 1993). Wavelet and more general multilevel methods have
been shown to be extremely efficient for the preconditioning of linear systems stemming
from Galerkin approximations for elliptic problems (Dahmen and Kunoth 1992; Jaffard
1992). At the same time wavelet methods prove of great worth for the compression of
full stiffness matrices arising in connection with integral or pseudodifferential operators
(Beylkin et al. 1991; Dahmen et al. 1993a). Both of these properties combined yield
a powerful tool for the numerical treatment of pseudodifferential operators (Dahmen
et al. 1993b; Dahlke et al. 1997). Wavelet methods have been used in approximation theory, both linear and non-linear (DeVore et al. 1992; de Boor et al. 1994; Sweldens and
Piessens to appear; Lei 1994; Fischer 1996) and for the characterization of Hardy spaces,
Hölder spaces, Besov spaces and others (Feichtinger and Gröchenig 1988; Meyer 1990;
Dahmen 1995). Of course, these few citations in no way do justice to the flood of excellent publications that have been turned out on these various subjects, or to the many fields
where wavelet techniques have been playing a role. In this chapter we will be concerned
with the interrelationship between wavelet analysis and sampling theory.

Let $\psi \in L^2(\mathbb{R})$ with $C_\psi := 2\pi \int_{\mathbb{R}} |\xi|^{-1} |\psi^\wedge(\xi)|^2 \, d\xi < \infty$. The *continuous wavelet
transform* (*CWT*) of a function $f \in L^2(\mathbb{R})$ with respect to ψ is given by

$$(W^\psi f)(a, b) := \int_{\mathbb{R}} f(x) |a|^{-1/2} \psi\left(\frac{x - b}{a}\right) dx, \qquad a, b \in \mathbb{R}, \, a \neq 0.$$

The CWT was first proposed by the geophysicist J. Morlet (Morlet et al. 1982; Morlet
1983) for the analysis of seismic data and then studied in more detail by A. Grossmann
and J. Morlet (Grossmann and Morlet 1984; Grossmann et al. 1985, 1986). One may
think of the scaling factor a as characterizing the "frequency" of the CWT, while b
stands for position. Without loss of generality let us suppose the function ψ has its
center of gravity close to the origin. For small values $|a| \ll 1$, the scaled and shifted
function $|a|^{-1/2} \psi\left((\cdot - b)/a\right)$ will then be concentrated around the point b with frequency
content mostly in the high-frequency range, while for $|a| \gg 1$ it will be spread out with
mostly low frequencies. Thus the CWT of a function f contains information about the
frequency behaviour of f locally in space (or time), comparable to the notes on a music

sheet. In higher dimensions the CWT even gathers directional features of a signal f, which is of importance, e.g. for edge detection in image analysis. Moreover, the CWT has the ability of zooming-in upon a point b for higher and higher frequencies $|a| \ll 1$. This is one of the characteristics that account for its importance.

In general, the information stored in the CWT of a signal $f \in L^2(\mathbb{R})$ is highly redundant, and the CWT may be discretized without loss. For fixed $a > 1, b > 0$ we choose a grid partition $\{(a^n, kba^n)\}_{n,k \in \mathbb{Z}}$ corresponding to equidistant sampling on successive scales $a^n, n \in \mathbb{Z}$, while adapting the translation steps to the width of the scaled function $|a|^{-1/2} \psi(\cdot/a)$. To permit numerically stable reconstruction of the signal f we need to require that the set $\{|a|^{-n/2} \psi((\cdot - kba^n)/a^n)\}_{k,n \in \mathbb{Z}}$ be a frame for $L^2(\mathbb{R})$. For suitable choices of ψ, a and b, this set actually is an orthonormal basis for $L^2(\mathbb{R})$. A common choice is $a := 2, b := 1$, corresponding to an analysis in octaves.

A *wavelet* on $\mathbb{R}$ is a function $\psi \in L^2(\mathbb{R})$ such that the set $\{2^{1/2} \psi(2^k \cdot -j)\}_{k,j \in \mathbb{Z}}$ is an orthonormal basis for $L^2(\mathbb{R})$. In higher dimensions one needs more than one wavelet to span the whole of $L^2(\mathbb{R}^d)$. Accordingly, a *wavelet set* is a finite set of functions $\{\psi_\nu\}_{\nu \in \mathcal{I}} \subset L^2(\mathbb{R}^d)$ such that the set $\{2^{kd/2} \psi_\nu(2^k \cdot -j)\}_{k \in \mathbb{Z}, j \in \mathbb{Z}^d, \nu \in \mathcal{I}}$, forms an orthonormal basis for $L^2(\mathbb{R}^d)$. Given a wavelet set $\{\psi_\nu\}_{\nu \in \mathcal{I}}$, any function $f \in L^2(\mathbb{R}^d)$ admits the *wavelet decomposition*

$$f = \sum_{k \in \mathbb{Z}} \sum_{j \in \mathbb{Z}^d} \langle f, 2^{dk/2} \psi_\nu(2^k \circ -j) \rangle 2^{dk/2} \psi_\nu(2^k \cdot -j).$$

Here $\langle \cdot, \cdot \rangle$ denotes the scalar product in $L^2(\mathbb{R}^d)$; concerning notation see also the end of this section.

For the construction of wavelets the concept of a multiresolution analysis (MRA) as introduced by S. Mallat (Mallat 1989a) has proved to be of value. An MRA is a nested sequence of closed subspaces of $L^2(\mathbb{R}^d)$

$$\cdots \subset V_{-1} \subset V_0 \subset V_1 \subset \cdots \subset L^2(\mathbb{R}^d)$$

arising from each other by subsequent scaling by the factor of 2. Their union $\bigcup_{n \in \mathbb{Z}} V_n$ is supposed to be dense in $L^2(\mathbb{R}^d)$. In addition, V_0 is assumed to be *shift-invariant*, i.e., invariant under translation by (multi-)integers, and to possess a Riesz basis consisting of the (multi-)integer shifts $\{\vartheta(\cdot - j)\}_{j \in \mathbb{Z}^d}$ of a function $\vartheta \in V_0$. ϑ is called a *scaling function* for the MRA. Given an MRA, we may define the difference space $W_n := V_{n+1} \ominus V_n$ as the orthogonal complement of V_n in $V_{n+1}, n \in \mathbb{Z}$. A finite set of functions $\{\psi_\nu\}_{\nu \in \mathcal{I}}$, whose integer translates form an orthonormal basis for W_0, turns out to be a wavelet set. One can show that for any MRA there exists such a corresponding wavelet set, and in many cases it can be explicitly constructed, see Section 7.4 for details.

To give an example of an MRA illustrating the intimate connections between sampling theory and wavelet theory, let V_0 be the well-known Paley–Wiener space $PW_\pi(\mathbb{R})$ consisting of all functions $f \in L^2(\mathbb{R})$ band-limited to the interval $[\pi, \pi]$. In view of the scaling property, for general $n \in \mathbb{Z}$ the space V_n then coincides with $PW_{2^n\pi}(\mathbb{R})$ consisting of all functions $f \in L^2(\mathbb{R}^d)$ band-limited to the interval $[-2^n\pi, 2^n\pi]$. Obviously, these spaces tend to $L^2(\mathbb{R}^d)$ for $n \to \infty$. One corresponding scaling function is given by the *Sinus cardinalis* sinc $x := \sin(\pi x)/(\pi x), x \neq 0$, and sinc $0 := 1$, whose integer translates form an orthonormal basis for $PW_\pi(\mathbb{R})$ (see Example 7.2 and Example 7.8 below). On the space $V_0 = PW_\pi(\mathbb{R})$, the Whittaker–Kotel'nikov–Shannon sampling theorem (Whittaker (1915), Kotel'nikov (1933), Shannon (1949)), is known to hold:

$$f(v) = \sum_{j \in \mathbb{Z}} f(j) \text{sinc}(v - j), \qquad v \in \mathbb{R}, f \in PW_\pi(\mathbb{R}). \qquad (7.1.1)$$

Similar interpolation formulae can be derived for more general shift-invariant spaces (see Theorem 7.5 below). In the instance of an MRA, such interpolation formulae can be "lifted" from the shift-invariant space V_0 to the general level V_n, $n \in \mathbb{Z}$, yielding representations of the form

$$f(v) = \sum_{j \in \mathbb{Z}} f(2^{-k}j) g(2^k v - j), \qquad v \in \mathbb{R}, f \in V_k, k \in \mathbb{Z}. \qquad (7.1.2)$$

One might ask whether anything can be said concerning the *aliasing error* arising when formula (7.1.2) is applied to a function f not contained in any of the spaces V_k, $k \in \mathbb{Z}$. As by assumption V_k tends to all of $L^2(\mathbb{R}^d)$ for $k \to \infty$, one might expect (7.1.2) to hold in some sense in the limit for $k \to \infty$. In fact, given a suitable MRA, it is possible to prove Jackson-type inequalities for sequences $\{T_n\}_{n \in \mathbb{Z}}$ of quite general bounded linear operators on $C_B(\mathbb{R}^d)$ possessing the reproducing property $T_n f \equiv f$ for all $f \in V_n \cap C_B(\mathbb{R}^d)$, $n \in \mathbb{Z}$ (see Theorem 7.13). One important class of applications is provided by sampling operators (see Theorem 7.15). A more classical method to prove Jackson-type inequalities for sampling operators is to require reproduction of polynomials up to a certain degree and then to approximate a sufficiently smooth function by a polynomial. Here we require reproduction of elements of $V_n \cap C_B(\mathbb{R}^d)$ instead, which is more natural in a way, since polynomials of positive degree do not belong to $C_B(\mathbb{R}^d)$.

For an MRA to be "suitable" in the above context, one basic assumption is that there exist a corresponding wavelet set $\{\psi_v\}_{v \in \mathcal{I}}$ satisfying the *vanishing moment condition* of some order $r \in \mathbb{N}$, i.e., $D^\alpha \psi_v^\wedge(0) = 0$ for all $|\alpha| < r$, $v \in \mathcal{I}$. This condition plays a big role in wavelet theory. It has an equally important counterpart in the theory of shift-invariant spaces and sampling theory, namely the *Strang–Fix condition* of order r. A function f is said to satisfy the Strang–Fix condition of order r, if $D^\alpha f^\wedge(2\pi j) = 0$ for all $j \in \mathbb{Z}^d \setminus \{0\}$, $|\alpha| < r$, while $f^\wedge(0) \neq 0$. For an MRA one can show that given a scaling function ϑ and a wavelet set $\{\psi_v\}_{v \in \mathcal{I}}$ with polynomial decay of sufficiently high order, then ϑ satisfies the Strang–Fix condition of order $r \in \mathbb{N}$ if and only if $\{\psi_v\}_{v \in \mathcal{I}}$ satisfies the vanishing moment condition of order r. See Section 7.5.2 for more details.

The chapter is organized as follows. Section 7.2 deals with the continuous wavelet transform and its discretization. Necessary and sufficient conditions for wavelet frames are given. In Section 7.3 some basic facts about finitely generated shift-invariant spaces are given. Several examples are discussed, and an interpolation theorem is stated. In Section 7.4 the notion of a multiresolution analysis is introduced, again several examples are given. Section 7.5 is concerned with the connections between sampling approximation and multiresolution analysis. A general approximation theorem is stated and several applications to sampling operators are given. Furthermore, the connections between the vanishing moment condition and the Strang–Fix condition are discussed.

Some notation

Throughout the chapter we will use standard multi-index notation: The order of a multi-index $\alpha = (\alpha_1, \ldots, \alpha_d)^{tr} \in \mathbb{N}_0^d$ is given by $|\alpha| := \alpha_1 + \cdots + \alpha_d$, and $\alpha! := \alpha_1! \cdot \alpha_2! \cdot \ldots \cdot \alpha_d!$. For any two vectors $x, y, \in \mathbb{R}^d$, $x \cdot y := x_1 y_1 + \cdots + x_d y_d$ is the scalar product of x and y, while $xy := (x_1 y_1, \ldots, x_d y_d)^{tr}$. Also $x^\alpha := \prod_{\nu=1}^d x_\nu^{\alpha_\nu}$, and we will use the Euclidean norm $\|x\| := (x \cdot x)^{1/2}$. $B_\rho(x)$ is the ball with radius ρ and centre $x \in \mathbb{R}^d$.

$B(\mathbb{R}^d)$ is the space of all bounded functions $f : \mathbb{R}^d \to \mathbb{C}$ endowed with the supremum norm $\|f\|_B := \sup_{x \in \mathbb{R}^d} |f(x)|$. $C_B(\mathbb{R}^d)$ is the space of all continuous and bounded functions on $\mathbb{R}^d$ with the norm $\|f\|_C := \|f\|_B$, and $C_{2\pi}$ is its subspace of 2π-periodic functions, i.e., functions $f \in C_B(\mathbb{R}^d)$ that are invariant under translation by elements of $2\pi\mathbb{Z}^d$. For $r \in \mathbb{N}$, $C_B^r(\mathbb{R}^d)$ is the space of all functions $f \in C_B(\mathbb{R}^d)$ with such that

$$D^\alpha f := \frac{\partial^{\alpha_1}}{\partial x_1^{\alpha_1}} \cdots \frac{\partial^{\alpha_d}}{\partial x_d^{\alpha_d}} \qquad f \in C_B(\mathbb{R}^d)$$

for all $\alpha \in \mathbb{N}_0$, $|\alpha| \le r$. Its subspace of 2π-periodic functions is denoted by $C_{2\pi}^r$.

For $1 \le p < \infty$ and a countable set $\mathcal{I}$, $l^p(\mathcal{I})$ (resp. $l^\infty(\mathcal{I})$) is the space of all complex-valued sequences over $\mathcal{I}$ such that $\|a\|_p := \left(\sum_{\ell \in \mathcal{I}} |a_\ell|^p\right)^{1/p} < \infty$ (resp. $\|a\|_\infty := \sup_{\ell \in \mathcal{I}} |a_\ell| < \infty$), and $L^p(\mathbb{R}^d)$ (resp. $L^\infty(\mathbb{R}^d)$) is the space of measurable functions $f : \mathbb{R}^d \longrightarrow \mathbb{C}$ with finite norm $\|f\|_p := (\int_{\mathbb{R}^d} |f(u)|^p du)^{1/p} < \infty$ (resp. $\|f\|_\infty := \operatorname{ess\,sup}_{x \in \mathbb{R}^d} |f(x)| < \infty$), with the usual identifications. $L_{2\pi}^p$ (resp. $L_{2\pi}^\infty$) is the space of 2π-periodic measurable functions f on $\mathbb{R}^d$ such that $\|f\|_p := \left(\int_{[-\pi,\pi]^d} |f(u)|^p du\right)^{1/p} < \infty$ (resp. $\|f\|_\infty := \operatorname{ess\,sup}_{x \in [-\pi,\pi]^d} |f(u)| < \infty$).

The Fourier transform of $f \in L^1(\mathbb{R}^d)$ is given by

$$f^\wedge(v) := \left(\sqrt{2\pi}\right)^{-d} \int_{\mathbb{R}^d} f(u) e^{-iv \cdot u} du, \qquad v \in \mathbb{R}^d,$$

while for $f \in L^2(\mathbb{R}^d)$ it is the $L^2(\mathbb{R}^d)$-limit function

$$\lim_{M \to \infty} \text{i. m.} (f \cdot \chi_{[-M,M]^d})^\wedge,$$

$\chi_{[-M,M]^d}$ being the characteristic function of the interval $[-M, M]^d$. The Fourier coefficients of $f \in L_{2\pi}^p$ are defined by

$$f^\wedge(k) := (2\pi)^{-d} \int_{[-\pi,\pi]^d} f(u) e^{-ik \cdot u} du, \qquad k \in \mathbb{Z}^d.$$

In all three cases the Fourier transform will be denoted by $f^\wedge$. The inverse Fourier transform of a function $f \in L^2(\mathbb{R}^d)$ is denoted by $f^\vee$. The scalar product of two functions $f, g \in L^2(\mathbb{R}^d)$ is given by $\langle f, g \rangle := \int_{\mathbb{R}^d} f(u) \overline{g(u)} du$.

7.2 The continuous wavelet transform

7.2.1 *Univariate theory*

Most of the material covered in this section is to be found in Daubechies (1990, 1992) or Heil and Walnut (1989). Another very good presentation of the material is given by Kölzow (1994). For $\psi \in L^2(\mathbb{R})$, let

$$\psi^{a,b} := |a|^{-1/2} \psi\left(\frac{\cdot - b}{a}\right), \qquad a, b \in \mathbb{R}, \ a \neq 0, \tag{7.2.1}$$

be the family of functions derived from ψ by translation and dilation. A function $\psi \in L^2(\mathbb{R})$ is called *admissible,* if

$$C_\psi := 2\pi \int_{\mathbb{R}} |\xi|^{-1} |\psi^\wedge(\xi)|^2 \, d\xi < \infty. \tag{7.2.2}$$

The *continuous wavelet transform (CWT)* of a function $f \in L^2(\mathbb{R})$ with respect to an admissible function $\psi \in L^2(\mathbb{R})$ is given by

$$(W^\psi f)(a, b) := \langle f, \psi^{a,b} \rangle = \int_{\mathbb{R}} f(x) |a|^{-1/2} \psi\left(\frac{x - b}{a}\right) dx, \qquad a, b \in \mathbb{R}, \ a \neq 0. \tag{7.2.3}$$

For an admissible function ψ one can show the Plancherel-type formula or *resolution of the identity* for the CWT

$$\int_{\mathbb{R}} \int_{\mathbb{R}} (W^\psi f)(a, b) \overline{(W^\psi f')(a, b)} \frac{da \, db}{a^2} = C_\psi \langle f, f' \rangle, \qquad f, f' \in L^2(\mathbb{R}). \tag{7.2.4}$$

This means that the continuous wavelet transform maps any function $f \in L^2(\mathbb{R})$ isometrically into $L^2(\mathbb{R}^2; C_\psi^{-1} a^{-2} da \, db)$, the space of all complex valued functions F on $\mathbb{R}^2$ with finite norm

$$\|F\| := \left(C_\psi^{-1} \int_{\mathbb{R}} \int_{\mathbb{R}} |F(a, b)|^2 \frac{da \, db}{a^2} \right)^{1/2}.$$

Equipped with the scalar product

$$\langle F, G \rangle_\psi := C_\psi^{-1} \int_{\mathbb{R}} \int_{\mathbb{R}} F(a, b) \overline{G(a, b)} \frac{da \, db}{a^2},$$

this space becomes a Hilbert space. The image $W^\psi(L^2(\mathbb{R})) =: \mathcal{H}^\psi$ is a closed subspace of $L^2(\mathbb{R}^2; C_\psi^{-1} a^{-2} da \, db)$. An immediate consequence of (7.2.4) is the following representation of an element $F \in \mathcal{H}^\psi$

$$\begin{aligned} F(a, b) = \langle f, \psi^{a,b} \rangle &= C_\psi^{-1} \int_{\mathbb{R}} \int_{\mathbb{R}} (W^\psi f)(a', b') \overline{(W^\psi \psi^{a,b})(a', b')} \frac{da' \, db'}{(a')^2} \\ &= C_\psi^{-1} \int_{\mathbb{R}} \int_{\mathbb{R}} K(a, b; a', b') F(a', b') \frac{da' \, db'}{(a')^2}, \end{aligned} \tag{7.2.5}$$

where $K(a, b; a', b') := \langle \psi^{a',b'}, \psi^{a,b} \rangle, a, b, a', b' \in \mathbb{R}; a, a' \neq 0$. Thus $\mathcal{H}^\psi$ is in fact a reproducing kernel Hilbert space.

Formula (7.2.4) may also be read as

$$f = C_\psi^{-1} \int_\mathbb{R} \int_\mathbb{R} (W^\psi f)(a, b) \, \psi^{a,b} \, \frac{da \, db}{a^2}, \qquad f \in L^2(\mathbb{R}), \qquad (7.2.6)$$

with convergence in the weak sense, i.e., taking scalar products on both sides. Several variations of this inversion formula have been shown under slightly more restrictive assumptions. There exist formulae involving only the values of $(W^\psi f)(a, b)$ for positive a, as well as pointwise versions of (7.2.6). The main requirement, however, that allows the recovery of a signal $f \in L^2(\mathbb{R})$ is the admissibility condition $C_\psi < \infty$. In case $\psi \in L^1(\mathbb{R})$ it is equivalent to the vanishing moment condition $\int_\mathbb{R} \psi(x) \, dx = 0$.

Typical admissible functions ψ used in vision analysis include the *Mexican hat function*

$$\psi_{Mex}(x) := \frac{2\pi}{\sqrt{3}} \pi^{-1/4} (1 - x^2) e^{-x^2/2}, \qquad x \in \mathbb{R}, \qquad (7.2.7)$$

which is the (normalized) second derivative of the Gaussian $e^{-x^2/2}$, or the so-called *Morlet-wavelet*

$$\psi_{Mor}(x) := \pi^{-1/4} \left(e^{-i\xi_0 x} - e^{-\xi_0^2/2} \right) e^{-x^2/2}, \qquad x \in \mathbb{R}, \qquad (7.2.8)$$

which is a modulated Gaussian.

7.2.2 *The continuous wavelet transform in higher dimensions*

There exist several possibilities to generalize the CWT to higher dimensions. One possibility is to choose one translation and one dilation parameter for each direction. The theory will then be very similar to the univariate theory. A different approach exploiting the underlying group theoretical structure was used by Murenzi (1989, 1990). He uses one "global" dilation parameter $a \in \mathbb{R}_+$ only and introduces as additional parameter a rotation $R \in SO(d)$. Here $SO(d)$ is the special orthogonal group in d dimensions. Let us understand this method.

Let G be a locally compact group, i.e., a locally compact topological space equipped with group operations such that the mappings $(g, g') \mapsto gg'$ from $G \times G \to G$ and $g \mapsto g^{-1}$ on G are continuous. See Nachbin (1965) and Reiter (1968, Chapter 3) for more details. On G there exists a measure μ, called the *left Haar measure*, which is invariant under multiplication on the left, i.e., $\int_G f(gg') \, d\mu(g') = \int_G f(g') \, d\mu(g')$ for every $g \in G$ and every integrable function f on G. The left Haar measure on G is uniquely determined up to a constant. These facts were established by A. Haar, J. von Neumann and A. Weil; analogous results hold for the *right Haar measure* defined in the obvious way. A mapping U from G into the group of automorphisms of a Hilbert space $\mathcal{H}$ with $U(g)U(g') = U(gg')$ for all $g, g' \in G$ is called a *representation of G on $\mathcal{H}$*. U is called *unitary*, if the map $U(g) : \mathcal{H} \to \mathcal{H}$ is unitary for each $g \in G$. It is *irreducible* if for any $h \in \mathcal{H}$, $h \neq 0$, the set span$\{U(g)h\}_{g \in G}$ is dense in $\mathcal{H}$, and it is called *square integrable* if it is irreducible and there exists an element $f \in \mathcal{H}$ such that

$$\int_G |\langle f, U(g)f \rangle|^2 \, d\mu < \infty. \qquad (7.2.9)$$

An element $f \in \mathcal{H}$ satisfying (7.2.9) is called *U–admissible*.

The first to realize the close connections of the CWT to representation theory was A. Grossmann. In fact, the family (7.2.1) corresponds to a unitary representation U of the affine or $(ax + b)$-group in $L^2(\mathbb{R})$: For each pair $(a, b) \in \mathbb{R}^2$, $a \neq 0$, let $U(a, b)$ be the operator

$$U(a, b)f := |a|^{-1/2} f\left(\frac{\cdot - b}{a}\right), \qquad f \in L^2(\mathbb{R}).$$

Then every $U(a, b)$ is a unitary operator on $L^2(\mathbb{R})$ and $U(a, b)U(a', b') = U(aa', ab' + b)$. One can also show that U is irreducible and square-integrable. In fact, a function $f \in L^2(\mathbb{R})$ is U–admissible if and only if it is admissible in the sense of (7.2.2).

It was shown by Grossmann et al. (1985, 1986) that the basic properties of the univariate CWT (7.2.4)–(7.2.6) hold in a much more general setting. Let U be a square-integrable unitary representation of g in a Hilbert space $\mathcal{H}$. Then there exists a dense set $\mathcal{D} \in \mathcal{H}$ such that (7.2.9) holds for any element $f' \in \mathcal{D}$, and there exists a (possibly unbounded) operator $A : \mathcal{D} \to \mathcal{H}$ such that for any $\psi \in \mathcal{D}$

$$\int_G \langle h_1, U(g)\psi \rangle \overline{\langle h_2, U(g)\psi \rangle} \, d\mu(g) = C_\psi \langle h_1, h_2 \rangle, \qquad h_1, h_2 \in \mathcal{H}, \qquad (7.2.10)$$

where $C_\psi = \langle A\psi, \psi \rangle$. This corresponds exactly to the resolution of the identity (7.2.4); in the wavelet case the operator A is given by $(Af)^\wedge(\xi) := 2\pi |\xi|^{-1} f^\wedge(\xi)$, a.e. on $\mathbb{R}$, $f \in L^2(\mathbb{R})$. Equivalent formulations of type (7.2.5) or (7.2.6) are possible as well.

The wavelet transform in higher dimensions is treated by Murenzi (1989, 1990); concerning the bivariate theory see also Antoine (1992), Antoine et al. (1993). The adequate choice here is the Euclidean group with dilations $IG(d)$, defined in the following way: For $d \geq 2$, let $G(d)$ be the set of real $d \times d$-matrices of the form

$$V = aR, \qquad a \in \mathbb{R}_+, \, R \in \mathrm{SO}(d),$$

with matrix multiplication as group operation. $G(d)$ has the structure of the direct product of the multiplicative group of positive real numbers $\mathbb{R}_+^*$ with $\mathrm{SO}(d)$ (dilations and rotations commute). An element V of $G(d)$ depends on $1 + d(d-1)/2$ parameters. The Euclidean d-dimensional group with dilations $IG(d)$ is the semidirect product of $G(d)$ with the additive group $\mathbb{R}^d$,

$$IG(d) = G(d) \otimes (\mathbb{R}^d) = \left[\mathbb{R}_+^* \times \mathrm{SO}(d)\right] \otimes \mathbb{R}^d,$$

i.e., the set of ordered pairs $(V, b) \in G(d) \times \mathbb{R}^d$ with group operation $(V, b) \cdot (V', b') := (VV', Vb' + b)$ and inverse $(V, b)^{-1} = (V^{-1}, -V^{-1}b)$. In the following, elements of $IG(d)$ will be denoted by a triplet (a, R, b) with $a \in \mathbb{R}_+$, $R \in G(d)$ and $b \in \mathbb{R}^d$.

$IG(d)$ is a locally compact group with left Haar measure $a^{-(d+1)} da \, dR \, db$ and right Haar measure $a^{-1} da \, dR \, db$, where dR is the Haar measure on $\mathrm{SO}(d)$, see Vilenkin (1968). For $\psi \in L^2(\mathbb{R}^d)$ let

$$\psi^{a, R, b} := a^{-d/2} \psi\left(a^{-1} R^{-1}(\cdot - b)\right), \qquad a \in \mathbb{R}_+, \, R \in \mathrm{SO}(d), \, b \in \mathbb{R}^d. \quad (7.2.11)$$

This defines a unitary square integrable representation of $IG(d)$ on $L^2(\mathbb{R}^d)$. The admissibility condition (7.2.9) in this setting is equivalent to

$$C_\psi := (2\pi)^d \int_{\mathbb{R}^d} \|\xi\|^{-d} |\psi^\wedge(\xi)|^2 \, d\xi < \infty. \tag{7.2.12}$$

Accordingly, a function $\psi \in L^2(\mathbb{R}^d)$ is called *admissible* if it satisfies (7.2.12). The CWT of a function $f \in L^2(\mathbb{R}^d)$ with respect to an admissible function $\psi \in L^2(\mathbb{R}^d)$ is given by

$$(W^\psi f)(a, R, b) := \langle f, \psi^{a,R,b} \rangle = \int_{\mathbb{R}^d} f(x)\, a^{-d/2} \psi\big(a^{-1} R^{-1}(x - b)\big) \, dx, \tag{7.2.13}$$

where $a \in \mathbb{R}_+$, $R \in SO(d)$ and $b \in \mathbb{R}^d$. The basic properties of the univariate CWT (7.2.4)–(7.2.6) can be transferred to the multivariate setting via the formula (7.2.10). Here the operator A is given by $(Af)^\wedge(\xi) := (2\pi)^d \|\xi\|^{-d} \psi^\wedge(\xi)$ a.e. on $\mathbb{R}^d$, $f \in L^2(\mathbb{R}^d)$. The Plancherel-type formula or resolution of the identity (7.2.4) turns into

$$\int_{\mathbb{R}_+} \int_{SO(d)} \int_{\mathbb{R}^d} (W^\psi f)(a, R, b) \overline{(W^\psi f')(a, R, b)} \, \frac{da \, dR \, db}{a^{d+1}} = C_\psi \, \langle f, f' \rangle \tag{7.2.14}$$

for any $f, f' \in L^2(\mathbb{R}^d)$, while the inversion formula (7.2.6) becomes

$$f = C_\psi^{-1} \int_{\mathbb{R}_+} \int_{SO(d)} \int_{\mathbb{R}^d} (W^\psi f)(a, R, b) \, \psi^{a,R,b} \, \frac{da \, dR \, db}{a^{d+1}}, \qquad f \in L^2(\mathbb{R}^d). \tag{7.2.15}$$

In case ψ is rotation-invariant, $W^\psi f$ is a function of the parameters $a \in \mathbb{R}_+$ and $b \in \mathbb{R}^d$ only, and the integral over $SO(d)$ in both (7.2.14) and (7.2.15) is absent. A typical example of this kind is the *two-dimensional Mexican hat function*, which is the (normalized) Laplacian of the Gaussian function

$$\psi_{Mex}(x) := \frac{(2\pi)^2}{3\sqrt{\pi}} \, (1 - \|x\|^2) \, e^{-\|x\|^2/2}, \qquad x \in \mathbb{R}^2. \tag{7.2.16}$$

It is a real-valued, rotation-invariant function with very good localization properties in both space and frequency, therefore it is efficient for a fine pointwise analysis, but not for detecting directions.

On the other hand, if ψ is not rotation-invariant, then $W^\psi f$ may be used for detecting oriented features of the signal f, for instance for edge detection in image analysis. An example of this sort is given by the *d-dimensional Morlet wavelet*

$$\psi_{Mor}(x) := \pi^{-1/2} \left(e^{-i\xi_0 \cdot x} - e^{-\xi_0^{tr} D \xi_0/2} \right) e^{-x^{tr} C x/2}, \qquad x \in \mathbb{R}^d, \tag{7.2.17}$$

where C is a positive definite $d \times d$-matrix, $D := C^{-1}$, and $\xi_0 \in \mathbb{R}^d$. The second term in the parentheses in (7.2.17) guarantees the admissibility condition; when $\|\xi_0\|$ is chosen large enough, it is negligible. In this case ψ_{Mor} is essentially a smoothed plane wave with wave vector ξ_0; its directivity may even be enhanced by a suitable choice of the matrix C. Therefore the two-dimensional Morlet wavelet is well adapted for edge detection (Antoine et al. (1993)).

7.2.3 Sampling of the continuous wavelet transform

By (7.2.5) and its multivariate counterpart, the image space $\mathcal{H}^{\psi}$ of the CWT is a reproducing kernel Hilbert space, so it is to be expected that the information stored in the CWT may actually be recovered from a discrete set of sampled values of the CWT. As it turns out, for suitable admissible functions ψ, the CWT can in fact be discretized without loss of information. This property of the CWT is essential for numerical purposes. Let us for simplicity restrict ourselves to the univariate case $d = 1$.

Fixing some $a > 1$, it seems reasonable to sample the CWT in "octaves", i.e., to use the scales a^n, $n \in \mathbb{Z}$. (If we choose $a = 2$, then we might in fact talk about an analysis in octaves.) Further, for $n = 0$ (i.e., no dilation of ψ) we choose to use equidistant sampling in the second variable, i.e., the translates of the function ψ by the integer multiples of some $b > 0$. For $n \neq 0$, the function $a^{-1/2}\psi(\cdot/a^n)$ has grown "wider" (or narrower, depending on the sign of $n \in \mathbb{Z}$) by the factor of a^n, so it seems natural to scale the distance between the sampling points by the same factor. What we end up with is the discrete (but infinite) set of data $(W^{\psi} f)(a^n, kba^n) = \langle f, \psi^{a^n, kba^n} \rangle$, $n, k \in \mathbb{Z}$ for some $a > 1$, $b > 0$. To be able to recover the signal f from these data in a numerically stable way we need to require that the set

$$\left\{ \psi^{a^n, kba^n} \right\}_{n,k \in \mathbb{Z}} \tag{7.2.18}$$

be a frame for $L^2(\mathbb{R})$, i.e., that there exist constants A, $B > 0$ such that

$$A\|f\|_2^2 \leq \sum_{n,k \in \mathbb{Z}} \left\langle f, \psi^{a^n, kba^n} \right\rangle \leq B\|f\|_2^2, \qquad f \in L^2(\mathbb{R}),$$

(see Higgins (1996) concerning basic facts about frames). A frame of this form is called a *wavelet frame*. Note that the corresponding dual frame $\{\tilde{\psi}_{n,k}\}_{n,k \in \mathbb{Z}}$ need not be a wavelet frame; a counterexample is given by Chui (1992, p. 13). Recall from Volume 1 that in this setting a signal $f \in L^2(\mathbb{R})$ can be recovered from the set of data $(W^{\psi} f)(a^n, kba^n) = \langle f, \psi^{a^n, kba^n} \rangle$, $n, k \in \mathbb{Z}$ via the reconstruction formula

$$f = \sum_{n,k \in \mathbb{Z}} \left\langle f, \psi^{a^n, kba^n} \right\rangle \tilde{\psi}_{n,k}. \tag{7.2.19}$$

An algorithm for the reconstruction of f that does not involve the explicit computation of the dual frame is given by Daubechies (1992, p. 63).

A common choice in (7.2.18) is $a := 2$, $b := 1$. Of particular interest is the case when the set $\{\psi^{a^n, kba^n}\}_{n,k \in \mathbb{Z}}$ is not only a frame, but an orthonormal basis for $L^2(\mathbb{R})$. A function ψ with this property is called a *wavelet*.[1] In this case the corresponding dual frame coincides with the frame itself, so the reconstruction formula (7.2.19) becomes particularly simple. A valuable tool for the construction of wavelets is the concept of

---

[1]To avoid confusion note that this is the terminology used by many authors dealing with the *discrete wavelet transform*, i.e., with discrete samples of the CWT. The Morlet wavelet (7.2.8) and (7.2.17) is not a wavelet in this sense. However, in the context of the *continuous* wavelet transform, the term "wavelet" is often used to denote *any* admissible function; this accounts for the name "Morlet wavelet".

a multiresolution analysis as introduced by Mallat (1989a), which will be discussed in more detail in Section 7.4.

Daubechies (1990, 1992) shows that if a function ψ generates a wavelet frame of the form (7.2.18) with frame bounds A, B, then it necessarily satisfies

$$\frac{b \log a}{2\pi} A \leq \int_0^\infty \frac{|\psi^\wedge(\xi)|^2}{\xi} d\xi \leq \frac{b \log a}{2\pi} B, \tag{7.2.20}$$

$$\frac{b \log a}{2\pi} A \leq \int_{-\infty}^0 \frac{|\psi^\wedge(\xi)|^2}{|\xi|} d\xi \leq \frac{b \log a}{2\pi} B. \tag{7.2.21}$$

This in turn implies the admissibility condition (7.2.2). She also shows (Daubechies 1990, 1992) that for an admissible function ψ with reasonable decay in both time and frequency there exists a whole range of values a, b such that the set (7.2.18) is a wavelet frame. This is the content of the next theorem.

Theorem 7.1 (Daubechies 1990; 1992) *Let* $\psi \in L^2(\mathbb{R})$, $a > 1$ *be such that*

$$\underset{1 \leq |\xi| \leq a}{\text{ess inf}} \sum_{m \in \mathbb{Z}} |\psi^\wedge(a^m \xi)|^2 > 0, \qquad \underset{1 \leq |\xi| \leq a}{\text{ess sup}} \sum_{m \in \mathbb{Z}} |\psi^\wedge(a^m \xi)|^2 < \infty.$$

Further suppose that for some $\varepsilon > 0$ *there exists a constant* K *such that*

$$\beta(s) := \underset{\xi \in \mathbb{R}}{\text{ess sup}} |\psi^\wedge(a^m \xi)||\psi^\wedge(a^m \xi + s)| \leq (1 + |s|)^{-(1+\varepsilon)}, \qquad a.e. \ on \ \mathbb{R}.$$

Then there exist $b_0 > 0$ *such that for all choices* $0 < b < b_0$ *the set* (7.2.18) *constitutes a frame for* $L^2(\mathbb{R})$ *with frame bounds*

$$A := \frac{2\pi}{b} \left\{ \underset{1 \leq |\xi| \leq a}{\text{ess inf}} \sum_{m \in \mathbb{Z}} |\psi^\wedge(a^m \xi)|^2 - \sum_{\substack{k \in \mathbb{Z} \\ k \neq 0}} \left[\beta\left(\frac{2\pi}{b}k\right) \beta\left(-\frac{2\pi}{b}k\right) \right]^{1/2} \right\},$$

$$B := \frac{2\pi}{b} \left\{ \underset{1 \leq |\xi| \leq a}{\text{ess sup}} \sum_{m \in \mathbb{Z}} |\psi^\wedge(a^m \xi)|^2 + \sum_{\substack{k \in \mathbb{Z} \\ k \neq 0}} \left[\beta\left(\frac{2\pi}{b}k\right) \beta\left(-\frac{2\pi}{b}k\right) \right]^{1/2} \right\}.$$

The decay conditions are satisfied if e.g. $|\psi^\wedge(\xi)| \leq C|\xi|^\alpha (1 + |\xi|)^\gamma$ *a.e. on* $\mathbb{R}$ *for some constants* $C, \alpha > 0$, $\gamma > \alpha + 1$.

Both the Mexican hat function (7.2.7) and the Morlet wavelet (7.2.8) satisfy the assumptions of Theorem 7.1 for any $a > 1$.

For $a = 2$, Ph. Tchamitchian (see Daubechies 1990) derived different expressions for the frame bounds A and B that take into account the phase of $\psi^\wedge$. Because of cancellation, these frame bounds may be better than those given above, unless $\psi^\wedge$ is a non-negative real function. Similar results for irregular sampling of the CWT with respect to a band-limited admissible function ψ as well as reconstruction algorithms were given by Gröchenig (1993) and Feichtinger and Gröchenig (1994). A multivariate analogue of Theorem 7.1 may be found in Murenzi (1990). A beautiful theory for the much more abstract group theoretical setting was developed by Feichtinger and Gröchenig (1988) and Gröchenig (1991). They obtained discrete expansions of vectors in a large class of Banach spaces called coorbit spaces.

7.3 Interpolation on shift-invariant spaces

7.3.1 *Shift-invariant spaces*

A closed space $V \subset L^2(\mathbb{R}^d)$ is called *shift-invariant*, if for any function $f \in V$ its *integer shifts* $f(\cdot - j)$, $j \in \mathbb{Z}^d$, are also contained in V. We say the shift-invariant space V is a *finitely generated shift-invariant space (FSI)*, if there exists a finite set of functions $\{\vartheta_\nu\}_{\nu \in \mathcal{I}} \subset L^2(\mathbb{R}^d)$ such that V is the $L^2(\mathbb{R}^d)$-closure of $\mathrm{span}\{\vartheta_\nu(\cdot - j); j \in \mathbb{Z}^d, \nu \in \mathcal{I}\}$. In case the set of generators consists of only one element ϑ, V is called a *principal shift-invariant space (PSI)*. We use the notation $V = S(\vartheta)$. A finite set of functions $\{\vartheta_\nu\}_{\nu \in \mathcal{I}}$ is called a *stable set of generators* for the FSI V, if their integer shifts $\{\vartheta_\nu(\cdot - j)\}_{j \in \mathbb{Z}^d, \nu \in \mathcal{I}}$ form a Riesz basis for V, i.e., if for any function $f \in V$ there exists a unique sequence $a = \{a_{j,\nu}\}_{j \in \mathbb{Z}^d, \nu \in \mathcal{I}} \in l^2(\mathbb{Z}^d \times \mathcal{I})$ such that f has a representation

$$f = \sum_{\nu \in \mathcal{I}} \sum_{j \in \mathbb{Z}^d} a_{j,\nu} \vartheta_\nu(\cdot - j),$$

and there exist constants $A, B > 0$ such that for any sequence $a \in l^2(\mathbb{Z}^d \times \mathcal{I})$

$$A \sum_{\nu \in \mathcal{I}} \sum_{j \in \mathbb{Z}^d} |a_{j,\nu}|^2 \leq \left\| \sum_{\nu \in \mathcal{I}} \sum_{j \in \mathbb{Z}^d} a_{j,\nu} \vartheta_\nu(\cdot - j) \right\|_2^2 \leq B \sum_{\nu \in \mathcal{I}} \sum_{j \in \mathbb{Z}^d} |a_{j,\nu}|^2. \tag{7.3.1}$$

In this case

$$V = \left\{ f = \sum_{\nu \in \mathcal{I}} \sum_{j \in \mathbb{Z}^d} a_{j,\nu} \vartheta_\nu(\cdot - j); \ a = \{a_{j,\nu}\}_{j \in \mathbb{Z}^d, \nu \in \mathcal{I}} \in l^2(\mathbb{Z}^d \times \mathcal{I}) \right\} \tag{7.3.2}$$

or, taking Fourier transforms,

$$V = \left\{ f = \left(\sum_{\nu \in \mathcal{I}} m_\nu \vartheta_\nu^\wedge \right)^\vee; \ m_\nu \in L^2_{2\pi}, \nu \in \mathcal{I} \right\}. \tag{7.3.3}$$

If $\{\vartheta_\nu(\cdot - j)\}_{j \in \mathbb{Z}^d, \nu \in \mathcal{I}}$ is an orthonormal basis for V, then the set $\{\vartheta_\nu\}_{\nu \in \mathcal{I}}$ is called an *orthonormal set of generators* for V. Note that by Parseval's identity, in this case $\{\vartheta_\nu(\cdot - j)\}_{j \in \mathbb{Z}^d, \nu \in \mathcal{I}}$ satisfies (7.3.1) with constants $A = B = 1$, so orthonormality implies stability. In fact, $\{\vartheta_\nu\}_{\nu \in \mathcal{I}}$ is an orthonormal set of generators for V if and only if it is a stable set of generators for V satisfying (7.3.1) with $A = B = 1$. This can be seen from (7.3.1) by choosing sequences $a \in l^2(\mathbb{Z}^d \times \mathcal{I})$ of the form $a_{j,\nu} := \delta_{j,k} \delta_{\nu,\mu} + \lambda \delta_{j,k'} \delta_{\nu,\mu'}$ for all $j \in \mathbb{Z}^d$, $\nu \in \mathcal{I}$ and some $k, k' \in \mathbb{Z}^d$, $\mu, \mu' \in \mathcal{I}$, $\lambda \in \{0, 1, i\}$.

We say a function $\vartheta \in L^2(\mathbb{R}^d)$ has *stable integer shifts* (or *orthonormal integer shifts*), if ϑ is a stable generator (or orthonormal generator) for the PSI $S(\vartheta)$, i.e., if (7.3.1) is satisfied (with constants $A = B = 1$ in the orthonormal case), where of course the summation over $\mathcal{I}$ is superfluous. In this setting there exists another characterization for stability which in many cases is more convenient. To see this, let $a = \{a_j\}_{j \in \mathbb{Z}^d} \in l^2(\mathbb{Z}^d)$ and define $m \in L^2(\mathbb{R}^d)$ by $m(x) := \sum_{j \in \mathbb{Z}^d} a_j e^{-ij \cdot x}$ a.e. on $\mathbb{R}^d$. Then

$$\sum_{j \in \mathbb{Z}^d} |a_j|^2 = (2\pi)^{-d} \|m\|_2^2, \tag{7.3.4}$$

and, by Parseval's formula,

$$
\left\| \sum_{j \in \mathbb{Z}^d} a_j \vartheta(\cdot - j) \right\|_2^2 = \int_{\mathbb{R}^d} \Big| \sum_{j \in \mathbb{Z}^d} a_j \vartheta(x - j) \Big|^2 \, dx = \int_{\mathbb{R}^d} |m(\xi) \vartheta^\wedge(\xi)|^2 \, d\xi
$$

$$
= \int_{[-\pi,\pi]^d} |m(\xi)|^2 \sum_{j \in \mathbb{Z}^d} |\vartheta^\wedge(\xi - 2\pi j)|^2 \, d\xi.
$$

In view of (7.3.1) and (7.3.4), an application of the Riesz representation theorem yields

A function $\vartheta \in L^2(\mathbb{R}^d)$ has stable integer translates if and only if there exist constants $A, B > 0$ such that

$$
A \le \sum_{j \in \mathbb{Z}^d} |\vartheta^\wedge(\xi - 2\pi j)|^2 \le B \qquad \text{a.e. on } \mathbb{R}^d; \tag{7.3.5}
$$

a function $\varphi \in L^2(\mathbb{R}^d)$ has orthonormal integer translates if and only if

$$
\sum_{j \in \mathbb{Z}^d} |\varphi^\wedge(\xi - 2\pi j)|^2 = (2\pi)^{-d} \qquad \text{a.e. on } \mathbb{R}^d. \tag{7.3.6}
$$

Note that if a function $\vartheta \in L^2(\mathbb{R}^d)$ has stable integer shifts, then in view of (7.3.5) one can define a function $\varphi \in L^2(\mathbb{R}^d)$ by

$$
\varphi^\wedge(\xi) := \frac{\vartheta^\wedge(\xi)}{\left((2\pi)^d \sum_{j \in \mathbb{Z}^d} |\vartheta^\wedge(\xi - 2\pi j)|^2 \right)^{1/2}} \qquad \text{a.e. on } \mathbb{R}^d. \tag{7.3.7}
$$

Combining (7.3.3), (7.3.5) and (7.3.6), we find that φ has orthonormal integer shifts, and $S(\vartheta) = S(\varphi)$. Thus we conclude that for any PSI with a stable generator there also exists an orthonormal generator.

In our setting it will be useful to define the following spaces: For $1 \le p \le \infty$ let $\mathcal{L}^p(\mathbb{R}^d)$ be the space introduced by Jia and Micchelli (1991), consisting of all measurable functions f on $\mathbb{R}^d$ such that $\sum_{j \in \mathbb{Z}^d} |f(\cdot - j)| \in L^p([0, 1]^d)$. Further let $\mathcal{B}(\mathbb{R}^d)$ be the space of all functions $f \in B(\mathbb{R}^d)$ such that $\sum_{j \in \mathbb{Z}^d} |f(\cdot - j)| \in B(\mathbb{R}^d)$, and let $\mathcal{C}(\mathbb{R}^d)$ be the space of all functions $f \in C_B(\mathbb{R}^d)$ such that the series $\sum_{j \in \mathbb{Z}^d} |f(\cdot - j)|$ converges uniformly on compact sets. Note that by periodicity $\mathcal{C}(\mathbb{R}^d) \subset \mathcal{B}(\mathbb{R}^d)$; also $\mathcal{L}^p(\mathbb{R}^d) \subset L^p(\mathbb{R}^d)$, $1 \le p \le \infty$, with $\mathcal{L}^1(\mathbb{R}^d) = L^1(\mathbb{R}^d)$ (Jia and Micchelli 1991). The following useful fact is easily seen from (7.3.2):

If $\{\vartheta_\nu\}_{\nu \in \mathcal{I}} \subset \mathcal{C}(\mathbb{R}^d)$ is a stable set of generators for the FSI V,
then $V \subset C_B(\mathbb{R}^d)$. $\tag{7.3.8}$

Further, for $r \in \mathbb{N}_0$, let

$$
\mathcal{L}_r^p(\mathbb{R}^d) := \{ f \in L^1(\mathbb{R}^d); \ (1 + \| \cdot \|)^r f \in \mathcal{L}^p(\mathbb{R}^d) \}, \quad 1 \le p \le \infty, \tag{7.3.9}
$$
$$
\mathcal{B}_r(\mathbb{R}^d) := \{ f \in B(\mathbb{R}^d); \ (1 + \| \cdot \|)^r f \in \mathcal{B}(\mathbb{R}^d) \}. \tag{7.3.10}
$$

see also Lei (1994b), Fischer (1996). If $1 \le p \le p' \le \infty$ and $r, r' \in \mathbb{N}_0$ with $r \le r'$, then $\mathcal{B}_r(\mathbb{R}^d) \subset \mathcal{L}_r^{p'}(\mathbb{R}^d) \subset \mathcal{L}_r^p(\mathbb{R}^d)$ and $\mathcal{L}_{r'}^p(\mathbb{R}^d) \subset \mathcal{L}_r^p(\mathbb{R}^d)$. Obviously,

$(1 + \| \cdot \|)^r f \in L^1(\mathbb{R}^d)$ for any function $f \in \mathcal{L}_r^p(\mathbb{R}^d)$, $1 \leq p \leq \infty$; in fact the elements of $\mathcal{L}_r^1(\mathbb{R}^d)$ are characterized by this property. Note that any function f satisfying $|f(x)| \leq K(1 + \|x\|)^{-r-d-\varepsilon}$, $x \in \mathbb{R}^d$, for some constant K and some $\varepsilon > 0$ is an element of $\mathcal{L}_r^p(\mathbb{R}^d)$ and $\mathcal{B}_r(\mathbb{R}^d)$. We will think of r as characterizing the polynomial decay rate of $f \in \mathcal{L}_r^p(\mathbb{R}^d)$.

Let us see some examples of functions having stable integer shifts.

7.3.2 Some examples

Example 7.2 Let the dimension $d = 1$; consider the *Sinus Cardinalis*

$$\operatorname{sinc} x := \begin{cases} 1, & x = 0, \\ \dfrac{\sin \pi x}{\pi x}, & \text{otherwise.} \end{cases} \tag{7.3.11}$$

This function has Fourier transform

$$\operatorname{sinc}^\wedge(\xi) = (2\pi)^{-1/2} \chi_{[-\pi,\pi]}(\xi), \qquad \xi \in \mathbb{R}^d, \tag{7.3.12}$$

where $\chi_{[-\pi,\pi]}$ is the characteristic function of the interval $[-\pi, \pi]$. By (7.3.6), the sinc-function has orthonormal integer translates. By (7.3.3), $S(\operatorname{sinc}) = PW_\pi(\mathbb{R})$ is the space of all functions $f \in L^2(\mathbb{R})$ band-limited to the interval $[-\pi, \pi]$, so the sinc-function is an orthonormal generator for $PW_\pi(\mathbb{R})$.

Example 7.3 As a second example let us consider univariate spline spaces. For $r \in \mathbb{N}$, let B_{r-1} be the cardinal B-spline of degree $r - 1$ defined via its Fourier transform

$$B_{r-1}^\wedge(v) := (2\pi)^{-1/2} \left(e^{-iv/2} \frac{\sin v/2}{v/2} \right)^r, \qquad v \in \mathbb{R}. \tag{7.3.13}$$

B_{r-1} has support on the interval $[0, r]$; it is a piecewise polynomial of degree $r - 1$ with knots at the integers, and it is an element of $C_B^{r-2}(\mathbb{R})$. The space $S(B_{r-1})$ generated by the cardinal B-spline B_{r-1} consists of all functions $f \in L^2(\mathbb{R}) \cap C_B^{r-2}(\mathbb{R})$ whose restriction $f|_{[m,m+1)}$ to the interval $[m, m + 1)$ is a polynomial of degree at most $r - 1$ for all $m \in \mathbb{Z}$. Schoenberg (1946) is generally considered the father of B-splines, a good survey of the history of the cardinal B-spline is given by Butzer, Schmidt and Stark (1988). Concerning the above results, see e.g. Schoenberg (1973) and de Boor (1987).

It can be shown that B_{r-1} satisfies (7.3.5) and thus has stable integer shifts (see e.g. Chui 1992, Chapter 4). This fact is even more easily seen using another criterion given by Jia and Micchelli (1991) for functions with some moderate decay:

A function $\vartheta \in \mathcal{L}^2(\mathbb{R}^d)$ has stable integer shifts if and only if

$$\sup_{j \in \mathbb{Z}^d} |\vartheta^\wedge(\xi - j)| > 0 \text{ for all } \xi \in \mathbb{R}^d. \tag{7.3.14}$$

An easy way to derive multivariate functions having stable or orthonormal integer translates from univariate ones is given by the tensor product: Let $\vartheta_\ell \in L^2(\mathbb{R})$, $1 \leq \ell \leq d$

be univariate functions; define their tensor product $\vartheta \in L^2(\mathbb{R}^d)$ by

$$\theta(x) := \prod_{\ell=1}^{d} \vartheta_\ell(x_\ell), \qquad x \in \mathbb{R}^d. \tag{7.3.15}$$

It is easily seen from (7.3.5) that the (multivariate) function θ has stable integer shifts if and only if all the (univariate) functions ϑ_ℓ, $1 \leq \ell \leq d$, have stable integer shifts, and θ has orthonormal integer shifts if and only if the ϑ_ℓ, $1 \leq \ell \leq d$, all have orthonormal integer shifts (up to normalization). One nice aspect of constructions of this type is that the properties of the tensor product θ can usually be derived from the univariate theory. However, they bring about a certain orientation of θ along the axes, which may be a disadvantage in some applications. Typical examples of genuinely multivariate functions having stable integer translates are provided by certain box splines:

Example 7.4 For $m \geq d$ let Ξ be a $d \times m$-matrix of rank d; denote its ℓ-th column by Ξ_ℓ, $1 \leq \ell \leq m$. The corresponding box spline M_Ξ as introduced by de Boor and Höllig (1982/83) is definded via its Fourier transform

$$M_\Xi^\wedge(\xi) := (2\pi)^{-d/2} \prod_{\ell=1}^{m} e^{-i\Xi_\ell \cdot \xi/2} \frac{\sin(\Xi_\ell \cdot \xi/2)}{\Xi_\ell \cdot \xi/2}, \qquad \xi \in \mathbb{R}^d. \tag{7.3.16}$$

The box spline M_Ξ has support on the (compact) set $\Xi \cdot [0,1]^m$; it is a piecewise polynomial function of degree at most $m - d$, and it is an element of $C_B^{\rho(\Xi)-1}(\mathbb{R}^d)$. Here $\rho(\Xi)$ is the maximal integer such that the removal of $\rho(\Xi)$ columns from Ξ does not reduce its rank (see e.g. de Boor, Höllig and Riemenschneider 1993). If in addition $\Xi \subset \mathbb{Z}^{d \times m}$ is *unimodular*, i.e., if any $d \times d$-submatrix of Ξ has determinant 0, 1 or -1, then one can show that M_Ξ has stable integer shifts (see e.g. de Boor, Höllig and Riemenschneider 1993, Chapter V).

Note that in the univariate case $d = 1$, if all entries of Ξ are chosen to be equal to 1, then the box spline M_Ξ is equal to the B-spline of degree $m - 1$. In case $d > 1$, if all columns of Ξ contain exactly one non-zero entry which is equal to 1, then M_Ξ is a tensor product of univariate B-splines. In general, however, the box-spline is not a tensor product of univariate functions.

7.3.3 Interpolation

Recall from Volume 1 that by the Whittaker–Kotel'nikov–Shannon sampling theorem (Whittaker 1915; Kotel'nikov 1933; Shannon 1949), any function $f \in PW_\pi(\mathbb{R})$ has a representation

$$f(v) = \sum_{j \in \mathbb{Z}} f(j)\text{sinc}(v - j), \qquad v \in \mathbb{R}. \tag{7.3.17}$$

As $\text{sinc}(j) = \delta_{0,j}$ for any $j \in \mathbb{Z}$, this is in fact an interpolation formula for the PSI $PW_\pi(\mathbb{R})$ of Example 7.2. Similar interpolation formulae can be derived in other PSI's as well. Interpolating splines were already studied by Schoenberg (1946, 1973); an extension to box spline spaces was given by Riemenschneider (1989). The following theorem as stated here can be found in Fischer (1995, 1996), but theorems of this type have been widely used, see e.g. Buhmann (1989), Walter (1992), Aldroubi and Unser (1992), Dahmen (1993), Steidl (1995), among many others.

Theorem 7.5 *Let $\vartheta \in C(\mathbb{R}^d)$ be a stable generator for the shift-invariant space V such that*

$$\vartheta^{\wedge *}(v) := \sum_{j \in \mathbb{Z}^d} \vartheta(j) e^{-ij \cdot v} \neq 0, \qquad v \in \mathbb{R}^d. \tag{7.3.18}$$

Then there exists a unique stable generator $g \in V \subset C_B(\mathbb{R}^d)$ such that for all $f \in V$

$$f(u) = \sum_{j \in \mathbb{Z}^d} f(j) g(u - j), \qquad u \in \mathbb{R}^d. \tag{7.3.19}$$

Moreover, $g \in \mathcal{B}(\mathbb{R}^d)$, its Fourier transform is given by

$$g^{\wedge}(v) = \vartheta^{\wedge}(v)/\vartheta^{\wedge *}(v), \qquad v \in \mathbb{R}^d, \tag{7.3.20}$$

and g has the interpolation property

$$g(j) = \delta_{0,j}, \qquad j \in \mathbb{Z}^d. \tag{7.3.21}$$

Remark 7.6 Since $\vartheta \in C(\mathbb{R}^d)$ it follows that $\vartheta^{\wedge *} \in C_{2\pi}$. Note that if ϑ itself satisfies $\vartheta(j) = \delta_{0,j}$, $j \in \mathbb{Z}^d$, then $\vartheta^{\wedge *} \equiv 1$ and $g \equiv \vartheta$ by (7.3.20).

By Poisson's formula, if $\vartheta^{\wedge} \in L^1(\mathbb{R}^d)$, then

$$\vartheta^{\wedge *}(v) := \sum_{j \in \mathbb{Z}^d} \vartheta(j) e^{-ij \cdot v} = (2\pi)^{d/2} \sum_{j \in \mathbb{Z}^d} \vartheta^{\wedge}(v + 2\pi j), \qquad v \in \mathbb{R}^d, \tag{7.3.22}$$

in case both series converge absolutely and uniformly on compact sets. Therefore it is easy to see that the B-spline of degree $r - 1$ introduced in Example 7.3 satisfies the assumptions of Theorem 7.5 for $r \in \mathbb{N}$ even. Further, if the matrix Ξ is as in Example 7.4 and in addition all the columns of Ξ have even multiplicity, then the corresponding box spline M_Ξ satisfies the assumptions of Theorem 7.5 as well. Some more examples will be introduced in the course of this chapter.

Note that the sinc-function is an orthonormal generator for the shift-invariant space $PW_\pi(\mathbb{R})$ satisfying the interpolation property (7.3.21), so at first sight one might suppose that the Whittaker–Kotel'nikov–Shannon sampling theorem (7.3.17) may be considered as a special case of Theorem 7.5. However, the assumption $\vartheta \in C(\mathbb{R}^d)$ is essential in the proof of Theorem 7.5 as given by Fischer (1995, 1996), but the sinc-function is not an element of $C(\mathbb{R}^d)$. Yet in some respects the Whittaker–Kotel'nikov–Shannon sampling theorem may be considered a limit case of Theorem 7.5. In fact, using as a starting point subsequent (weighted) convolutions ϑ_n, $n \in \mathbb{N}$, of a suitable generator ϑ with itself, then for $n \to \infty$ the sampling representation (7.3.19) turns into the Whittaker–Kotel'nikov–Shannon sampling theorem in a certain sense. In the instance of B-splines this was first proved by Schoenberg (1972, 1973, Lecture 9) and independently by Golitschek (1972). The theory was extended to box-splines by de Boor, Höllig and Riemenschneider (1986, 1993) and may be discussed in a much more general setting (Aldroubi and Unser 1992, 1993, 1994).

We will later-on see that "scaled versions" of formulae of type (7.3.19) can be used to efficiently approximate smooth functions. Let us first turn our attention to the notion of a multiresolution analysis, which is basically a sequence of closed subspaces of $L^2(\mathbb{R}^d)$ resulting from subsequent scaling of a PSI by the factor of 2.

7.4 Multiresolution analysis

7.4.1 Scaling functions

A *multiresolution analysis (MRA)* as introduced by Mallat (1989a) is a nested sequence of closed subspaces of $L^2(\mathbb{R}^d)$ satisfying

$$\cdots \subset V_{-1} \subset V_0 \subset V_1 \subset \cdots \subset L^2(\mathbb{R}^d); \tag{7.4.1}$$

$$\bigcap_{n \in \mathbb{Z}} V_n = \{0\}; \qquad \overline{\bigcup_{n \in \mathbb{Z}} V_n} = L^2(\mathbb{R}^d); \tag{7.4.2}$$

$$\text{for all } n \in \mathbb{Z}, \ f \in V_n \text{ if and only if } f(2^{-n}\cdot) \in V_0; \tag{7.4.3}$$

$$V_0 \text{ is a shift-invariant space}; \tag{7.4.4}$$

$$\text{there exists a stable generator } \vartheta \text{ for } V_0. \tag{7.4.5}$$

The function ϑ is called a *scaling function* for the MRA. If ϑ is an orthonormal generator for V_0, then it is called an *orthonormal scaling function*. Recall from Section 7.3 that the existence of a stable generator for V_0 implies the existence of an orthonormal generator, hence for any MRA there exists an orthonormal scaling function.

Let us see how we can define an MRA. Suppose $\vartheta \in L^2(\mathbb{R}^d)$ has stable integer shifts; define the spaces

$$V_n := \{f(2^n\cdot); \ f \in S(\vartheta)\}, \qquad n \in \mathbb{Z}. \tag{7.4.6}$$

These spaces obviously form a family of closed subspaces of $L^2(\mathbb{R}^d)$ satisfying (7.4.3), (7.4.4) and (7.4.5). As de Boor, DeVore and Ron (1993) show, the first condition in (7.4.2) is satisfied as well. They also show that in case the nestedness condition (7.4.1) is satisfied the second condition in (7.4.2) is equivalent to $\bigcup_{n \in \mathbb{Z}} \mathrm{supp}\, \vartheta^\wedge(2^n\cdot) = \mathbb{R}^d$ up to a set of measure zero. This condition in turn is satisfied for instance if $\vartheta^\wedge$ is continuous in a neighbourhood of zero with $\vartheta^\wedge(0) \neq 0$ or, more generally, if $\vartheta^\wedge$ is non-zero almost everywhere in a neighbourhood of zero.

Concerning the nestedness (7.4.1), note that by the definition (7.4.6) it is sufficient to show

$$V_0 \subset V_1. \tag{7.4.7}$$

From (7.4.6) it follows that V_1 is invariant under translation by half-integers, and by assumption the functions $\vartheta(\cdot - k), k \in \mathbb{Z}^d$, form a Riesz basis for $V_0 = S(\vartheta)$. Therefore (7.4.7) holds if and only if $\vartheta \in V_1$. Further, the scaled functions $\vartheta(2 \cdot -k), k \in \mathbb{Z}^d$, form a Riesz basis for the scaled space V_1, so $\vartheta \in V_1$ if and only if there exists a sequence $\{a_k\}_{k \in \mathbb{Z}^d} \in l^2(\mathbb{Z}^d)$ such that

$$\vartheta = \sum_{k \in \mathbb{Z}^d} a_j \vartheta(2 \cdot -j). \tag{7.4.8}$$

Taking Fourier transforms on both sides, this means that (7.4.7) holds if and only if there exists a function $q \in L^2_{2\pi}$ such that

$$\vartheta^\wedge = 2^{-d} q(\cdot/2) \, \vartheta^\wedge(\cdot/2). \qquad (7.4.9)$$

Any function satisfying a *refinement equation* of type (7.4.8) or, equivalently, (7.4.9) is said to be *refinable*. The function $q \in L^2_{2\pi}$ is called the *symbol* of the refinement equation, and the sequence $\{a_j\}_{j \in \mathbb{Z}^d} \in l^2(\mathbb{Z}^d)$ is the *refinement mask* of ϑ. They are related by

$$q(x) = \sum_{j \in \mathbb{Z}^d} a_j e^{-ij \cdot x} \qquad \text{a.e. on } \mathbb{R}^d. \qquad (7.4.10)$$

So far we have seen the following:

Theorem 7.7 *Let $\vartheta \in L^2(\mathbb{R}^d)$ have stable integer shifts, further suppose $\vartheta^\wedge$ is non-zero almost everywhere in a neighbourhood of zero, and let there exist a function $q \in L^2_{2\pi}$ such that (7.4.9) holds. Then the spaces $\{V_n\}_{n \in \mathbb{Z}}$ as given by (7.4.6) form an MRA with scaling function ϑ.*

7.4.2 Wavelets

Given an MRA, we may introduce the "difference space" $W_n := V_{n+1} \ominus V_n$ as the orthogonal complement of V_n in V_{n+1}, $n \in \mathbb{Z}$. By (7.4.2),

$$L^2(\mathbb{R}^d) = V_n \oplus \bigoplus_{k \geq n} W_k = \bigoplus_{k \in \mathbb{Z}} W_k, \qquad n \in \mathbb{Z}. \qquad (7.4.11)$$

It is easily checked that W_0 is a shift-invariant space. Furthermore, the scaling property (7.4.3) for the family $\{V_n\}_{n \in \mathbb{Z}}$ implies the same property for the spaces $\{W_k\}_{k \in \mathbb{Z}}$, namely $f \in W_k$ if and only if $f(2^{-k} \cdot) \in W_0$ for all $k \in \mathbb{Z}$. Therefore, given an orthonormal set of generators $\{\psi_v\}_{v \in \mathcal{I}}$ for the space W_0, the set $\{\psi_v(2^k \cdot - j)\}_{j \in \mathbb{Z}^d, v \in \mathcal{I}}$ forms an orthonormal basis for the space W_k, $k \in \mathbb{Z}$. In view of (7.4.11) this means, that the set $\{\psi_v(2^k \cdot - j)\}_{j \in \mathbb{Z}^d, k \in \mathbb{Z}, v \in \mathcal{I}}$ forms an orthonormal basis for $L^2(\mathbb{R}^d)$, so $\{\psi_v\}_{v \in \mathcal{I}}$ is a *wavelet set*. For practical purposes, orthonormality of the integer shifts of $\{\psi_v\}_{v \in \mathcal{I}}$ is often too strong a requirement that excludes certain other desirable properties. Relaxing the orthonormality condition, given a finite set of functions $\{\psi_v\}_{v \in \mathcal{I}}$ that is (only) a stable set of generators for W_0, then one can check that the set $\{\psi_v(2^k \cdot - j)\}_{j \in \mathbb{Z}^d, k \in \mathbb{Z}, v \in \mathcal{I}}$ is no longer an orthonormal basis for $L^2(\mathbb{R}^d)$, but it is still a Riesz basis for $L^2(\mathbb{R}^d)$. In this case, $\{\psi_v\}_{v \in \mathcal{I}}$ is called a *prewavelet set*. As was shown by de Boor, DeVore and Ron (1993), any prewavelet set necessarily consists of $2^d - 1$ elements. It will be convenient to index a prewavelet set by the set $E^* := E \backslash \{0\}$, where $E := \{0, 1\}^d$ is the set of vertices of the unit cube $[0, 1]^d$.

In the univariate case, $2^1 - 1 = 1$, hence the wavelet space W_0 is generated by a single wavelet ψ. In this case there exists a simple characterization of all possible wavelets corresponding to a given MRA (see e.g. Daubechies 1992, Chapter 5): Let φ be an orthonormal scaling function for an MRA with symbol $q \in L^2_{2\pi}$ given by (7.4.9). The function $\psi \in L^2(\mathbb{R}^d)$ defined by

$$\psi^\wedge(\xi) := \frac{1}{2} e^{-i\xi/2} \overline{q(\xi/2 + \pi)} \, \varphi^\wedge(\xi/2) \qquad \text{a.e. on } \mathbb{R} \qquad (7.4.12)$$

is a wavelet for this MRA, and any wavelet $\tilde{\psi}$ for the same MRA is of the form

$$\tilde{\psi}^{\wedge} = h\psi^{\wedge}, \tag{7.4.13}$$

for some function $h \in L^2_{2\pi}$ satisfying $|h(\xi)| = 1$ a.e. on $\mathbb{R}^d$. One important feature of the construction (7.4.12) is that the resulting wavelet inherits many "nice" properties of the original scaling function, for instance regularity (Meyer 1990, p. 70) and decay properties like compact support (Daubechies 1992, p. 167), exponential decay (Jia and Micchelli 1991), rapid (Meyer 1990) or polynomial decay (Fischer 1995, 1996).

In the multivariate setting, the easiest case is the tensor product case, i.e., when the scaling function φ is of the form

$$\varphi(\xi) = \prod_{\ell=1}^{d} \phi_0^{\ell}(\xi_{\ell}), \qquad \xi \in \mathbb{R}^d, \tag{7.4.14}$$

where the functions $\phi_0^{\ell} \in L^2(\mathbb{R})$, $1 \leq \ell \leq d$, are orthonormal scaling functions for a univariate MRA. Let ϕ_1^{ℓ} be corresponding (univariate) wavelets. The choice

$$\psi_{\nu}(\xi) := \prod_{\ell=1}^{d} \phi_{\nu_{\ell}}^{\ell}(\xi_{\ell}), \qquad \xi \in \mathbb{R}^d, \ \nu \in E^*, \tag{7.4.15}$$

then yields a wavelet set corresponding to the (multivariate) scaling function φ (see Meyer 1990, p. 79).

The general case, however, is not as easily solved. Several theoretical results have been obtained. Gröchenig (1987) was the first to show that the existence of a scaling function with some regularity implies the existence of a wavelet set with the same regularity (see also Meyer 1990, p. 90). The same method can be used to prove the existence of a wavelet set for an arbitrary MRA (Rückforth 1994) or to show that given a scaling function with exponential decay, rapid decay or some sort of polynomial decay there also exists a wavelet set with the same sort of decay (Fischer 1996, 1997). Using the Quillen–Suslin theorem, Jia and Micchelli (1991) prove that the existence of a scaling function with compact support implies the existence of a prewavelet set with compact support.

Although these approaches do not yield explicit formulae, we will see in Section 7.5 that these results can be useful for theoretical purposes. For numerical purposes, however, explicit formulae are necessary. Many authors have been concerned with this problem. In dimension $d \leq 3$, starting from a scaling function ϑ skew-symmetric about some centre $c \in \frac{1}{2}\mathbb{Z}^d$, Riemenschneider and Shen (1990, 1991) give a representation of a prewavelet set that is very similar to the univariate one (7.4.12). As in the univariate case, it preserves compact support, regularity and decay properties (see also Fischer, to appear, 1996). If the scaling function is orthonormal, then it yields a wavelet set. A similar construction is given by Chui, Stöckler and Ward (1992) (see also de Boor, DeVore and Ron 1993; de Boor, Höllig and Riemenschneider 1993; Stöckler 1992). These methods, however, cannot be generalized to higher dimensions. de Boor, DeVore and Ron (1993) give a construction that can be used in arbitrary dimensions; they assume that the Fourier transform of the scaling function has support on all of $\mathbb{R}^d$. Note that for compactly supported scaling functions this condition is satisfied.

7.4.3   Some examples

Example 7.8 (Shannon MRA) *Suppose the dimension $d = 1$; let $V_n := PW_{2^n\pi}(\mathbb{R})$ be the space of functions $f \in L^2(\mathbb{R})$ band-limited to the interval $[-2^n\pi, 2^n\pi]$, $n \in \mathbb{Z}$. This defines a nested sequence of closed subspaces of $L^2(\mathbb{R})$ satisfying (7.4.1), (7.4.2), (7.4.3) and (7.4.4). As for (7.4.5), we have already seen in Section 7.3 that the sinc-function is an orthonormal generator for the space V_0. Thus we have found a first example of an MRA. An application of (7.4.12) yields the Shannon wavelet $\psi(x) := 2\mathrm{sinc}(2x - 1) - \mathrm{sinc}(x - 1/2)$.*

Note that no scaling function or prewavelet for the Shannon MRA can be an element of $L^1(\mathbb{R})$: The Fourier transform of the sinc-function $\mathrm{sinc}^\wedge = (2\pi)^{-1/2}\chi_{[-\pi,\pi]}$ has support on the interval $[-\pi, \pi]$. By (7.3.3), the Fourier transform of any element of V_0 vanishes outside of this interval. On the other hand, any scaling function must also satisfy the stability condition (7.3.5), hence its Fourier transform must have discontinuities at least at one of the points $-\pi, \pi$. A function having discontinuous Fourier transform, however, cannot be an element of $L^1(\mathbb{R})$. The same argument can be applied to prewavelets, noting that the Fourier transform of the Shannon wavelet $\psi^\wedge(\xi) = (2\pi)^{-1/2}e^{-i\xi/2}(\chi_{[-2\pi,2\pi]}(\xi) - \chi_{[-\pi,\pi]}(\xi))$, $\xi \in \mathbb{R}$, has support on the intervals $[-2\pi, -\pi] \cup [\pi, 2\pi]$. A generalization avoiding this problem was given by Lemarié-Rieusset and Meyer (1986), see also Meyer (1990, p. 23). Here we will discuss a slightly more general version.

Example 7.9 (Meyer-type MRA) *Let $\theta \in L^2(\mathbb{R}^d)$ be such that $\theta^\wedge$ is non-negative a.e. on $\mathbb{R}^d$ with*

$$\theta^\wedge(y) \geq \varepsilon \text{ a.e. on } [-2\pi/3, 2\pi/3]^d \text{ for some } \varepsilon > 0; \tag{7.4.16}$$

$$\theta^\wedge \equiv 0 \text{ outside } [-4\pi/3, 4\pi/3]^d; \tag{7.4.17}$$

$$\underset{y \in [-\pi,\pi]^d}{\mathrm{ess\ inf}} \sum_{v \in E} \theta^{\wedge 2}(y + 2\pi v) > 0. \tag{7.4.18}$$

In view of (7.3.5), relation (7.4.18) implies that θ has stable integer shifts. Setting

$$q^0(y + 2j\pi) := \begin{cases} 2^d \theta^\wedge(2y)/\theta^\wedge(y), & \text{if } y \in [-2\pi/3, 2\pi/3]^d, j \in \mathbb{Z}^d, \\ 0, & \text{otherwise}, \end{cases} \tag{7.4.19}$$

then $q^0 \in L^2_{2\pi}$ satisfies the refinement equation (7.4.9). Thus by Theorem 7.7, the spaces V_n, $n \in \mathbb{Z}$, as given by (7.4.6) form an MRA.

Note that the sinc-function satisfies the assumptions upon θ of Example 7.9, thus the Shannon MRA may also be considered as an MRA of Meyer type. In general, however, for a Meyer-type MRA there may exist scaling functions with good decay. For instance, choosing $\theta^\wedge \in C_B^\infty(\mathbb{R}^d)$ then by (7.4.17) θ will have rapid decay.

Example 7.10 As a further example let us consider univariate spline spaces. For $r \in \mathbb{N}$, $n \in \mathbb{Z}$, let V_n be the space of all functions $f \in L^2(\mathbb{R}) \cap C_B^{r-2}(\mathbb{R})$ whose restriction $f \mid_{[2^{-n}m, 2^{-n}(m+1))}$ is a polynomial of degree at most $r - 1$ for all $m \in \mathbb{Z}$. This defines a nested sequence of closed subspaces of $L^2(\mathbb{R})$ satisfying (7.4.1), (7.4.2), (7.4.3) and (7.4.4). As we have seen in Example 7.3, a stable generator for V_0 is given by the cardinal B-spline B_{r-1} of degree $r - 1$, so we have found another type of MRA.

In case $r = 1$, the space V_n consists of all functions $f \in L^2(\mathbb{R})$ that are constant if restricted to an interval of the form $[2^{-n}m, 2^{-n}(m + 1))$, $m \in \mathbb{Z}$. The corresponding scaling function $B_0 = \chi_{[0,1)}$ turns out to be the characteristic function of the interval $[0, 1)$ which has orthonormal shifts. Applying (7.4.12) yields the *Haar wavelet* $\psi_H :=$ $\chi_{[0,1/2)} - \chi_{[1/2,1)}$ already known to Haar (1910).

Analogous multivariate examples may be constructed by a tensor product approach or, more generally, using box splines.

Example 7.11 Let the matrix Ξ be as in Example 7.4. The corresponding box spline M_Ξ has stable integer shifts, and by (7.3.16)

$$M_\Xi^\wedge(\xi) = \left(\prod_{\ell=1}^{m} \frac{1 + e^{-i\Xi_\ell \cdot \xi/2}}{2} \right) M_\Xi^\wedge(\xi/2), \qquad \xi \in \mathbb{R}^d. \qquad (7.4.20)$$

Thus by Theorem 7.7, the spaces V_n defined by (7.4.6) with $\vartheta = M_\Xi$ form an MRA.

Note that the box spline M_Ξ is symmetric about its centre $c_\Xi := \sum_{\ell=1}^{m} \Xi_\ell/2 \in \frac{1}{2}\mathbb{Z}^d$, therefore in dimension $d \leq 3$, prewavelet sets can be constructed using the elegant methods of Riemenschneider and Shen (1990, 1991) or Chui, Stöckler and Ward (1992) mentioned above.

Remark 7.12 Of particular interest in numerical analysis are compactly supported wavelets. For the univariate case these were studied in detail by Daubechies (1988). Her construction relies on the following observation: Let $d = 1$ again. Iteration of the refinement relation (7.4.9) leads to

$$\vartheta^\wedge(v) = \prod_{\ell=1}^{m} \frac{1}{2} q(2^{-\ell}v)\vartheta^\wedge(2^{-m}v), \qquad v \in \mathbb{R}, \ m \in \mathbb{N}. \qquad (7.4.21)$$

If the product converges for $m \to \infty$ and if $\vartheta^\wedge$ is continuous in a neighbourhood of zero, then ϑ can be described in terms of its symbol (up to the constant $\vartheta^\wedge(0)$)

$$\vartheta^\wedge(v) = \prod_{\ell=1}^{\infty} \frac{1}{2} q(2^{-\ell}v)\vartheta^\wedge(0), \qquad v \in \mathbb{R}. \qquad (7.4.22)$$

Conversely, one may use as starting point a trigonometric function $q \in L^2_{2\pi}$ such that the right-hand side of (7.4.22) converges and then proceed to *define* a function ϑ by (7.4.22). This function will automatically satisfy the refinement relation (7.4.9). As I. Daubechies observed (see also Daubechies (1992, p. 193)), choosing a suitable trigonometric *polynomial* q, this method leads to a compactly supported orthonormal scaling function ϑ. A compactly supported wavelet ψ may then be constructed using the standard method (7.4.12). I. Daubechies also derived a useful algorithm for the numerical calculation of ψ. We will not go into the details here.

7.5 Sampling approximation

7.5.1 *Motivation and general theory*

Let us return to the interpolation formulae of Section 7.3, this time focussing on MRA's. Suppose $\vartheta \in C(\mathbb{R}^d)$ is a scaling function for an MRA such that

$$\vartheta^{\wedge *}(v) = \sum_{j \in \mathbb{Z}^d} \vartheta(j) e^{-ij \cdot v} \neq 0, \qquad v \in \mathbb{R}^d, \tag{7.5.1}$$

then by Theorem 7.5 there exists a further scaling function $g \in C_B(\mathbb{R}^d) \cap \mathcal{B}(\mathbb{R}^d)$ for the same MRA, uniquely determined by

$$g^{\wedge}(v) := \vartheta^{\wedge}(v) / \vartheta^{\wedge *}(v), \qquad v \in \mathbb{R}^d, \tag{7.5.2}$$

such that for all $f \in V_0$

$$f(u) = \sum_{j \in \mathbb{Z}^d} f(j) g(u - j), \qquad u \in \mathbb{R}^d. \tag{7.5.3}$$

Note the analogy of this formula to the Whittaker–Kotel'nikov–Shannon sampling theorem (7.3.17): V_0 may be viewed as a space of functions band-limited in the wavelet sense, i.e., for which the wavelet coefficients $\langle f, 2^{kd/2} \psi(2^k \cdot -j) \rangle$ vanish for all "frequencies" $k \geq 0$.

Using the scaling property (7.4.3), formula (7.5.3) can be "lifted" to the general level V_n, $n \in \mathbb{N}$: For $f \in V_n$ arbitrary, the function $f(2^{-n} \cdot)$ is an element of V_0, so an application of (7.5.3) yields

$$f(u) = \sum_{j \in \mathbb{Z}^d} f(2^{-n} j) g(2^n u - j), \qquad u \in \mathbb{R}^d, \; f \in V_n. \tag{7.5.4}$$

Thus we have found sampling representations on all the levels V_n, $n \in \mathbb{N}$. Now $\{V_n\}_{n \in \mathbb{Z}}$ is a nested sequence of spaces whose union is dense in $L^2(\mathbb{R}^d)$, so it seems reasonable to ask whether operators as defined by the right-hand side of (7.5.4) may be used for the approximation of arbitrary elements of $L^2(\mathbb{R}^d) \cap C_B(\mathbb{R}^d)$. This problem has been studied by Fischer (1995, 1996); in fact, it is possible to prove Jackson-type inequalities for these operators exploiting known results concerning the approximation properties of an MRA. As it turns out, formulae of type (7.5.4) can even be used to approximate functions $f \in C_B(\mathbb{R}^d)$ that are not elements of $L^2(\mathbb{R}^d)$. We will first give a rather general result and then check that it applies to our sampling operators. We will need the following condition: A finite set of functions $\{\vartheta_v\}_{v \in \mathcal{I}} \subset \mathcal{L}^1_{r-1}(\mathbb{R}^d)$ is said to satisfy the *vanishing moment condition of order* $r \in \mathbb{N}$ iff

$$D^\alpha \vartheta_v^\wedge(0) = 0, \qquad |\alpha| < r, \; v \in \mathcal{I}. \tag{7.5.5}$$

Note that this is equivalent to $\int_{\mathbb{R}^d} u^\alpha \vartheta_v(u) \, du = 0, |\alpha| < r, \; v \in \mathcal{I}$, i.e., the vanishing of the integral moments of the functions ϑ_v, $v \in \mathcal{I}$, up to the order $r - 1$.

Theorem 7.13 *Let* $r \in \mathbb{N}$ *and let* $\{V_n\}_{n \in \mathbb{Z}}$ *be an MRA either of Meyer type or containing a prewavelet set* $\{\psi_v\}_{v \in E^*} \subset \mathcal{B}_r(\mathbb{R}^d) \cap C_B(\mathbb{R}^d)$ *satisfying the vanishing moment condition of order* r. *Suppose* $\{T_n\}_{n \in \mathbb{Z}}$ *is a uniformly bounded sequence of linear operators from* $C_B(\mathbb{R}^d)$ *into* $B(\mathbb{R}^d)$ *such that* T_n *reproduces identically any function* $f \in V_n \cap C_B(\mathbb{R}^d)$, $n \in \mathbb{Z}$.

(a) *There exists a constant* $K > 0$ *such that for any* $f \in L^2(\mathbb{R}^d) \cap C_B^r(\mathbb{R}^d)$,

$$\|f - T_n f\|_B \leq K \max_{|\alpha| = r} \|D^\alpha f\|_C \cdot 2^{-nr}, \qquad n \in \mathbb{Z}. \tag{7.5.6}$$

(b) *Suppose the sequence $\{T_n\}_{n\in\mathbb{Z}}$ also satisfies the following condition:*
For any (fixed) $x \in \mathbb{R}^d$ and any uniformly bounded sequence $\{q_k\}_{k\in\mathbb{N}} \subset C_B^r(\mathbb{R}^d)$
with $q_k(u) \equiv 0$ on $B_k(x)$, $k \in \mathbb{N}$, there holds

$$\lim_{k\to\infty}(T_n q_k)(x) = 0, \qquad n \in \mathbb{Z}. \tag{7.5.7}$$

Then (7.5.6) holds for all $f \in C_B^r(\mathbb{R}^d)$.

Let us first discuss the meaning of (7.5.7). This condition ensures that each of the operators T_n is local in the sense that disturbances far enough away from a point $x \in \mathbb{R}^d$ will not change $(T_n f)(x)$ very much; in other words it describes a very weak version of local continuity. As we will see it is satisfied for most sampling operators bounded on $C_B(\mathbb{R}^d)$.

With the help of Theorem 7.13, several classes of operators including quasi-interpolation type operators are studied by Fischer (1995, 1996). Our main interest here is in sampling operators; before turning to an application of Theorem 7.13 to these let us study MRAs satisfying its assumptions.

7.5.2 Suitable MRAs

Let us examine which of the MRA's introduced in Section 7.4 satisfy the assumptions of Theorem 7.13. Obviously the Meyer-type MRA of Example 7.9 does, and so does the Shannon MRA of Example 7.8, which may be considered to be a special case of a Meyer-type MRA. To evaluate the remaining examples it is helpful to first discuss the relationship between the vanishing moment condition and another condition frequently used in this context: A function $\vartheta \in \mathcal{L}_{r-1}^1(\mathbb{R}^d)$ satisfies the *Strang–Fix condition of order* $r \in \mathbb{N}$ iff

$$D^\alpha\vartheta^\wedge(2j\pi) = 0, \quad j \in \mathbb{Z}^d \setminus \{0\}, \ |\alpha| < r; \qquad \vartheta^\wedge(0) \neq 0. \tag{7.5.8}$$

Both the Strang–Fix condition and the vanishing moment condition play an important role in approximation theory. The Strang–Fix condition was used by Strang and Fix (1973) to study "controlled approximation" by FSIs with compactly supported generators in $L^2(\mathbb{R}^d)$. The theory was later extended to L^p-approximation and spaces with generators satisfying weaker decay requirements (Dahmen and Micchelli 1984; de Boor and Jia 1985; de Boor 1990; Light and Cheney 1992; Halton and Light 1993; Jia and Lei 1993; Lei 1994b). The Strang–Fix condition is also important in the theory of interpolation, sampling and quasi-interpolation (Schoenberg 1973; Ries and Stens 1984; Butzer, Splettstößer and Stens 1988; Buhmann 1989; Butzer et al. 1992, 1990; Dyn et al. 1992; Burchard et al. to appear; Burchard and Lei 1995, among many others). In this context it was already used by Schoenberg (1946) long before G. Strang und G. Fix.

Wavelets or prewavelets satisfying the vanishing moment condition were used e.g. by Sweldens and Piessens (to appear) and Fischer (1995, 1996) to characterize the approximation behaviour of an MRA or associated operators (compare Theorem 7.13); Beylkin, Coifman and Rokhlin (1991) use wavelets with vanishing moments to compress large matrices.

In the context of an MRA the two concepts are basically equivalent. This was first shown for r-regular functions, i.e. functions with rapid decay or at least polynomial decay of sufficiently high order and continuous derivatives up to order $r - 1$ with the same decay properties (Meyer 1990). In fact, r-regularity of a scaling function implies that it satisfies the Strang–Fix condition of order r (Meyer 1990, p. 56; Cavaretta et al. 1991, p. 158), while r-regularity of a wavelet set implies that it satisfies the vanishing moment condition of order r (Meyer 1990, p. 93; Daubechies 1992, p. 153). Furthermore, one can show that the existence of an r-regular scaling function implies the existence of an r-regular wavelet set (Gröchenig 1987; Meyer 1990).

However, one can show that the two concepts are equivalent even when no regularity is present, assuming only sufficient decay of the functions involved to ensure that the corresponding Fourier transforms are smooth enough to give meaning to the definitions (7.5.5) and (7.5.8). Some of the implications involved have been treated by several authors under various assumptions (Mallat 1989a; Strang 1989; Meyer 1990; Daubechies 1992; Stöckler 1992). The univariate case is rather easily solved (Butzer et al. 1994) exploiting the explicit representation (7.4.12) of a wavelet, while the multivariate case is more complicated since in general a comparable explicit representation of a wavelet set is not known. The following rather general multivariate result is established by Fischer (to appear, 1996) using an approximation theory approach.

Theorem 7.14 *Let $r \in \mathbb{N}$, and suppose $\vartheta \in \mathcal{B}_r(\mathbb{R}^d)$ is a scaling function for an MRA. The following are equivalent:*

$$\vartheta \text{ satisfies the Strang–Fix condition of order } r; \tag{7.5.9}$$

there exists an orthonormal scaling function $\varphi \in \mathcal{B}_r(\mathbb{R}^d)$ *satisfying*

$$\text{the Strang–Fix condition of order } r; \tag{7.5.10}$$

any function $f \in V_0 \cap \mathcal{B}_r(\mathbb{R}^d)$ *satisfies*

$$D^\alpha f^\wedge(2j\pi) = 0, \qquad j \in \mathbb{Z}^d \setminus \{0\}, \ \alpha \in \mathbb{N}_0^d, \ |\alpha| < r; \tag{7.5.11}$$

there exists a prewavelet set $\{\psi_\nu\}_{\nu \in E^*} \subset \mathcal{B}_r(\mathbb{R}^d)$ *satisfying the*

$$\text{vanishing moment condition of order } r; \tag{7.5.12}$$

there exists a wavelet set $\{\psi_\nu\}_{\nu \in E^*} \subset \mathcal{B}_r(\mathbb{R}^d)$ *satisfying the*

$$\text{vanishing moment condition of order } r; \tag{7.5.13}$$

any function $f \in W_0 \cap \mathcal{B}_r(\mathbb{R}^d)$ *satisfies*

$$D^\alpha f^\wedge(4j\pi) = 0, \qquad j \in \mathbb{Z}^d, \ \alpha \in \mathbb{N}_0^d, \ |\alpha| < r. \tag{7.5.14}$$

Note that we postulate the existence of a scaling function $\vartheta \in \mathcal{B}_r(\mathbb{R}^d)$, so that assertion (7.5.11) is not vacuous. One can show that the existence of a scaling function $\vartheta \in \mathcal{B}_r(\mathbb{R}^d)$ implies the existence of a wavelet set $\{\psi_\nu\}_{\nu \in E^*} \subset \mathcal{B}_r(\mathbb{R}^d)$ (Fischer, to appear, 1996); hence assertion (7.5.14) is not vacuous either and in fact trivially implies (7.5.13). We do not know at present whether the converse is true, i.e., whether the existence of a prewavelet set $\{\psi_\nu\}_{\nu \in E^*} \subset \mathcal{B}_r(\mathbb{R}^d)$ implies the existence of a scaling function $\vartheta \in \mathcal{B}_r(\mathbb{R}^d)$. Some results in this respect were obtained by Lemarié (1991a, 1991b); he shows that in the univariate case the existence of a wavelet with compact support implies the existence of a scaling function with compact support. We do not know of any results concerning weaker decay properties, such as exponential or polynomial decay. Therefore to formulate a meaningful theorem we cannot renounce the a priori assumption $\vartheta \in B_r(\mathbb{R}^d)$.

Observing that $B_r(\mathbb{R}^d) \cap C_B(\mathbb{R}^d) \subset C(\mathbb{R}^d)$ (Fischer 1996), Theorem 7.14 together with (7.3.8) show that the existence of a scaling function $\vartheta \in B_r(\mathbb{R}^d) \cap C_B(\mathbb{R}^d)$ satisfying the Strang–Fix-condition of order r suffices to ensure the existence of a prewavelet set $\{\psi_v\}_{v \in E^*} \subset B_r(\mathbb{R}^d) \cap C_B(\mathbb{R}^d)$ satisfying the vanishing moment condition of order r. Returning to our original question we are now ready to tackle the remaining examples of Section 7.4. Concerning Example 7.10, given $r \in \mathbb{N}$, then for any $s \geq \max\{1, r-1\}$ the B-spline B_s of degree s is a compactly supported and continuous function which by (7.3.13) satisfies the Strang–Fix condition of order $s+1$. So this MRA can be used in Theorem 7.13. Similarly, if Ξ is a unimodular $d \times m$-matrix with $\rho(\Xi) \geq r-1$, then the corresponding box spline M_Ξ of (7.3.16) satisfies the Strang–Fix condition of order r (see Butzer et al. 1990), so the corresponding MRA of Example 7.11 may also be used. As to the MRA of Daubechies in Remark 7.12, it can be shown that the scaling function ϑ satisfies the Strang–Fix condition of order r iff its symbol q satisfies $D^\alpha q(\pi) = 0$, $0 \leq \alpha < r$ (see e.g. Butzer et al. 1994); in other words, the trigonometric polynomial q must contain the factor $(1 + e^{-i\xi})^r$.

7.5.3 Sampling operators

Let us now derive a "sampling version" of Theorem 7.13.

Theorem 7.15 *Let $r \in \mathbb{N}$ and let $\{V_n\}_{n \in \mathbb{Z}}$ be an MRA either of Meyer type or containing a prewavelet set $\{\psi_v\}_{v \in E^*} \subset B_r(\mathbb{R}^d) \cap C_B(\mathbb{R}^d)$ satisfying the vanishing moment condition of order r. Further let there exist for each $n \in \mathbb{Z}$ a sequence $\{t_{n,j}\}_{j \in \mathbb{Z}^d} \subset \mathbb{R}^d$ and functions $g_{n,j}$, $j \in \mathbb{Z}^d$, satisfying*

$$\sup_{n \in \mathbb{Z}^d} \sup_{u \in \mathbb{R}^d} \sum_{j \in \mathbb{Z}^d} |g_{n,j}(u)| < \infty, \tag{7.5.15}$$

such that for any $f \in V_n \cap C_B(\mathbb{R}^d)$ there holds the pointwise sampling representation

$$f(u) = \sum_{j \in \mathbb{Z}^d} f(t_{n,j}) g_{n,j}(u), \qquad u \in \mathbb{R}^d. \tag{7.5.16}$$

For $n \in \mathbb{Z}$, define the operator $S_n : C_B(\mathbb{R}^d) \longrightarrow B(\mathbb{R}^d)$ by

$$(S_n f)(u) := \sum_{j \in \mathbb{Z}^d} f(t_{n,j}) g_{n,j}(u), \qquad u \in \mathbb{R}^d, \ f \in C_B(\mathbb{R}^d). \tag{7.5.17}$$

Then there exists a constant $K > 0$ such that

$$\|f - S_n f\|_B \leq K \max_{|\alpha| = r} \|D^\alpha f\|_C \cdot 2^{-nr}, \qquad n \in \mathbb{Z}, \ f \in C_B^r(\mathbb{R}^d). \tag{7.5.18}$$

Proof It is easy to check that the operators S_n as given by (7.5.17) are bounded linear operators from $C_B(\mathbb{R}^d)$ to $B(\mathbb{R}^d)$ satisfying $\|S_n\| \leq \sup_{u \in \mathbb{R}^d} \sum_{j \in \mathbb{Z}^d} |g_{n,j}(u)|$, $n \in \mathbb{Z}$. Therefore by (7.5.15) and (7.5.16) the sampling operators $\{S_n\}_{n \in \mathbb{Z}}$ form a uniformly bounded sequence of operators from $C_B(\mathbb{R}^d)$ to $B(\mathbb{R}^d)$ such that S_n reproduces identically any function $f \in V_n \cap C_B(\mathbb{R}^d)$, $n \in \mathbb{Z}$. Thus, in view of Theorem 7.13 we only need to check (7.5.7). Choose $x \in \mathbb{R}^d$, and let $\{q_k\}_{k=1}^{\infty} \subset C_B(\mathbb{R}^d)$ be uniformly bounded with $q_k(u) = 0$ for all $u \in B_k(x)$. Then for any (fixed) $n \in \mathbb{Z}$,

$$|(S_n q_k)(x)| \le \sum_{j \in \mathbb{Z}^d} |q_k(t_{n,j})||g_{n,j}(x)| \le \|q_k\|_C \sum_{\substack{j \in \mathbb{Z}^d \\ t_{n,j} \notin B_k(x)}} |g_{n,j}(x)|,$$

which by (7.5.15) and the uniform boundedness of $\{q_k\}_{k=1}^{\infty}$ tends to zero as $k \to \infty$.

Remark 7.16 Note that in case $\{t_{n,j}\}_{j \in \mathbb{Z}^d}$ does not have any cluster points and $t_{n,\nu} \ne t_{n,\mu}$ for all $\nu, \mu \in \mathbb{Z}^d$ with $\nu \ne \mu$ there actually holds $\|S_n\| = \sup_{u \in \mathbb{R}^d} \sum_{j \in \mathbb{Z}^d} |g_{n,j}(u)|$, $n \in \mathbb{Z}$. Thus for "reasonable" sequences, assumption (7.5.15) is necessary for the uniform boundedness of the sequence $\{S_n\}_{n \in \mathbb{Z}}$ and we get the additional assumption (7.5.7) of Theorem 7.13 for free.

As we have seen, the Shannon MRA $\{V_n\}_{n \in \mathbb{Z}} = \{PW_{2^n\pi}(\mathbb{R})\}_{n \in \mathbb{Z}}$ of Example 7.8 is of Meyer type, so we might first try to apply Theorem 7.15 to representations derived from the Whittaker–Kotel'nikov–Shannon sampling theorem

$$f(x) = \sum_{j \in \mathbb{Z}} f(2^{-n}j)\text{sinc}(2^n x - j), \qquad x \in \mathbb{R}, \ f \in PW_{2^n\pi}(\mathbb{R}), \ n \in \mathbb{Z}. \quad (7.5.19)$$

This is a representation of type (7.5.16) with $g_{n,j} := \text{sinc}(2^n \cdot -j)$ and $t_{n,j} := 2^{-n}j$, $n, j \in \mathbb{Z}$. Unfortunately the sinc-function is not an element of $B(\mathbb{R}^d)$, therefore (7.5.15) cannot be fulfilled and there is no natural extension of these representations to bounded linear operators on $C_B(\mathbb{R}^d)$. However, Benedetto and Heller (1990) and Benedetto (1992, 1994) give similar sampling representations for functions in $PW_\pi(\mathbb{R})$ that do yield applications of Theorem 7.15. The boundedness of the corresponding operators is paid for with *oversampling*, namely a denser set of sampling points than in the Shannon case. The following theorem was established by J.J. Benedetto and W. Heller in a slightly more general form, but for the univariate case only; their proof is easily transferred to the multivariate version as given here. All the necessary tools may be found in Benedetto and Walnut (1994).

Theorem 7.17 (Benedetto 1994) *Let* $\pi < R < \infty$, $\omega \in PW_R(\mathbb{R}^d)$ *such that* $\omega^\wedge$ *is continuous on* $\mathbb{R}^d$ *with* $\omega^\wedge \equiv 1$ *on the interval* $[-\pi, \pi]^d$, *and* $\omega^\wedge(x) > 0$ *for all* $x \in [-R, R]^d$. *Further suppose* $\pi + R \le b < 2R$. *Setting*

$$\Omega(x) := \sum_{j \in \mathbb{Z}^d} |\omega^\wedge(x - jb)|^2, \qquad x \in \mathbb{R}^d,$$

$$g^\wedge := (2\pi)^{-d/2}\omega^\wedge/\Omega, \qquad\qquad\qquad (7.5.20)$$

then $g \in PW_R(\mathbb{R}^d)$, $g^\wedge \equiv (2\pi)^{-d/2}$ *on* $[-\pi, \pi]^d$, *and every function* $f \in PW_\pi(\mathbb{R}^d)$ *has a representation*

$$f = \frac{\pi^d}{R^d} \sum_{j \in \mathbb{Z}^d} f\left(\frac{j\pi}{R}\right) g\left(\cdot - \frac{j\pi}{R}\right) \qquad\qquad (7.5.21)$$

with convergence of the series in $L^2(\mathbb{R}^d)$.

Note that for $n \in \mathbb{Z}$, scaling by the factor of 2^n yields the representations

$$f = \frac{\pi^d}{R^d} \sum_{j \in \mathbb{Z}^d} f\left(\frac{2^{-n} j\pi}{R}\right) g\left(2^n \cdot - \frac{j\pi}{R}\right), \qquad f \in PW_{2^n\pi}(\mathbb{R}), \qquad (7.5.22)$$

in the $L^2(\mathbb{R}^d)$-sense. Obviously these representations are of type (7.5.16) with sampling points $t_{n,j} := 2^{-n} j\pi/R$ and "weights" $g_{n,j} := g(2^n \cdot -j\pi/R)$. One can show (Fischer 1996) that the additional assumption $(1 + \|\cdot\|)^{d+\epsilon} \omega \in B(\mathbb{R}^d)$ for some $\varepsilon > 0$ suffices to guarantee that these representations in fact hold in the pointwise sense and that the corresponding operators

$$S_n f := \frac{\pi^d}{R^d} \sum_{j \in \mathbb{Z}^d} f\left(\frac{2^{-n} j\pi}{R}\right) g\left(2^n \cdot - \frac{j\pi}{R}\right), \qquad n \in \mathbb{Z}, f \in C_B(\mathbb{R}^d), \qquad (7.5.23)$$

form a bounded family of operators on $C_B(\mathbb{R}^d)$. Note that this additional requirement upon ω can be met by requiring sufficient smoothness of $\omega^\wedge$. An application of Theorem 7.15 yields

Corollary 7.18 Let R, ω, g be as in Theorem 7.17; further suppose $(1 + \|\cdot\|)^{d+\epsilon} \omega \in B(\mathbb{R}^d)$ for some $\varepsilon > 0$, and let the operators S_n, $n \in \mathbb{Z}$, be given by (7.5.23). For any $r \in \mathbb{N}$ there exists a constant K such that

$$\|f - S_n f\|_B \leq K \max_{|\alpha|=r} \|D^\alpha f\|_C \cdot 2^{-nr}, \qquad n \in \mathbb{Z}, \ f \in C_B^r(\mathbb{R}^d). \qquad (7.5.24)$$

A suitable sampling representation for functions in a more general Meyer-type MRA is given by Walter (1993).

7.5.4 *Interpolating sampling operators*

Returning to our starting point we might also try to use Theorem 7.15 to investigate the approximation behaviour of operators derived from the interpolation formula (7.5.4). Obviously, (7.5.4) is a representation of type (7.5.16) with sampling points $t_{n,j} := 2^{-n} j$ and "weights" $g_{n,j} := g(2^n \cdot -j)$, $j \in \mathbb{Z}^d, n \in \mathbb{Z}$. The corresponding sampling operators (7.5.17) are given by

$$(S_n f)(u) := \sum_{j \in \mathbb{Z}^d} f(2^{-n} j) g(2^n u - j), \qquad u \in \mathbb{R}^d, \ f \in C_B(\mathbb{R}^d). \qquad (7.5.25)$$

To be able to apply Theorem 7.15 we need to check (7.5.15). In this respect,

$$\sup_{n \in \mathbb{Z}} \sup_{u \in \mathbb{R}^d} \sum_{j \in \mathbb{Z}^d} |g_{n,j}(u)| = \sup_{n \in \mathbb{Z}} \sup_{u \in \mathbb{R}^d} \sum_{j \in \mathbb{Z}^d} |g(2^n u - j)|$$

$$= \sup_{u \in \mathbb{R}^d} \sum_{j \in \mathbb{Z}^d} |g(u - j)| < \infty$$

since $g \in \mathcal{B}(\mathbb{R}^d)$ by Theorem 7.5, so (7.5.15) is satisfied. From (7.5.4) and Theorem 7.15 we conclude

Corollary 7.19 *Let* $r \in \mathbb{N}$ *and let* $\{V_n\}_{n \in \mathbb{Z}}$ *be an MRA either of Meyer type or containing a prewavelet set* $\{\psi_v\}_{v \in E^*} \subset \mathcal{B}_r(\mathbb{R}^d) \cap C_B(\mathbb{R}^d)$ *satisfying the vanishing moment condition of order* r. *Further suppose* $\theta \in C(\mathbb{R}^d)$ *is a corresponding scaling function satisfying* (7.5.1), *and define the sampling operators* S_n, $n \in \mathbb{Z}$, *by* (7.5.25) *with* g *given by* (7.5.2). *Then there exists a constant* $K > 0$ *such that*

$$\|f - S_n f\|_B \leq K \max_{|\alpha|=r} \|D^\alpha f\|_C \cdot 2^{-nr}, \qquad n \in \mathbb{Z}, \ f \in C_B^r(\mathbb{R}^d). \qquad (7.5.26)$$

As mentioned in Section 7.3 (compare also Section 7.5.2), the scaling function ϑ may be chosen to be a B-spline of odd degree or a box spline M_Ξ corresponding to a matrix Ξ satisfying the assumptions of Example 7.4, all of whose columns have even multiplicity. In view of (7.3.22), any scaling function $\vartheta \in C(\mathbb{R}^d)$ generating a Meyer-type MRA also satisfies the assumptions of Corollary 7.19.

Recall from Remark 7.6 that if ϑ itself satisfies $\vartheta(j) = \delta_{0,j}$, $j \in \mathbb{Z}^d$, then the interpolating scaling function g agrees with ϑ. In the univariate setting such interpolating scaling functions have been much studied. By a tensor product approach one may also construct multivariate interpolating scaling functions. Using a method based on the approach of I. Daubechies as sketched in Remark 7.12, Rückforth (1994) constructs compactly supported interpolating scaling functions satisfying the Strang–Fix condition (7.5.8) of some order $r \in \mathbb{N}$. By Theorem 7.14 these yield an application of Corollary 7.19.

Of particular interest in the literature have been *orthonormal* interpolating scaling functions. Lemarié (1990) gives orthonormal interpolating scaling functions with rapid decay. Xia and Zhang (1993) even give constructions of orthonormal interpolating scaling functions with exponential decay; by a different method Lewis (1994) constructs orthonormal interpolating scaling functions with exponential decay and corresponding wavelets with vanishing moments up to some degree $r \in \mathbb{N}$. He also discusses their regularity and symmetry properties. It can be shown, however (Xia and Zhang 1993; Lewis 1994; Goodman and Micchelli 1994), that there only exists one orthonormal interpolating scaling function with compact support, namely the generator $\chi_{[0,1)}$ of the Haar MRA (see Example 7.10). Goodman and Micchelli (1994) give a characterization of orthonormal interpolating scaling functions that are symmetric about a point x. They show that the symbol of the corresponding refinement equation is necessarily discontinuous. This in turn implies the non-existence of any scaling function $\vartheta \in C(\mathbb{R}^d)$ possessing these three properties (Fischer 1996).

Note that the operators occurring in both Corollary 7.18 and Corollary 7.19 are quasi-interpolation type operators, i.e., operators of the form

$$(T_W^{\lambda,g} f)(x) := \sum_{j \in \mathbb{Z}^d} \lambda\left(f\left(\frac{\cdot + j}{W}\right)\right) g(Wx - j), \qquad x \in \mathbb{R}^d, \ f \in C_B(\mathbb{R}^d), \quad (7.5.27)$$

where $\lambda : C_B(\mathbb{R}^d) \longrightarrow \mathbb{C}$ is a bounded linear functional, $g \in \mathcal{B}(\mathbb{R}^d)$ and $W \in \mathbb{R}_+^d$. Recall that for $x, y \in \mathbb{R}^d$, x/y denotes the vector $(x_1/y_1, \ldots, x_d/y_d)^{tr}$, while $xy := (x_1 y_1, \ldots, x_d y_d)^{tr}$. In our setting, λ is the point functional defined by $\lambda(f) := f(0)$, $f \in C_B(\mathbb{R}^d)$, and $W := 2^n(1, \ldots, 1)^{tr}$. For this type of operator one can show

(Fischer 1996) that the Jackson-type inequalities (7.5.26) or (7.5.24) imply the more general version

$$\| S_W^g f - f \|_B \le K \max_{|\alpha|=r} \frac{\| D^\alpha f \|_C}{\alpha!} W^{-\alpha}, \quad f \in C_B^r(\mathbb{R}^d),\ W \in \mathbb{R}_+^d, \tag{7.5.28}$$

with the *generalized sampling series*

$$S_W^g f(x) := \sum_{j \in \mathbb{Z}^d} f\left(\frac{j}{W}\right) g(Wx - j), \quad x \in \mathbb{R}^d,\ f \in C_B(\mathbb{R}^d),\ W \in \mathbb{R}_+^d. \tag{7.5.29}$$

Sampling operators of this form constitute an important type of quasi-interpolation operator and have been studied by mathematicians since about 1975. A good survey especially of the univariate theory is given by Butzer, Splettstößer and Stens (1988); concerning multivariate results see e.g. Buhmann (1989), Jackson (1989), Butzer et al. (1992, 1990), Fischer and Stens (1990), Dyn et al. (1992). A common procedure to prove estimates of type (7.5.28) or (7.5.26) is to require polynomial reproduction of some order $r \in \mathbb{N}$ and approximate the function f by suitable polynomials. However, polynomials p of positive degree do not belong to $C_B(\mathbb{R}^d)$, hence $S_n p$ is not a priori well defined, and unless $g \in \mathcal{B}_r(\mathbb{R}^d)$ there does not exist an obvious extension. Our approach avoids this problem by requiring reproduction of elements in $V_n \cap C_B(\mathbb{R}^d)$ instead and approximating f locally by such functions. Let us study an example where this is of significance.

Let F_d be Fejér's kernel

$$F_d(x) := \frac{1}{(\sqrt{2\pi})^d} \prod_{\ell=1}^d \left(\frac{\sin(x_\ell/2)}{x_\ell/2} \right)^2, \quad x \in \mathbb{R}^d,$$

with Fourier transform

$$F_1^\wedge(v) = \begin{cases} 1 - |v|, & |v| \le 1 \\ 0, & |v| > 1,\ v \in \mathbb{R}, \end{cases}$$

$$F_d^\wedge(y) = \prod_{\ell=1}^d F_1^\wedge(y_\ell), \quad y \in \mathbb{R}^d,$$

(Butzer and Nessel 1971, p. 516). Here the index d denotes the dimension. Setting

$$F_{d,\rho} := F_d(\rho \cdot), \quad \rho \in \mathbb{R}_+,$$

one easily checks that for $\pi < \rho \le 4\pi/3$, this scaled version $F_{d,\rho} \in C(\mathbb{R}^d)$ of Fejér's kernel is a scaling function for an MRA of Meyer's type satisfying (7.5.1), so Corollary 7.19 can be applied. The corresponding interpolating scaling function $g_{d,\rho}$ as defined by (7.5.2) turns out to be

$$g_{1,\rho} = \frac{1}{\sqrt{2\pi}} \left(\frac{\rho^2}{2(\rho - \pi)} F_{1,\rho} - \frac{(2\pi - \rho)^2}{2(\rho - \pi)} F_{1,(2\pi-\rho)} \right),$$

$$g_{d,\rho}(x) = \prod_{\ell=1}^d g_{1,\rho}(x_\ell), \quad x \in \mathbb{R}^d,\ \pi < \rho \le 4\pi/3. \tag{7.5.30}$$

An application of Corollary 7.19 now yields

Corollary 7.20 *For* $\pi < \rho \leq 4\pi/3$ *let* $g_{d,\rho}$ *be given by (7.5.30). Further let* $W^{(n)} :=$ $2^n(1,\ldots,1)^{tr}$ *and define the operators* $S_n := S_{W^{(n)}}^{g_{d,\rho}}$, $n \in \mathbb{N}$. *For any* $r \in \mathbb{N}$ *and* $f \in C_B^r(\mathbb{R}^d)$ *there exists a constant* K *such that*

$$\|S_n f - f\|_B \leq K \max_{|\alpha|=r} \|D^\alpha f\|_C \cdot 2^{-nr}, \qquad n \in \mathbb{Z},$$

$$\|S_W^{g_{d,\rho}} f - f\|_B \leq K \max_{|\alpha|=r} \frac{\|D^\alpha f\|_C}{\alpha!} W^{-\alpha}, \qquad W \in \mathbb{R}_+^d.$$

Note that in the limit case $\rho = 4\pi/3$ we have in particular

$$g_{d,4\pi/3} = \frac{(2\pi)^{d/2}}{3^d} \Theta_{d,2\pi/3},$$

with de la Vallée Poussin's kernel

$$\Theta_d(x) := \left(\frac{4}{\sqrt{2\pi}}\right)^d \prod_{\ell=1}^d \left(\frac{\sin(x_\ell/2)\sin(3x_\ell/2)}{x_\ell^2}\right), \qquad x \in \mathbb{R}^d,$$

and

$$\Theta_{d,\rho} := \Theta_d(\rho \cdot), \qquad \rho \in \mathbb{R}_+.$$

Note that both Fejér's kernel and de la Vallée Poussin's kernel are elements of $\mathcal{B}_0(\mathbb{R}^d)$ but not of $\mathcal{B}_r(\mathbb{R}^d)$ for any $r > 0$; therefore Corollary 7.20 cannot be derived by any argument involving polynomial reproduction. However, both kernels are band-limited, and in this particular instance it is possible to estimate the approximation error of the generalized sampling series using known results concerning the approximation behaviour of the corresponding singular integral of Fejér's type (Splettstößer 1978, 1979; Stens 1980; Splettstößer et al. 1981; Butzer and Stens 1983; Butzer, Splettstößer and Stens 1988; Butzer et al. 1990).

APPROXIMATION BY TRANSLATES OF
A RADIAL FUNCTION

8.1 Introduction

The study of approximation schemes based on translates of a radial function to a set of points (centres) in $\mathbb{R}^d$ is a central subject in multivariate approximation theory. Here we study the L_∞-approximation orders of such schemes, first for centres constituting a regular grid, and then for quasi-uniformly scattered centres.

8.1.1 *Radial functions and multivariate interpolation*

The interest in radial function approximation was initiated by applications. Radial functions provide a convenient and simple tool for global interpolation of scattered multivariate data. Given a univariate function $g(t) : \mathbb{R}_+ \to \mathbb{R}$, and a non-negative integer m, the interpolation problem to the scattered data

$$(x^i, f_i), \quad x^i \in \mathbb{R}^d, \quad f_i \in \mathbb{R}, \quad i = 1, \ldots, N, \tag{8.1.1}$$

based on the radial function $\phi(x) = g(\|x\|)$, consists of finding a function of the form

$$S(x) = \sum_{i=1}^{N} v_i \phi(x - x^i) + p_m(x),$$

$$p_m \in \pi_m, \quad \sum_{i=1}^{N} v_i q(x^i) = 0, \quad q \in \pi_m, \tag{8.1.2}$$

satisfying

$$S(x^i) = f_i, \quad i = 1, \ldots, N. \tag{8.1.3}$$

Here π_m is the space of all algebraic polynomials of degree at most m on $\mathbb{R}^d$, and $\|\cdot\|$ is the Euclidean norm on $\mathbb{R}^d$. This method of interpolation reproduces polynomials in π_m, whenever (8.1.3) is uniquely solvable. Classes of functions $g(t)$ which are well known in the literature include:

(i) "Surface splines"

$$g(t) = \begin{cases} t^{2k-d} \log t, & d \text{ even}, \\ t^{2k-d}, & d \text{ odd}, \end{cases} \tag{8.1.4}$$

with k an integer satisfying $2k > d$ and with $m = k - 1$, studied in a series of papers (see e.g. Duchon 1976, 1977; Meinguet 1979). The corresponding interpolant (8.1.2) minimizes the functional

$$\mathcal{R}_k(f) = \int_{\mathbb{R}^d} \sum_{|\alpha|=k} (D^\alpha f)^2 dx, \tag{8.1.5}$$

among all functions interpolating the data in the space

$$\chi_k = \left\{ f \in C(\mathbb{R}^d), \quad D^\alpha f \in L_2(\mathbb{R}^d), \quad |\alpha| = k \right\}. \tag{8.1.6}$$

Here and hereafter we use the notation

$$\alpha = (\alpha_1, \ldots, \alpha_d) \in \mathbb{Z}_+^d, \quad D^\alpha = \frac{\partial^{|\alpha|}}{\partial x_1^{\alpha_1} \ldots \partial x_d^{\alpha_d}}, \quad |\alpha| = \sum_{i=1}^d \alpha_i.$$

The functional $\mathcal{R}_k(f)$ is rotation-invariant, and this choice reflects the assumption that there are no preferable directions in the scattered data (8.1.1). The functions (8.1.4) are fundamental solutions of the k'th iterated Laplacian,

$$\Delta^k \phi = c\delta. \tag{8.1.7}$$

For $d = 1$ the surface spline coincides with the natural spline of order $2k$.

The variational formulation yields the unique solvability of (8.1.2), (8.1.3) whenever the set $X = \{x^1, \ldots, x^N\}$ satisfies the geometric condition

$$\dim \pi_m \Big|_X = \dim \pi_m. \tag{8.1.8}$$

In (Duchon 1977) the class of surface splines is extended by considering functionals of the type (8.1.5) corresponding to derivatives of fractional orders,

$$\mathcal{R}_{s,k}(f) = \int_{\mathbb{R}^d} \|\omega\|^{2s} \sum_{|\alpha|=k} (\widehat{D^\alpha f})^2(\omega) d\omega, \quad 0 < s < 1, \quad |\alpha| = k, \tag{8.1.9}$$

where $\widehat{f}$ denotes the Fourier transform of f. For $2k + 2s > d$, the solution to the variational problem determined by (8.1.9) and the scattered data (8.1.1) is given by the solution to (8.1.2), (8.1.3) with $m = k - 1$ and

$$g(t) = \begin{cases} t^{2k+2s-d} \log t, & 2k + 2s - d \text{ even,} \\ t^{2k-2s-d}, & \text{otherwise.} \end{cases} \tag{8.1.10}$$

(ii) "Multiquadrics"

$$g(t) = (t^2 + c^2)^\beta, \quad \beta = \pm\frac{1}{2}, \quad c > 0, \quad d = 2, \quad m = -1, \tag{8.1.11}$$

introduced in (Hardy 1971) (see also (Hardy 1990)), for geophysical applications. The value $m = -1$ in (8.1.11) corresponds to π_m being the empty set in (8.1.2).

(iii) "Shifted surface splines"

$$g(t) = \begin{cases} (t^2 + c^2)^{(2k-d)/2} \log(t^2 + c^2)^{\frac{1}{2}}, & 2k \geq d, \quad d \text{ even}, \\ (t^2 + c^2)^{(2k-d)/2}, & \text{otherwise}, \ k \geq 1, \end{cases} \tag{8.1.12}$$

with $c > 0$ and $m = k - 1$, introduced in (Dyn et al. 1986) for $d = 2$. These functions are the "shifted" version of the fundamental solutions of the iterated Laplacian of order $k \geq 1$. With the choice $c > 0$, $g(0)$ is well defined also for $1 \leq k \leq d/2$, in contrast to the case $c = 0$ of the surface splines.

In a comparative study (Franke 1982), the quality of interpolation of scattered data in $\mathbb{R}^2$ by radial functions of classes (i), (ii) is found to be superior to other methods of interpolation.

The classes of radial functions for which the interpolation problem (8.1.2), (8.1.3) is uniquely solvable under condition (8.1.8) in all $\mathbb{R}^d$ is studied in (Micchelli 1986; Madych and Nelson 1990; Guo et al. 1993). This class is characterized by the strict complete monotonicity of order $m + 1$ of $g(\sqrt{t})$, namely

$$\frac{d^m}{dt^m} g(\sqrt{t}) \not\equiv \text{const}, \quad \varepsilon(-1)^j \frac{d^j}{dt^j} g(\sqrt{t}) \geq 0, \quad t > 0, \quad j \geq m + 1 \tag{8.1.13}$$

with $\varepsilon = 1$ or $\varepsilon = -1$.

This allows one to extend further the class of radial functions of interest to

$$g(t) = \begin{cases} t^\gamma, & \gamma \in \mathbb{R}_+ \backslash 2\mathbb{Z}_+, \\ t^\gamma \log t, & \gamma \in 2\mathbb{Z}_+, \end{cases} \tag{8.1.14}$$

in any $\mathbb{R}^d$ independent of the parity of d, with $m > \gamma/2 - 1$ in (8.1.2). The corresponding "shifted" version of the class (8.1.14) is even wider

$$g(t) = \begin{cases} (t^2 + c^2)^{\gamma/2}, & \gamma > -d, \quad \gamma \notin 2\mathbb{Z}_+, \\ (t^2 + c^2)^\gamma \log(t^2 + c^2)^{1/2}, & \gamma \in 2\mathbb{Z}_+, \end{cases} \tag{8.1.15}$$

with $c > 0$ and where the corresponding m satisfies $m > \gamma/2 - 1$ for $\gamma \geq 0$, and $m = -1$ otherwise.

For review papers on various aspects of the theory of radial functions see e.g. (Dyn 1987, 1989; Powell 1992; Buhmann 1993).

8.1.2 *The distributional Fourier transform of ϕ*

It is the properties of the distributional Fourier transform of $\phi(x) = g(\|x\|)$, considered as a tempered distribution (Gelfand and Shilov 1964), which are relevant to the solvability of the interpolation problem as well as to the theory of approximation orders of schemes based on translates of such radial functions. While for the solvability of the interpolation problem many more radial functions can be considered, such as the Gaussian function $g(t) = e^{-t^2/a}$, $a > 0$, and the radial functions of compact support introduced by Wu (1994), the analysis in this chapter is confined to radial functions of the class (8.1.14), (8.1.15) and to related functions with similar properties of their distributional Fourier transform, such as fundamental solutions of homogeneous elliptic operators, which are not necessarily radial.

The distributional Fourier transform of $\phi = g(\|\cdot\|)$, with g as in (8.1.14) (8.1.15), coincides away from the origin with a function $\hat{\phi}$ of the form (Gelfand and Sl: :· 1964)

$$\hat{\phi}(\omega) = a_{\gamma,c}\|\omega\|^{-\gamma-d}F_{\gamma,c}(\omega), \tag{8.1.16}$$

where $a_{\gamma,c}$ is a positive constant which depends on γ and c, and where

$$F_{\gamma,c}(\omega) = \begin{cases} 1, & c = 0, \\ \tilde{K}_{(d+\gamma)/2}(c\|\omega\|), & c > 0. \end{cases} \tag{8.1.17}$$

Here $\tilde{K}_\nu(t) = t^\nu K_\nu(t)$, with K_ν the modified Bessel function. Relevant properties of $\tilde{K}_\nu$ to our analysis are (Dyn 1989)

$$\begin{aligned}
&\tilde{K}_\nu \in C(\mathbb{R}), \quad \tilde{K}_\nu(t) > 0, \quad t \geq 0, \quad \nu > 0, \\
&\lim_{t\to\infty} \tilde{K}_\nu(t) = 0 \text{ exponentially}, \\
&\tilde{K}_n \in C^{2n-1}(\mathbb{R}) \cap C^\infty(\mathbb{R}\backslash 0), \\
&\tilde{K}_n^{(2n)}(t) = 0(\log t), \quad t \to 0^+, \\
&\tilde{K}_{n+\frac{1}{2}} \in C^\infty(\mathbb{R}), \quad n \in \mathbb{Z}_+.
\end{aligned} \tag{8.1.18}$$

A fundamental solution of the homogeneous elliptic operator $G(D)$ of order $2m$ has a generalized Fourier transform which coincides on $\mathbb{R}^d\backslash 0$ with $1/G(\omega)$ up to a multiplicative constant.

The important features of $\hat{\phi}$ are the order m' of its singularity at the origin, namely as $\|\omega\| \to 0$, and the rate of its decay as $\|\omega\| \to \infty$. For the fundamental solutions of homogeneous elliptic operators and for the class (8.1.14), the order of the singularity at the origin of $\hat{\phi}$ equals the rate of its decay at infinity. For the class of radial functions (8.1.15) the decay of $\hat{\phi}$ at infinity is exponential. For all these functions the distributional Fourier transform at the origin is a distribution of order less than m'.

8.1.3   Outline of the chapter

All the sections in this chapter are concerned with estimating the error measured in the L_∞-norm, incurred by approximation schemes based on translates of a radial function and its scales.

Sections 8.2 and 8.3 study two different types of approximation schemes, based on shifts of $\phi(h^{-1}\cdot)$ to the points of $h\mathbb{Z}^d$. The first type, analysed in Section 8.2, consists of quasi-interpolation schemes of the form

$$Q_{\psi,h}f = \sum_{\alpha\in\mathbb{Z}^d} f(h\alpha)\psi(h^{-1}\cdot-\alpha), \tag{8.1.19}$$

where ψ is a finite linear combination of shifts of ϕ to points of $\mathbb{Z}^d$ near the origin. The approximation order is the power of h in the error $\|f - Q_{\psi,h}f\|_\infty$ for all f in an admissible set of functions W. Section 8.2 presents the results in (Dyn et al. 1992), which apply to a wide class of functions ϕ. Other results in this direction for specific radial functions are presented in (Jackson 1989) and (Buhmann 1988). In (Beatson and

Light 1992) approximation orders by qausi-interpolation schemes based on the Gaussian radial function are derived.

In Section 8.3 the approximation scheme is an optimal one which achieves the optimal approximation orders possible. This scheme introduced in (de Boor and Ron 1992) uses global information on the approximated function and has the form

$$L_h f = \sum_{\alpha \in \mathbb{Z}^d} \Lambda_h f(h\alpha) \psi(h^{-1} \cdot -\alpha) . \tag{8.1.20}$$

Here $\psi = \sum_{\alpha \in \mathbb{Z}^d} \mu_\alpha \phi(\cdot - \alpha)$, with $\{\mu_\alpha : \alpha \in \mathbb{Z}^d\}$ not necessarily of finite support, and

$$\widehat{\Lambda_h f} = \hat{\lambda}(h \cdot) \hat{f}, \quad \hat{\lambda} = \eta/\hat{\psi} , \tag{8.1.21}$$

with η any smooth function of compact support which is 1 on a ball centred at the origin. Approximation orders of optimal schemes in the $L_2(\mathbb{R}^d)$ setting are studied in (Ron 1992).

Section 8.4 extends the results of the previous sections to analogous approximation schemes based on translates of ϕ to quasi-uniformly scattered centres. For the quasi-interpolation schemes, the required information on the approximated function is confined to the same set of centres. Section 8.4 is mainly based on (Dyn and Ron 1995), with some results taken from (Buhmann et al. 1995), where the notion of quasi-uniformly scattered centres is first introduced. Analogous results to those in (Dyn and Ron 1995) for approximation in the L_p–norms are derived in (Buhmann and Ron 1994).

All the results presented in this chapter deal with approximation orders defined by scaling the function ψ (and therefore ϕ) by h^{-1}, and then translating it to a set of centres with distances of order h between neighbouring centres. There are other notions of approximation orders, corresponding to different types of scaling of the function ϕ (see e.g. Beatson and Light 1992; Ron 1992).

Another important type of approximation order is based on translates of ϕ, without any scaling, to sets of centres with increasing density. For "homogeneous" radial functions this approach yields the same orders as scaling by h^{-1}, while for the others, qualitatively different results are obtained (see e.g. Buhmann and Dyn 1993; Madych and Nelson 1992; Schaback and Wu 1993).

8.2 Quasi-interpolation on regular grids

In this section, we analyse the approximation order of schemes based on function values on a regular grid $h\mathbb{Z}^d$ and on the $h\mathbb{Z}^d$ translates of a scaled basis function ψ, consisting of a finite linear combination of multi-integer translates of a radial function or a related function ϕ. The schemes we study are quasi-interpolatory of the form

$$Q_{\psi,h} f = \sum_{\alpha \in \mathbb{Z}^d} f(\alpha h) \psi(h^{-1} \cdot -\alpha).$$

The analysis employs two important ingredients of the scheme: the decay of $\psi(x)$ as $\|x\| \to \infty$ and polynomial reproduction. In case ψ is of compact support or decays fast enough the argument for getting the approximation order is standard and we present it here. We denote by Q_ψ the operator $Q_{\psi,1}$ and by A any constant appearing in the various bounds.

Theorem 8.1 *Assume that*

$$|\psi(x)| \le A(1 + \|x\|)^{-(d+k)}, \quad k > \ell + 1, \qquad (8.2.1)$$

and that

$$Q_\psi p = p, \quad p \in \pi_\ell. \qquad (8.2.2)$$

Then for every f with bounded derivatives of order $\ell + 1$

$$\|Q_{\psi,h} f - f\|_\infty \le A\|f\|_{\infty,\ell+1} h^{\ell+1}, \qquad (8.2.3)$$

where $\|f\|_{\infty,\ell+1} = \sum_{|\alpha|=\ell+1} \|D^\alpha f\|_\infty < \infty$.

Proof First we note that (8.2.2) also holds for $Q_{\psi,h}$. This is easily seen for the basis of powers of π_ℓ, and hence holds for all π_ℓ.

Let $T_x f$ be the Taylor polynomial of f of degree ℓ at x, namely $D^\alpha(f - T_x f)(x) = 0$, $|\alpha| \le \ell$. Then for $g = f - T_x f$ we get by (8.2.2) that

$$Q_{\psi,h} f - f = Q_{\psi,h} g - g, \qquad (8.2.4)$$

while by the definition of g,

$$|g(z)| \le A\|f\|_{\infty,\ell+1} \|z - x\|^{\ell+1}. \qquad (8.2.5)$$

Thus

$$|(Q_{\psi,h} f - f)(x)| = |Q_{\psi,h} g(x)| = \sum_{\alpha \in \mathbb{Z}^d} |\psi(h^{-1}x - \alpha)| \, |g(\alpha h)|$$

$$\le A\|f\|_{\infty,\ell+1} \sum_{\alpha \in \mathbb{Z}^d} \|\alpha h - x\|^{\ell+1} |\psi(h^{-1}x - \alpha)|$$

$$\le A h^{\ell+1} \|f\|_{\infty,\ell+1} \sum_{\alpha \in \mathbb{Z}^d} \|h^{-1}x - alp\|^{\ell+1} |\psi(h^{-1}x - \alpha)|$$

$$\le A h^{\ell+1} \|f\|_{\infty,\ell+1}, \qquad (8.2.6)$$

where in the last inequality we used (8.2.1) to bound the sum above by a constant independent of h and x.

For many of the radial functions and related functions that we consider in this paper Theorem 8.1 does not yield the optimal approximation orders. In fact we can show (8.2.1) for $k = \ell + 1$ at most, where ℓ is the maximal value possible in (8.2.2). A finer analysis is needed then to obtain approximation orders $O(h^{\ell+1} \log |h|)$ or sometimes even $O(h^{\ell+1})$.

8.2.1 The general setting

The first step in the presentation of the approximation scheme is the construction of ψ satisfying (8.2.1) with $k = \ell + 1$. This rate of decay is sufficient for Q_ψ to be well defined on π_ℓ. The next stage is to show that (8.2.2) holds. Finally the sum $\sum_{\alpha \in \mathbb{Z}^d} |\psi(h^{-1}x - \alpha)| \, |g(\alpha h)|$ has to be estimated.

The class of approximating spaces under investigation is of the form

$$S_h(\phi) = \text{span}\{\phi(h^{-1}x - \alpha) : \alpha \in \mathbb{Z}^d\} \tag{8.2.7}$$

with the span standing for the closure of the algebraic span under the topology of uniform convergence on compact sets. Here ϕ is a function which grows at most as a power of $\|x\|$ as $\|x\| \to \infty$, and whose distributional Fourier transforms $\widehat{\phi}$ satisfies the equation

$$G\widehat{\phi} = F. \tag{8.2.8}$$

The distributional Fourier transform of ϕ as a tempered distribution is defined by the equality

$$\int_{\mathbb{R}^d} \phi(\omega)s(\omega)d\omega = \int_{\mathbb{R}^d} \widehat{\phi}(\omega)\widehat{s}(\omega)d\omega, \quad s \in S,$$

where S is the space of all C^∞ rapidly decaying test functions (Gelfand and Shilov 1964).

There are several assumptions on F and G typical of the class of functions ϕ under investigation:

(a) $G(\omega) \neq 0$ if $\omega \neq 0$,

(b) $G(\omega)$ is a homogeneous polynomial of degree $2m$,

(c) $F(0) \neq 0$, $F(x) - \displaystyle\sum_{|\alpha| \leq m_0} \frac{D^\alpha F(0)}{\alpha!} x^\alpha \in \mathcal{F}_{m_0+\theta}$ for some $\theta > 0$, $\qquad$ (8.2.9)

(d) $F \in C^\infty(\mathbb{R}^d \backslash 0)$,

(e) $|D^\alpha(F/G)(\omega)| \leq \dfrac{A_\alpha}{\|\omega\|^{d+\alpha+\varepsilon}}$ for $\|\omega\| \geq 1$, $\varepsilon > 0$, $\alpha \in \mathbb{Z}_+^d$,

where in (c) we use the notation

$$\mathcal{F}_r = \left\{ f \in C^\infty(\mathbb{R}^d \backslash 0) : D^\alpha f(x) = 0(\|x\|^{r-|\alpha|}) \text{ as } \|x\| \to 0, \ \alpha \in \mathbb{Z}_+^d \right\}. \tag{8.2.10}$$

In the case of fundamental solutions of homogeneous elliptic operators, $F \equiv 1$.

Condition (a) in (8.2.9) guarantees that $\widehat{\phi} = F/G$ as functions on $\mathbb{R}^d \backslash 0$. The behaviour of $\widehat{\phi}$ at the origin is defined in a distributional sense by

$$\widehat{\phi}[s] = \int_{\mathbb{R}^d} s(\omega) \frac{F(\omega)}{G(\omega)} d\omega, \quad s \in S_{2m-1}, \tag{8.2.11}$$

where

$$S_{2m-1} = \{s \in S : D^\alpha s(0) = 0, \|\alpha\| \leq 2m - 1\}. \tag{8.2.12}$$

8.2.2   The construction of ψ

By (8.2.9) the behaviour of the singularity of $\widehat{\phi}$ near the origin is as the reciprocal of a polynomial of degree $2m$, hence by taking a finite linear combination of shifts of ϕ

$$\psi = \sum_{\alpha \in I} \mu_\alpha \phi(\cdot - \alpha), \quad I \subset \mathbb{Z}^d, \tag{8.2.13}$$

one can get $\widehat{\psi}$ to be defined as an ordinary Fourier transform of ψ, which is well defined on $\mathbb{R}^d$ and is of the form

$$\widehat{\psi} = \widehat{\phi} e, \quad e(\omega) = \sum_{\alpha \in I} \mu_\alpha e^{-i\alpha \cdot \omega}. \tag{8.2.14}$$

The coefficients $\{\mu_\alpha : \alpha \in I\}$ are chosen to satisfy

$$D^\alpha(e - G/F)(0) = 0, \quad |\alpha| \le 2m + \ell, \tag{8.2.15}$$

for some $\ell \in [0, m_0] \cap \mathbb{Z}$. Thus $D^\alpha e(0) = 0$ for $\|\alpha\| \le 2m - 1$, and the zero of e at the origin cancels the singularity of $\widehat{\phi}$ there. Note that as ℓ increases in (8.2.15) the set I that supports the sequence $\{\mu_\alpha\}$ is bigger, since more conditions in (8.2.15) have to be satisfied. Conditions (8.2.15) together with the $(2\pi)^d$-periodicity of $e(\omega)$ and assumptions (b), (c) of (8.2.9), lead to

Proposition 8.2 Let ψ be defined by (8.2.13)–(8.2.15). Then $\widehat{\psi} \in C^\ell(\mathbb{R}^d)$ and satisfies

$$\widehat{\psi}(0) = 1, \quad D^\alpha \widehat{\psi}(0) = 0, \quad 1 \le |\alpha| \le \ell, \tag{8.2.16}$$

$$p(-iD)\widehat{\psi}(2\pi\beta) = 0, \quad \beta \in \mathbb{Z}^d \backslash 0, \quad p \in \mathcal{P}_G \cap \pi_{\ell+2m}, \tag{8.2.17}$$

where $\mathcal{P}_G$ is the kernel of the operator $G(D)$. In particular

$$D^\alpha \widehat{\psi}(2\pi\beta) = 0, \quad \beta \in \mathbb{Z}^d \backslash 0, \quad 0 \le |\alpha| \le 2m - 1, \tag{8.2.18}$$

since $\pi_{2m-1} \in \mathcal{P}_G$.

Proposition 8.2 is the key to both the decay of ψ and the polynomial reproduction property of Q_ψ. Using the behaviour of $\widehat{\psi}$ near zero and its decay as $\|x\| \to \infty$ it is possible to estimate the decay rate of $|\psi(x)|$ as $\|x\| \to \infty$.

Theorem 8.3 Under assumptions (8.2.9), (8.2.11), (8.2.13) and (8.2.15) with $0 \le \ell < m_0$,

$$|\psi(x)| \le A(1 + \|x\|^{-d-\ell-1}) \quad as \quad \|x\| \to \infty, \tag{8.2.19}$$

and

$$Q_\psi p = p, \quad p \in \pi_\ell \cap \mathcal{P}_G \tag{8.2.20}$$

with uniform convergence of $Q_\psi p$ to p on compact sets.

The polynomial reproduction property (8.2.20) follows from the extended version (8.2.16) and (8.2.17) of the "Strang–Fix conditions" (Strang and Fix 1973), and from the uniform convergence of $Q_\psi p$ on compact sets, which allows the use of the Poisson summation formula.

Since the decay of ϕ in (8.2.19) is due to the choice of the sequence $\{\mu_\alpha\}$ in (8.2.15), such a sequence is termed hereafter a "localization sequence". It defines a difference operator $\sum_{\alpha \in I} \mu_\alpha f(\cdot - \alpha)$, which vanishes on $\mathcal{P}_G \cap \pi_{2m+\ell}$.

We conclude from Theorem 8.3 that $\mathcal{P}_G \cap \pi_{m_0-1} \subset S_h(\phi)$, and that $\ell = \min(2m - 1, m_0 - 1)$ is the maximal ℓ such that π_ℓ is reproduced by Q_ψ, and therefore by $Q_{\psi,h}$.

For two important classes of radial functions these consequences can be strengthened.

Corollary 8.4 *In case $F = 1$, namely ϕ is a fundamental solution of the elliptic operator $G(D)$, $m_0 = \infty$ in Theorem 8.3 and for any $\ell \in \mathbb{Z}_+$ there exists ψ such that (8.2.19) and (8.2.20) hold. Thus $\mathcal{P}_G \subset S_h(\phi)$. In particular $\pi_{2m-1} \subset S_h(\phi)$, and for $\ell = 2m - 1$, Q_ψ reproduces π_{2m-1}, which is the maximal total degree polynomial space contained in $\mathcal{P}_G$.*

For later reference, we call the class of functions discussed in Corollary 8.4 class A. A second specific class of interest is that of the shifted fundamental solutions of the iterated Laplacian, referred to hereafter as class B. For this class $m_0 = 2m - 1$, and near the origin $F(\omega) = H(\|\omega\|)$ has an expansion of the form

$$H(r) = \sum_{k=0}^{\infty} a_k r^{2k} + (\log r) r^{2m} \sum_{k=0}^{\infty} b_k r^{2k} \quad \text{as } r \to 0^+ \tag{8.2.21}$$

with $a_0 > 0$ and $b_0 \neq 0$.

The inverse Fourier transform of the first homogeneous $2m + 1$ terms in the expansion of

$$\hat{\psi}(\omega) = 1 + \frac{F(\omega)}{G(\omega)}\left[e(\omega) - \frac{G(\omega)}{F(\omega)}\right],$$

near the origin can be obtained explicitly, in view of (8.2.15), from which it is possible to conclude (8.2.19) with $\ell = m_0 = 2m - 1$.

Corollary 8.5 *For ϕ in class B, condition (8.2.15) with $\ell = 2m - 1 = m_0$ generates ψ which satisfies (8.2.19) and (8.2.20) with $\ell = 2m - 1$. Thus π_{2m-1} is reproduced by Q_ψ, implying that $\pi_{2m-1} \subset S_h(\phi)$.*

An important observation about class A, which does not hold for class B, is that $Q_{\psi,h}f \in \text{span}\{\phi(x - \alpha) : \alpha \in h\mathbb{Z}^d\}$. This follows since any ϕ in class A is a homogeneous function up to a polynomial of degree $\leq 2m - 1$, which is cancelled in (8.2.13) since $\sum_{\alpha \in I} \mu_\alpha p(\cdot - \alpha) = 0$, for $p \in \pi_{2m-1}$.

8.2.3 The approximation orders on $\mathbb{R}^d$

The approximation order of $Q_{\psi,h}$ is now obtained from Theorem 8.3 and Corollaries 8.4, 8.5. As a first step we use Theorem 8.1 with (8.2.19) and (8.2.20).

Theorem 8.6 *Under the assumptions of Theorem 8.3 with any* $0 \leq \ell \leq \min(2m, m_0 - 1)$

$$\|f - Q_{\psi,h} f\|_\infty \leq A \|f\|_{\infty,\ell} h^\ell , \tag{8.2.22}$$

for any $f \in C^\ell(\mathbb{R}^d)$ *with bounded derivatives of order* ℓ. *If* ϕ *is in class A then (8.2.22) holds for* $\ell \leq 2m$, *while if* ϕ *is in class B, then (8.2.22) holds for* $\ell \leq 2m - 1 = m_0$.

Under the same conditions as in Theorem 8.6, higher approximation orders than (8.2.22) for $Q_{\psi,h}$ can be achieved. This requires a finer analysis of the sum $Q_{\psi,h} g$ appearing in the proof of Theorem 8.1.

Theorem 8.7 *Let* ψ *satisfy (8.2.19) for some* $\ell \geq 0$ *and let*

$$Q_\psi p = p, \quad p \in \pi_\ell. \tag{8.2.23}$$

Then for $f \in C^{\ell+1}(\mathbb{R}^d)$ *with bounded derivatives of orders* ℓ *and* $\ell + 1$

$$\|f - Q_{\psi,h} f\|_\infty \leq A(\|f\|_{\infty,\ell} + \|f\|_{\infty,\ell+1}) h^{\ell+1} |\log h|. \tag{8.2.24}$$

Proof Let g be as in the proof of Theorem 8.1. Then by (8.2.23)

$$(f - Q_{\psi,h} f)(x) = Q_{\psi,h} g(x). \tag{8.2.25}$$

To estimate the error in the approximation, we partition $Q_{\psi,h} g(x)$ into two sums:

$$Q_{\psi,h} g(x) = \sum_{\alpha \in S_{x,h}} g(\alpha h) \psi(h^{-1} x - \alpha) + \sum_{\alpha \in \mathbb{Z}^d \setminus S_{x,h}} g(\alpha h) \psi(h^{-1} x - \alpha), \tag{8.2.26}$$

where $S_{x,h} = \mathbb{Z}^d \cap h^{-1}(x + [-1, 1]^d)$.

In the first sum we use (8.2.5), while in the second we use a bound as (8.2.5) but with ℓ there replaced by $\ell - 1$, which also holds for g by the assumptions on f. Thus we get

$$|Q_{\psi,h} g(x)| \leq A \|g\|_{\infty,\ell+1} h^{\ell+1} \sum_{\alpha \in S_{x,h}} (\|h^{-1} x - \alpha\| + 1)^{-d}$$

$$+ A \|g\|_{\infty,\ell} h^\ell \sum_{\alpha \in \mathbb{Z}^d \setminus S_{x,h}} (\|h^{-1} x - \alpha\| + 1)^{-d-1}. \tag{8.2.27}$$

Applying Lemma 4.2 of (Dyn et al. 1992), stating that the first sum is bounded by $A |\log h|$ while the second sum by Ah, we finally obtain (8.2.24).

Theorem 8.7 and Corollary 8.5, applied to the radial functions of class B, yield approximation order $O(h^{2m} |\log h|)$ for the choice $\ell = m_0 = 2m - 1$. Similar approximation orders are obtained for the radial functions of class A by taking $\ell = 2m - 1$ in Theorem 8.7. Yet by Theorem 8.6 with $\ell = 2m$, which is a proper choice for functions in class A (ℓ in (8.2.15) can be any positive integer), one gets the better approximation order $O(h^{2m})$.

It is shown in (Dyn et al. 1992) by quite involved analysis that the $|\log h|$ factor in (8.2.24) can be removed for class A also for the choice $\ell = 2m - 1$ in (8.2.15), but cannot be removed for class B. For the latter class there are other approximation schemes which achieve the approximation order $O(h^{2m})$, such as the cardinal interpolation scheme (Buhmann 1990) or the optimal approximation scheme of de Boor and Ron (1992). The cardinal interpolation scheme also uses the values of f on $h\mathbb{Z}^d$, but uses the shifts of a function ψ which is an infinite linear combination of shifts of ϕ. On the other hand, the optimal approximation scheme uses global information on the approximated function f, but ψ is much simpler than in the quasi-interpolation case, with a very mild decay, independent of the approximation order (see Section 8.3).

8.2.4   The approximation orders on finite domains

The fast decay of ψ in the quasi-interpolation schemes presented here has a very important consequence: given function values on $h\mathbb{Z}^d \cap \Omega$, where Ω is an open bounded region of $\mathbb{R}^d$, it is possible to get the same approximation order that $Q_{\psi,h}$ achieves on $\mathbb{R}^d$, on any closed subdomain of Ω by the restricted scheme

$$Q_{\psi,h,\Omega} f = \sum_{\alpha \in \mathbb{Z}^d \cap h^{-1}\Omega} f(h\alpha)\psi(h^{-1}x - \alpha). \tag{8.2.28}$$

In fact, we can get a somewhat stronger result. We denote

$$\|f\|_{\infty,\Omega} = \sup_{x \in \Omega} |f(x)|, \quad \|f\|_{\infty,\ell,\Omega} = \sum_{|\alpha|=\ell} \|D^\alpha f\|_{\infty,\Omega},$$

and

$$\Omega_\delta = \{y \in \Omega : \|y - z\|_\infty \leq \delta \Rightarrow z \in \Omega\}. \tag{8.2.29}$$

Theorem 8.8 *Let $\Omega \subset \mathbb{R}^d$ be open and bounded, let $f \in C^{\ell+1}(\overline{\Omega})$ and let ψ and ℓ be as in Theorem 8.7. Then*

$$\|f - Q_{\psi,h,\Omega} f\|_{\infty,\Omega_{\delta(h)}} \leq A(\|f\|_{\infty,\ell,\Omega} + \|f\|_{\Omega,\ell+1,\Omega})h^{\ell+1}|\log h|, \tag{8.2.30}$$

where

$$\delta(h) \geq A|\log h|^{-1/(\ell+1)}. \tag{8.2.31}$$

Also

$$\|f - Q_{\psi,h,\Omega} f\|_{\infty,\Omega_{\delta(h)}} \leq A\|f\|_{\infty,\ell,\Omega}h^\ell \tag{8.2.32}$$

where

$$\delta(h) \geq Ah^{1/(\ell+1)}. \tag{8.2.33}$$

The idea of the proof is first to show that the main part of the approximation at $x \in \Omega_\delta$ is obtained by the local sum

$$Q_{\psi,h,\Omega,\delta} f(x) = \sum_{\{\alpha \in \mathbb{Z}^d : \|h\alpha - x\|_\infty \leq \delta\}} f(\alpha h)\psi(h^{-1}x - \alpha), \tag{8.2.34}$$

and that all the other terms in (8.2.28) contribute to the sum a magnitude of the order $O((h/\delta)^{\ell+1})$. The second important step is to show that the local scheme (8.2.34) for $f \in \pi_\ell$ approximates $f(x)$ with error of the order $O((h/\delta)^{\ell+1})$.

8.3   Optimal approximation schemes on regular grids

The construction of the optimal approximation schemes in (de Boor and Ron 1992) is aimed at providing schemes which achieve the maximal possible approximation order from spaces generated by shifts of the h^{-1}-scales of a basis function to $h\mathbb{Z}^d$. The setting in (de Boor and Ron 1992) is quite general, and includes also other scales of the basis function. Here we present the results related to the shift-invariant spaces

$$S(\phi) = \text{span}\{\phi(\cdot - \alpha) : \alpha \in \mathbb{Z}^d\}, \qquad (8.3.1)$$

and their scales

$$S_h(\phi) = \text{span}\{\phi(h^{-1} \cdot -\alpha) : \alpha \in \mathbb{Z}^d\}, \qquad (8.3.2)$$

where ϕ is a radial or a related function.

An important ingredient in the analysis is the determination of an upper bound for the approximation order from (8.3.2). Then an approximation scheme is termed optimal, if the approximation order provided by it is equal to the upper bound.

8.3.1   *The general setting*

The assumptions on ϕ in this section are such that all the radial functions and the related functions presented in the Introduction are included. The function ϕ grows at most as a power of $\|x\|$ as $\|x\| \to \infty$, and its distributional Fourier transform $\hat{\phi}$ satisfies the three conditions

(a) $\hat{\phi} \in C(\mathbb{R}^d \backslash 0)$, $\quad \hat{\phi}(\omega) > 0$, $\quad \omega \in \mathbb{R}^d \backslash 0$,

(b) for some $\delta > 0$, $\quad \|\hat{\phi}(\omega)\|_\infty = 0(\|\omega\|^{-d-\delta})$, $\quad \|\omega\| \to \infty$, $\qquad (8.3.3)$

(c) for some $m' \geq 0$, $\quad 0 < A_1 \leq \|\omega\|^{m'} |\hat{\phi}(\omega)| \leq A_2 < \infty$, $\|\omega\| < \rho$.

Under these conditions there exists a continuous 2π-periodic function $\mu(\omega)$, $\mu(\cdot) = \sum_{\alpha \in \mathbb{Z}^d} \mu_\alpha e^{-i\alpha \cdot \omega}$ such that the function

$$\hat{\psi}(\omega) = \mu(\omega)\hat{\phi}(\omega) \qquad (8.3.4)$$

is the proper Fourier transform of the $L_1(\mathbb{R}^d)$ function

$$\psi(x) = \sum_{\alpha \in \mathbb{Z}^d} \mu_\alpha \phi(x - \alpha). \qquad (8.3.5)$$

Moreover,

$$\hat{\psi}(0) \neq 0 \quad \text{and} \quad \sum_{\alpha \in \mathbb{Z}^d} |\psi(\cdot - \alpha)| \in L_\infty(\mathbb{R}^d). \qquad (8.3.6)$$

The optimal approximation order from the scales $S_h(\psi)$ of the shift-invariant space

$$S(\psi) = \text{span}\{\psi(x - \alpha) : \alpha \in \mathbb{Z}^d\}$$

is found to be independent of the localization sequence $\{\mu_\alpha\}$ in (8.3.5), and hence is attributed to the spaces (8.3.2).

The approximation orders in this setting are derived for the following classes of admissible functions, defined in terms of their distributional Fourier transforms:

Definition 8.9 *A function f of at most polynomial growth at infinity is termed k- admissible if $(1 + \| \cdot \|^k) \widehat{f}$ is a Radon measure such that*

$$\| f \|_k' = \int_{\mathbb{R}^d} (1 + \|x\|^k) |\widehat{f}(x)| dx < \infty. \tag{8.3.7}$$

The collection of these functions is denoted by $\widetilde{W}_k^\infty(\mathbb{R}^d)$.

It can be shown that any admissible f is bounded. In particular $f(x) = e^{-i\theta \cdot x}$, $\theta \in \mathbb{R}^d$, is admissible of any order k, since $\widehat{f}(\omega) = \delta_{-\theta}(\omega)$, and thus $\| f \|_k' = 1 + \|\theta\|^k$. If $\widehat{f}$ is a function then f is admissible if $(1 + \| \cdot \|^k) \widehat{f} \in L_1(\mathbb{R}^d)$. In case $k \in \mathbb{Z}_+$, f is k-admissible if and only if the distributional Fourier transforms of f and all its k'th-order derivatives are measures of finite total mass. In this case $\widetilde{W}_k^\infty(\mathbb{R}^d)$ is continuously embedded into the space of functions with continuous bounded derivatives of order $\leq k$, namely into $W_\infty^k(\mathbb{R}^d) \cap C^k(\mathbb{R}^d)$. In the results on approximation orders, the order of the admissibility class of f is related to the value m' in (8.3.3)(c).

8.3.2   The upper bound

The main necessary condition on the upper bound of the approximation order is obtained from the approximation of the exponential function $e^{-i\theta \cdot x}$, $\theta \in \mathbb{R}^d$ by $S_h(\psi)$, for quite a general class of functions ψ.

Theorem 8.10 *The space $S_h(\psi)$ with ψ satisfying (8.3.6), provides approximation order $k \geq 0$ to $f(x) = e^{-i\theta \cdot x}$, $\theta \in \mathbb{R}^d$ only if*

$$\widehat{\psi}(h\theta + \beta) = O(h^k), \quad \beta \in 2\pi \mathbb{Z}^d \backslash 0, \ \theta \in \mathbb{R}^d. \tag{8.3.8}$$

The upper bound for the approximation order from $S_h(\psi)$, for ψ given by (8.3.5), is obtained from Theorem 8.10 and assumptions (8.3.3).

Theorem 8.11 *The approximation order provided by the space $S_h(\psi)$ with ψ given by (8.3.5), is at most the order of the singularity of $\widehat{\phi}$ at the origin, as defined by m' of (8.3.3)(c).*

Proof Properties (8.3.3) of $\widehat{\phi}$ imply that for small enough h there is a constant A dependent on β such that

$$\left| \frac{h^{-m'} \widehat{\phi}(h\theta + \beta)}{\widehat{\phi}(h\theta)} \right| \geq A > 0, \quad \beta \in 2\pi \mathbb{Z}^d \backslash 0. \tag{8.3.9}$$

Multiplying numerator and denominator by $\mu(h\theta) = \mu(h\theta + \beta)$, for $\beta \in 2\pi \mathbb{Z}^d \backslash 0$, one obtains

$$\left| \frac{h^{-m'} \widehat{\psi}(h\theta + \beta)}{\widehat{\psi}(h\theta)} \right| \geq A > 0, \tag{8.3.10}$$

and since $\widehat{\psi}(0) \neq 0$, $|\widehat{\psi}(h\theta + \beta)| \geq A h^{m'}$ for small enough h, implying that the approximation order from $S_h(\psi)$ is at most m', independent of $\{\mu_\alpha\}$.

8.3.3 The optimal scheme

The scheme, which is shown later to be optimal, has the following form

$$L_h f = \sum_{\alpha \in \mathbb{Z}^d} \psi(h^{-1} \cdot -\alpha) \Lambda_h f(h\alpha), \qquad (8.3.11)$$

where

$$\widehat{\Lambda_h f} = \widehat{\lambda}(h \cdot)\widehat{f}, \quad \widehat{\lambda} = \sigma/\widehat{\psi}, \qquad (8.3.12)$$

with σ any smooth function of compact support Ω, which is 1 on a ball B_ρ of radius ρ centred at the origin. The information on f required by the scheme is global, as it depends on the values of $\widehat{f}$ in the support of $\widehat{\lambda}(h \cdot)$.

The general result which yields the approximation order of the scheme L_h is

Theorem 8.12 *Assume that ψ satisfies (8.3.6), and that for some $k \geq 0$*

$$\sum_{\beta \in 2\pi \mathbb{Z}^d \setminus 0} \left\| \frac{\widehat{\psi}(\cdot + \beta)}{\| \cdot \|^k \widehat{\psi}(\cdot)} \right\|_{\infty, \Omega} < \infty. \qquad (8.3.13)$$

Then for every $f \in \widetilde{W}_k^\infty(\mathbb{R}^d)$

$$\| f - L_h f \|_\infty \leq A \| f \|_k' h^k + o(h^k). \qquad (8.3.14)$$

Proof The main idea of the proof of Theorem 8.12 is to represent the error $f - L_h f$ in terms of $\widehat{f}$, and then to decompose it into two components, namely

$$(f - L_h f)(x) = (2\pi)^{-d} \int_{\mathbb{R}^d} \left[1 - \sigma(h\omega) \right] \widehat{f}(\omega) e^{i\omega \cdot x} d\omega$$

$$+ (2\pi)^{-d} \int_{\mathbb{R}^d} \sigma(h\omega) \left[1 - \frac{E_h(\omega, x)}{\widehat{\psi}(h\omega)} \right] \widehat{f}(\omega) e^{i\omega \cdot x} d\omega, \qquad (8.3.15)$$

with

$$E_h(\omega, x) = \sum_{\alpha \in \mathbb{Z}^d} \psi(h^{-1}x - \alpha) e^{i\omega \cdot (\alpha h - x)}.$$

The first integral in (8.3.15) vanishes for ω such that $\|\omega\| \leq h^{-1}\rho$. Hence

$$(h^{-1}\rho)^k \left| \int_{\mathbb{R}^d} \left[1 - \sigma(h\omega) \right] \widehat{f}(\omega) e^{i\omega \cdot x} d\omega \right|$$

$$\leq \int_{\|\omega\| \geq h^{-1}\rho} \|\omega\|^k |\widehat{f}(\omega)| \, |1 - \sigma(h\omega)| d\omega \to 0, \quad h \to 0. \qquad (8.3.16)$$

This proves that the first integral in (8.3.15) is $o(h^k)$.

To bound the second integral in (8.3.15), we note that the function $E_h(\omega, x)$ is h-periodic in the x variable, and can be written in terms of its Fourier series as

$$E_h(\omega, x) = \sum_{\beta \in 2\pi \mathbb{Z}^d} \widehat{\psi}(h\omega + \beta) e^{-i\beta \cdot h^{-1} x}, \qquad (8.3.17)$$

since (8.3.17) is uniformly convergent by (8.3.6). Thus

$$\frac{E_h(\omega, x)}{\widehat{\psi}(h\omega)} - 1 = \sum_{\beta \in 2\pi \mathbb{Z}^d \backslash 0} \frac{\widehat{\psi}(h\omega + \beta)}{\widehat{\psi}(h\omega)} e^{-i\beta \cdot h^{-1} x}, \qquad (8.3.18)$$

which together with (8.3.13) and the k-admissibility of f implies that the second integral in (8.3.15) is bounded by $A\|f\|'_k h^k$.

As a direct consequence of Theorem 8.12, we get

Corollary 8.13 *Let ψ be given by (8.3.5) with ϕ satisfying (8.3.3). Then the approximation scheme (8.3.11)–(8.3.12) provides the optimal approximation order m' for m'-admissible functions.*

Proof The first assumption in Theorem 8.12 is satisfied by ψ. To show (8.3.13) with k replaced by m', we use the relation $\widehat{\psi} = \mu\widehat{\phi}$ with μ 2π-periodic, and assumption (8.3.3)(c). Thus for $\beta \in 2\pi \mathbb{Z}^d \backslash 0$

$$\left\| \frac{\widehat{\psi}(\cdot + \beta)}{\|\cdot\|^m \widehat{\psi}(\cdot)} \right\|_{\infty,\Omega} = \left\| \frac{\widehat{\phi}(\cdot + \beta)}{\|\cdot\|^m \widehat{\phi}(\cdot)} \right\|_{\infty,\Omega} \leq A\|\widehat{\phi}(\cdot + \beta)\|_{\infty,\Omega}. \qquad (8.3.19)$$

This together with (8.3.3)(b) yields (8.3.13), and hence (8.3.14) with k replaced by m'.

8.4   Approximation on quasi-uniformly scattered centres

In this section we construct analogous approximation schemes to those studied in the previous sections, based on the translates of $\phi(h^{-1}\cdot)$ to sets of scatterd points $\{\Xi_h\}_{h>0}$. These schemes achieve the same approximation orders as their regular-grid analogues, provided that the sets $\{\Xi_h\}_{h>0}$ satisfy certain quasi-uniformity conditions.

8.4.1   *Quasi-uniform sets of points*

We start by stating the quasi-uniformity conditions on the sets of points $\{\Xi_h\}_{h>0}$.

Definition 8.14 *A set Ξ_h of points (centres) is called quasi-uniform of type ρ at level h if*

$$\{y : \|y - x\| \leq \rho\} \cap h^{-1}\Xi_h \neq \emptyset \quad \text{for all } x \in \mathbb{R}^d. \qquad (8.4.1)$$

In fact a weaker implicit condition on Ξ_h is the key property needed in the forthcoming theory.

Definition 8.15 *A set Ξ_h is called k-approximating of type R, E at level h, if there exists a matrix $\{K(\alpha, \xi) : \alpha \in \mathbb{Z}^d, \ \xi \in \Xi_h\}$ with the properties*

(a) $\displaystyle\sum_{\xi \in \Xi_h} |K(\alpha, \xi)| < E, \quad \alpha \in \mathbb{Z}^d,$

(b) $K(\alpha, \xi) = 0 \quad \text{if } \|\alpha - h^{-1}\xi\| > R, \ \alpha \in \mathbb{Z}^d, \ \xi \in \Xi_h,$ \hfill (8.4.2)

(c) $\displaystyle\sum_{\xi \in \Xi_h} K(\alpha, \xi) p(h^{-1}\xi) = p(\alpha), \quad p \in \pi_k, \ \alpha \in \mathbb{Z}^d.$

It is shown by Buhmann et al. (1995) that a quasi-uniform set of points of type $\underset{\sim}{\rho}$ at level h is also k-approximating of type R, E at level $\tilde{h}$ for any k, with R, E, and $\tilde{h}$ depending on ρ and k.

In (Dyn and Ron 1995) condition (b) is replaced by the weaker condition

$$\sum_{\xi \in \Xi_h} |K(\alpha, \xi)|(1 + \|h^{-1}\xi - \alpha\|^j) < E_j, \quad \alpha \in \mathbb{Z}^d, \ j = 1, \ldots, s, \qquad (8.4.3)$$

with $s > k$.

In the following we assume that the set Ξ_h under investigation is k-approximating of type R, E at level h for the required k, with fixed R and E. We denote the "active" subset of Ξ_h

$$\{\xi \in \Xi_h : K(\alpha, \xi) \neq 0 \ \text{for some } \alpha \in \mathbb{Z}^d\}, \qquad (8.4.4)$$

as our set Ξ_h.

8.4.2 *The approximation scheme*

The approximation scheme to be constructed is of the form

$$\mathcal{L}_{\Xi_h} f(x) = \sum_{\xi \in \Xi_h} \psi_\xi(h^{-1}x)\Gamma_h f(\xi), \qquad (8.4.5)$$

where Γ_h is either the identity for the quasi-interpolatory schemes, or $\Gamma_h = \Lambda_h$ for the optimal schemes, and where

$$\psi_\xi(x) = \sum_{\eta \in \Xi_h} N(\xi, \eta)\phi(x - h^{-1}\eta), \quad \xi \in \Xi_h. \qquad (8.4.6)$$

The matrix N is defined in terms of the matrix K as

$$N(\xi, \eta) = \sum_{\alpha, \beta \in \mathbb{Z}^d} K(\alpha, \xi)\mu_{\beta - \alpha} K(\beta, \eta), \quad \xi, \eta \in \Xi_h, \qquad (8.4.7)$$

with $\{\mu_\alpha\}$ an appropriate localization sequence, used in the analogous scheme on the regular grid.

The method for deriving the approximation order provided by $\mathcal{L}_{\Xi_h}$ is by comparison to the corresponding scheme on the regular grid $\mathcal{L}_h$ given by

$$\mathcal{L}_h f(x) = \sum_{\alpha \in \mathbb{Z}^d} \psi(h^{-1}x - \alpha) \Gamma_h f(h\alpha), \qquad (8.4.8)$$

with Γ_h as above, and with

$$\psi = \sum_{\alpha \in \mathbb{Z}^d} \mu_\alpha \phi(\cdot - \alpha). \qquad (8.4.9)$$

The comparison is aimed at showing that

$$\|\mathcal{L}_{\Xi_h} f - \mathcal{L}_h f\|_\infty \le A h^\ell, \qquad (8.4.10)$$

with ℓ not smaller than the known approximation order provided by $\mathcal{L}_h$. In many cases ℓ is greater than or equal to the optimal approximation order provided by the given ϕ (see Section 8.3). In fact this method of comparison works for any scheme derived from one on a regular grid, if the schemes (8.4.5) and (8.4.8) are defined by the same localization sequence $\{\mu_\alpha\}$ and by the same operator Γ_h. This is the general setting in which the results in (Dyn and Ron 1995) are derived.

8.4.3   The pseudo-shifts

To reveal the similarity between $\mathcal{L}_h$ and $\mathcal{L}_{\Xi_h}$ we first introduce the "pseudo-shifts" $\{\phi_\alpha : \alpha \in \mathbb{Z}^d\}$, which approximate the shifts $\{\phi(\cdot - \alpha) : \alpha \in \mathbb{Z}^d\}$ in a relevant sense to the comparison (8.4.10). The pseudo-shifts have the form

$$\phi_\alpha = \sum_{\xi \in \Xi_h} K(\alpha, \xi) \phi(\cdot - h^{-1}\xi), \qquad \alpha \in \mathbb{Z}^d, \qquad (8.4.11)$$

and their relation to the shifts is of the following nature.

Theorem 8.16 *Let ϕ be one of the radial or related functions introduced in the Introduction, and denote by $m' \in \mathbb{R}_+$ the order of the singularity of $\hat{\phi}$ at the origin. Let K be the matrix satisfying (8.4.2). Then*

$$|\Phi_\alpha(x)| = |\phi_\alpha(x) - \phi(x - \alpha)| \le A(1 + \|x - \alpha\|)^{-n_\Phi}, \qquad \alpha \in \mathbb{Z}^d, \qquad (8.4.12)$$

with A dependent on E, R, k but not on α, and with

$$n_\Phi = k - m' + d + 1. \qquad (8.4.13)$$

The proof of this theorem is based on the explicit form of the expansion of the Fourier transform of Φ_α near the origin and its decay behaviour as $\|\omega\| \to \infty$. Since

$$\widehat{\Phi}_\alpha(\omega) = e_\alpha(\omega)\widehat{\phi}(\omega), \quad e_\alpha(\omega) = \sum_{\xi \in \Xi_h} K(\alpha, \xi) e^{-i\omega \cdot h^{-1}\xi} - e^{-i\omega \cdot \alpha}, \qquad (8.4.14)$$

it is the zero of order $k + 1$ of $e_\alpha(\omega)$ at the origin which cancels the singularity of $\hat{\phi}$ there in case $k \ge [m']$. For the comparison analysis we assume that in (8.4.2) $k = [m']$ so that $n_\Phi > d$.

With the pseudo-shifts of ϕ we also define the pseudo-shifts of ψ, in an analogous form to (8.4.9)

$$\psi_\alpha = \sum_{\beta \in \mathbb{Z}^d} \mu_{\beta-\alpha} \phi_\beta. \tag{8.4.15}$$

In case the localization sequence $\{\mu_\alpha\}$ is of finite support, as in Section 8.2, we conclude from (8.4.12) and (8.4.15) that

$$|\psi_\alpha(x) - \psi(x - \alpha)| \le A(1 + \|x - \alpha\|)^{-n_\Phi}, \quad \alpha \in \mathbb{Z}^d. \tag{8.4.16}$$

A bound as in (8.4.16) also holds for more general localization sequences $\{\mu_\alpha\}$.

Theorem 8.17 *Assume*

$$|\mu_\alpha| \le A(1 + \|\alpha\|)^{-n_\mu}, \quad n_\mu > d, \tag{8.4.17}$$

$$|\psi(x)| \le A(1 + \|x\|)^{-n_\psi}, \quad n_\psi > d, \tag{8.4.18}$$

and that the sum in (8.4.9) is uniformly convergent on compact sets of $\mathbb{R}^d$. Then

$$|\Psi_\alpha(x)| = |\psi_\alpha(x) - \psi(x - \alpha)| \le A(1 + \|x - \alpha\|)^{-n_\Psi} \tag{8.4.19}$$

with $n_\Psi = \min\{n_\mu, n_\Phi\} > d$. Also

$$|\psi_\alpha(x)| \le A(1 + \|x - \alpha\|)^{-n'_\psi}, \tag{8.4.20}$$

with $n'_\psi = \min\{n_\mu, n_\Phi, n_\psi\} > d$.

In the proof of (8.4.19), the sum defining Ψ_α

$$\Psi_\alpha(x) = \psi_\alpha(x) - \psi(x - \alpha) = \sum_{\beta \in \mathbb{Z}^d} \mu_{\beta-\alpha} [\phi_\beta(x) - \phi(x - \beta)], \tag{8.4.21}$$

is estimated, in view of (8.4.12), by the discrete convolution

$$\sum_{\beta \in \mathbb{Z}^d} |\mu_\beta| (1 + \|(y + \alpha') - \beta\|)^{-n_\Phi}, \quad \|y\|_\infty \le \frac{1}{2}, \quad \alpha' \in \mathbb{Z}^d.$$

The estimate (8.4.20) is a direct consequence of (8.4.18) and (8.4.19).

Using the pseudo-shifts $\{\psi_\alpha\}$, we rewrite ψ_ξ of (8.4.6) as

$$\psi_\xi = \sum_{\alpha \in \mathbb{Z}^d} K(\alpha, \xi) \psi_\alpha, \tag{8.4.22}$$

which together with (8.4.20) leads to

Corollary 8.18 *Under the conditions of Theorem 8.17,*

$$|\psi_\xi(x)| \le A(1 + \|x - h^{-1}\xi\|)^{-n'_\psi}, \tag{8.4.23}$$

and the approximation scheme (8.4.5) is well defined for $f \in C(\mathbb{R}^d) \cap L_\infty(\mathbb{R}^d)$.

8.4.4 Comparison theorem for $C^\ell(\mathbb{R}^d) \cap W_\infty^\ell(\mathbb{R}^d)$

One more observation is needed before we can prove the first comparison result.

Lemma 8.19 *Let $\{\mu_\alpha\}$ be a localization sequence such that ψ satisfies (8.4.18), and such that the linear functional*

$$\mu p = \sum_{\alpha \in \mathbb{Z}^d} \mu_\alpha p(-\alpha), \tag{8.4.24}$$

is well defined for $p \in \pi_\ell$. Then $\mu p = 0$, $p \in \pi_\ell$ whenever $\ell < m'$. Moreover if μ is well defined on $\pi_{\ell+1}$, then for $f \in C^{\ell+1}(\mathbb{R}^d) \cap W_\infty^{\ell+1}(\mathbb{R}^d)$

$$\left|\mu f(h(\beta + \cdot))\right| \le A h^{\ell+1} \|f\|_{\infty,\ell+1}, \quad \beta \in \mathbb{Z}^d. \tag{8.4.25}$$

Proof Assume $\ell < m'$. By the conditions on μ, $\widehat{\mu}(\omega) = \sum_{\alpha \in \mathbb{Z}^d} \mu_\alpha e^{-i\omega \cdot \alpha}$ is ℓ-times differentiable. It has a zero of order $\ell + 1$ at the origin, since $\psi = \widehat{\mu}\widehat{\phi}$ is continuous everywhere. This implies that $\mu p = 0$ for $p \in \pi_\ell$, from which (8.4.25) follows by considering the operation of μ on $(f - T_{h\beta}f)(h(\beta+\cdot))$, with $T_x f$ the Taylor polynomial of f at x of degree ℓ.

With the results of Lemma 8.19, it is possible to compare $\mathcal{L}_{\Xi_h} f$ with $\mathcal{L}_h f$ in the case where $\mathcal{L}_h$ is a quasi-interpolation scheme or a cardinal interpolation scheme. For the schemes considered in Section 8.2 $m' = 2m$, $\{\mu_\alpha\}$ is of compact support and Lemma 8.19 holds for $0 \le \ell \le 2m - 1$. For the case of cardinal interpolation schemes see (Buhmann 1990; Dyn and Ron 1995) for the relevant properties of $\{\mu_\alpha\}$, depending on the properties of ϕ. For these two types of schemes the relevant comparison theorem is

Theorem 8.20 *Let $\{\mu_\alpha\}$ be a localization scheme satisfying the conditions of Lemma 8.19, and let $\Gamma_h f = f$. Then for $f \in C^{\ell+1}(\mathbb{R}^d) \cap W_\infty^{\ell+1}(\mathbb{R}^d)$*

$$\|(\mathcal{L}_{\Xi_h} - \mathcal{L}_h)f\|_\infty \le A\|f\|_{\infty,\ell+1} h^{\ell+1}. \tag{8.4.26}$$

In case ℓ is the maximal integer smaller than m', $\mathcal{L}_{\Xi_h}$ provides the same approximation order as $\mathcal{L}_h$.

Proof Let us introduce an intermediate scheme between $\mathcal{L}_h$ and $\mathcal{L}_{\Xi_h}$:

$$\widetilde{\mathcal{L}}_h f = \sum_{\alpha \in \mathbb{Z}^d} f(\alpha h)\psi_\alpha(h^{-1}\cdot). \tag{8.4.27}$$

Then

$$
\begin{aligned}
(\widetilde{\mathcal{L}}_h - \mathcal{L}_h)f(x) &= \sum_{\alpha \in \mathbb{Z}^d} f(\alpha h)\Psi_\alpha(h^{-1}x) \\
&= \sum_{\alpha \in \mathbb{Z}^d} f(\alpha h) \sum_{\beta \in \mathbb{Z}^d} \mu_{\beta-\alpha}\Phi_\beta(h^{-1}x).
\end{aligned}
\tag{8.4.28}
$$

Since Φ_β satisfies (8.4.12), we can change the order of summation in the right-hand side of (8.4.28) and apply (8.4.25) to obtain

$$|(\widetilde{\mathcal{L}}_h - \mathcal{L}_h)f(x)| = \left| \sum_{\beta \in \mathbb{Z}^d} \Phi_\beta(h^{-1}x)\mu f(h(\beta + \cdot)) \right| \leq Ah^{\ell+1}\|f\|_{\infty,\ell+1}. \quad (8.4.29)$$

To conclude (8.4.26) we have still to estimate

$$(\mathcal{L}_{\Xi_h} - \widetilde{\mathcal{L}}_h)f(x) = \sum_{\xi \in \Xi_h} f(\xi) \sum_{\alpha \in \mathbb{Z}^d} K(\alpha, \xi)\psi_\alpha(h^{-1}x)$$
$$- \sum_{\alpha \in \mathbb{Z}^d} f(\alpha h)\psi_\alpha(h^{-1}x). \quad (8.4.30)$$

Again, since ψ_α satisfies (8.4.20) with $n'_\psi > d$, we can change the order of summation in the first sum in (8.4.30) to obtain

$$(\mathcal{L}_{\Xi_h} - \widetilde{\mathcal{L}}_h)f(x) = \sum_{\alpha \in \mathbb{Z}^d} \left[\sum_{\xi \in \Xi_h} K(\alpha, \xi)f(\xi) - f(\alpha h) \right] \psi_\alpha(h^{-1}x). \quad (8.4.31)$$

Using properties (8.4.2) of the matrix K with $k = [m']$, we obtain in analogy to (8.4.25)

$$\left| f(\alpha h) - \sum_{\xi \in \Xi_h} K(\alpha, \xi)f(\xi) \right| \leq A\|f\|_{\infty,\ell+1}h^{\ell+1}, \quad \ell \leq [m']. \quad (8.4.32)$$

This leads to the estimate

$$\|(\mathcal{L}_{\Xi_h} - \widetilde{\mathcal{L}}_h)f\|_\infty \leq A\|f\|_{\infty,\ell+1}h^{\ell+1}, \quad (8.4.33)$$

which together with (8.4.29) yields (8.4.26). In case ℓ is the greatest integer less than m' then $\ell + 1$ is greater than or equal to the possible optimal approximation order m', and the last claim of the theorem follows.

The proof of Theorem 8.20 gives another interpretation to the scheme $\mathcal{L}_{\Xi_h}f$, namely

$$\mathcal{L}_{\Xi_h}f = \sum_{\xi \in \Xi_h} f(\xi)\psi_\xi(h^{-1}x) = \sum_{\alpha \in \mathbb{Z}^d} \widetilde{f}(h\alpha)\psi_\alpha(h^{-1}x), \quad (8.4.34)$$

where $\widetilde{f}(\alpha h) = \sum_{\xi \in \Xi_h} K(\alpha, \xi)f(\xi)$ approximates $f(\alpha h)$ with an appropriate order.

Remarks

(i) For the quasi-interpolation schemes in Section 8.2, $\{\mu_\alpha\}$ is of compact support and ℓ in Theorem 8.20 coincides with ℓ of Section 8.2, thus guaranteeing that $\mathcal{L}_{\Xi_h}$ provides the same approximation order as $\mathcal{L}_h$ for $0 \leq \ell \leq m' - 1$. More complicated situations, where $\ell + 1$ in (8.4.26) is not the optimal approximation order but equals or exceeds the one achieved by $\mathcal{L}_h$, are analysed in (Dyn and Ron 1995) in the context of cardinal interpolation.

(ii) The class of approximated functions f in Theorem 8.20 is smaller than that considered in the regular grid case, when polynomial reproduction arguments are employed. This is due to the mild requirement on the decay of Φ_α (and hence on ψ_α and ψ_ξ) in the comparison analysis. It is possible to obtain by similar arguments the results of Theorem 8.20 for unbounded $f \in C^{\ell+1}(\mathbb{R}^d)$, with bounded derivatives of order ℓ and $\ell + 1$, if Φ_α is assumed to decay as fast as ψ in (8.4.9). This can be obtained by taking a large enough k in condition (8.4.2) on the matrix K. In this setting, $\mathcal{L}_{\Xi_h}$ reproduces polynomials in π_ℓ for $\ell < m'$, although this property is not needed in the comparison analysis (see details in Dyn and Ron 1995).

(iii) The polynomial reproduction property of $\mathcal{L}_{\Xi_h}$ is the key property used in (Buhmann et al. 1995) for obtaining the approximation order provided by quasi-interpolation schemes of the form $\mathcal{L}_{\Xi_h} f = \sum_{\xi \in \Xi_h} f(\xi) \psi_\xi(h^{-1}\cdot)$, for ϕ satisfying the assumptions of Section 8.2. There the decay of ψ_ξ is obtained directly from properties of the matrix N in (8.4.6), (8.4.7) and from similar arguments to those in (Dyn et al. 1992). The proof of the polynomial reproduction property in (Buhmann et al. 1995) considers a sequence of sets $\{\Xi_{h,M} : M \in \mathbb{Z}_+\}$ such that $\Xi_{h,M}$ coincides with Ξ_h for $\|h^{-1}\xi\|_\infty \le M$ and with an appropriate regular grid for $\|h^{-1}\xi\|_\infty \ge M + A$, and shows that $\mathcal{L}_{\Xi_h} p$, $p \in \pi_\ell$, depends mostly on that part of the set of points which is regular.

8.4.5 Comparison theorem for $\widetilde{W}_\infty^m(\mathbb{R}^d)$

A second comparison theorem for schemes defined on the spaces $\widetilde{W}_\infty^m(\mathbb{R}^d)$, as in Section 8.3, is based on the following analogue of Lemma 8.19.

Lemma 8.21 *Assume $\{\mu_\alpha\}$ is such that for some $m \in \mathbb{R}_+$, $\|\cdot\|^{-m}|\hat{\mu}|$ is bounded. Let $\hat{\lambda} \in L_\infty(\mathbb{R}^d)$, and define $\widehat{\Gamma_h f} = \hat{f}\,\hat{\lambda}$. Then for any $f \in \widetilde{W}_\infty^m(\mathbb{R}^d)$*

$$\left|\mu\Gamma_h f\big(h(\beta + \cdot)\big)\right| \le Ah^m |f|'_{\infty,m}, \quad \beta \in \mathbb{Z}^d, \tag{8.4.35}$$

where $|f|'_m = \int_{\mathbb{R}^d} \|\omega\|^m |\hat{f}(\omega)| d\omega < \infty$.

The proof of (8.4.35) is carried in the Fourier domain and is quite involved. Lemma 8.21 leads to

Theorem 8.22 *Assume $\{\mu_\alpha\}$ and λ satisfy the requirements of Lemma 8.21. Then for every $f \in \widetilde{W}_\infty^m(\mathbb{R}^d)$, $m \le m'$*

$$\|(\mathcal{L}_{\Xi_h} - \mathcal{L}_h)f\|_\infty \le Ah^m |f|'_{\infty,m}. \tag{8.4.36}$$

In particular for $f \in \widetilde{W}_\infty^{m'}(\mathbb{R}^d)$, $\mathcal{L}_{\Xi_h}$ provides the same approximation order as $\mathcal{L}_h$.

Proof The proof is similar to that of Theorem 8.20, with Lemma 8.21 replacing Lemma 8.19. One has also to show that for $m \le m'$

$$\left|\sum_{\xi \in \Xi_h} K(\alpha, \xi) f(\xi) - f(\alpha h)\right| \le Ah^m |f|'_{\infty,m}, \quad \alpha \in \mathbb{Z}^d, \tag{8.4.37}$$

for the estimation of $(\mathcal{L}_{\Xi_h} - \widetilde{\mathcal{L}}_h)f$. This follows from the observation that

$$\sum_{\xi \in \Xi_h} K(\alpha, \xi) f(\xi) - f(\alpha h) = \frac{1}{(2\pi)^d} \int_{\mathbb{R}^d} e_\alpha(h\omega) \widehat{f}(\omega) d\omega, \qquad (8.4.38)$$

with $e_\alpha(\omega)$ defined as in (8.4.14). Since $e_\alpha \in C^{[m']+1}(\mathbb{R}^d)$ and has a zero of order $[m']+1$ at the origin (by conditions (8.4.2) on K with $k = [m']$), then $\|\omega\|^{-m} e_\alpha(\omega)$ is bounded for $m \leq m'$ and (8.4.37) follows from (8.4.38).

9

ALMOST SURE SAMPLING RESTORATION OF
BAND-LIMITED STOCHASTIC SIGNALS

9.1 Historical overview and oversampling

The origin of sampling principles for stochastic signals is not exactly known. Work of Kotel'nikov (1933) in this area was first made known to Western mathematicians by Kolmogorov, who mentioned it in a brief manner at the end of his talk at the Symposium on Information Theory at MIT, held on September 10–12, 1956 (Kolmogorov 1956). This involves the well-known sampling series, usually attributed to Whittaker, Kotel'nikov and Shannon in the Western literature, but often called the Kotel'nikov formula in Eastern Europe and the former Soviet Union countries.

In France, Oswald used this formula for weakly stationary stochastic processes (denoted by WSSP throughout this chapter) and stationary Gaussian processes (Laplace processes in his terminology) but without any remark as to its origin (see Oswald 1951). Also, Ville (1953) considered the sampling expansion of white noise. Yaglom (1962, p. 204) attributes this formula for band-limited (BL in the sequel) WSSPs in part to himself (see also Yaglom 1949). Similarly, Belyaev (1959) also attributes the formula to Yaglom (1955) and to Harkevich (1955). However, the formula for homogeneous random fields (HRF in the sequel) seems to originate in the early 1950s in the work of Obukhov (Skorokhod 1990).[1] Here we point out the famous paper of Balakrishnan (1957) where he initiated the sampling restoration of BL WSSPs with full mathematical rigour. In his paper the mean-square (m.s. in the sequel) restoration is established if the masses of the spectral distribution function of the observed process are equal to zero at the end-points of the frequency spectrum support interval $[-w, w]$. This sufficient condition appears in several subsequent studies; for example Balakrishnan (1957) established the exact value of the m.s. sampling cardinal series (Kotel'nikov series) at $\pm w$ for processes, and Pogány (1991) computed the value of multiple Kotel'nikov series at the vertices of the frequency spectrum support rectangle for BL HRFs.

Throughout this chapter we shall continue to use the name Kotel'nikov series, or formula, thereby emphasizing the stochastic setting in which the discussion takes place.

However, the first two papers which consider sampling reconstruction in the almost sure sense (a.s. in the sequel), or restoration "with probability 1", were Belyaev (1959) and Lloyd (1959). (In fact, the stochastic signal is defined on a probability space $(\Omega, \mathcal{A}, \mathbf{P})$. Its sampling restoration in the a.s. sense means that the Kotel'nikov formula is valid for almost all sample functions of the considered stochastic signal, i.e. the Kotel'nikov formula is valid almost everywhere with respect to the probability measure

---

[1] A search of certain papers by Obukhov, published between 1941 and 1954, did not confirm this assertion. However, it is known that Parzen (1956) derived the sampling formula for HRF's, and that Miyakawa was also one of the first authors of this formula in 1959, see Jerri (1977).

P$(d\omega)$ on Ω.) In this context we draw the reader's attention to the paper of Belyaev (1959). He reported that:

" ... the Kotel'nikov formula appeared often in the literature (e.g. Harkevich 1955; Yaglom 1955), but without rigorous mathematical treatment in proof for the sample functions of the considered BL stochastic process ... ".

Using an oversampling approach, Belyaev proved that if $\{\xi(t) := \xi(t, \omega); t \in \mathbf{R}, \omega \in \Omega\}$ is the zero-mean WSSP, BL to $\tilde{w} > 0$, i.e., its covariance function $B(t) = \mathbf{E}\xi(t)\xi^*(0)^2$ has the form:

$$B(t) = \int_{-\tilde{w}}^{\tilde{w}} e^{it\lambda} F(d\lambda); \qquad (9.1.1)$$

and

$$\xi(t) = \int_{-\tilde{w}}^{\tilde{w}} e^{it\lambda} \zeta(d\lambda), \qquad (9.1.2)$$

then

$$\xi(t) = \sum_{n \in \mathbf{Z}} \xi(n\pi/w) \frac{\sin(wt - n\pi)}{wt - n\pi}, \qquad (9.1.3)$$

in the almost sure sense for all $w > \tilde{w}$, uniformly with respect to t on all bounded subsets of $\mathbf{R}$.[3] Here $\zeta(\Delta)$ is the random spectral measure with orthogonal increments on $\mathbf{R}$ and $F(\Delta_1 \cap \Delta_2) = \mathbf{E}\zeta(\Delta_1)\zeta^*(\Delta_2)$ is the spectral measure of $\xi(t)$; $\Delta_1, \Delta_2 \subset \mathbf{R}$. At the same time, oversampling allows the inconvenient requirement that F be continuous at $\pm w$, i.e., the necessity that $F(\{\pm w\}) = 0$, to be avoided.

Remark 1 *Throughout this chapter Kolmogorov's Hilbert space approach will be exploited. Let $L_2(\Omega)$ be the Hilbert space of random variables with finite second moment. Denote by $L_2(\mathbf{R}, F)$ the Hilbert space of square-integrable functions $\phi(\lambda)$ with respect to the measure $F(d\lambda)$ on $\mathbf{R}$. If $H(\xi) \subset L_2(\Omega)$ is the L_2-closed linear span of the WSSP $\xi(t)$ (its Hilbert space), then there is an isometry $\xi(t) \longleftrightarrow e^{it\lambda}$ between $H(\xi)$ and $L_2(\mathbf{R}, F)$, realized by the mathematical expectation $\mathbf{E}$.*

The basic idea in Belyaev's oversampling approach was the following. It is easy to see that if the stochastic signal (scalar or vectorial process, or field) is BL to the given bandwidth $\tilde{w}$, then it is also BL to any greater bandwidth w. Denote by $\xi_N(t)$ the so-called *truncated Kotel'nikov series*:

$$\xi_N(t) := \sum_{n=-N}^{N} \xi(n\pi/w) \frac{\sin(wt - n\pi)}{wt - n\pi}. \qquad (9.1.4)$$

Since the m.s. truncation error $\mathcal{E}_N(\xi, t) := \mathbf{E}|\xi(t) - \xi_N(t)|^2$ possesses the upper bound

$$\mathcal{E}_N(\xi, t) \leq \frac{16B(0)(2\pi + |t|w)^2}{\pi^4(1 - \tilde{w}/w)^2 N^2} \qquad (9.1.5)$$

---

[2]α^* denotes the complex conjugate of α.

[3]Without loss of generality, we consider zero-mean stochastic signals throughout this chapter.

for all $t \in \mathbf{R}$, it follows by the Chebyshev inequality that the series

$$\sum_{n \geq N} \mathbf{P}\{|\xi(t) - \xi_n(t)| \geq \varepsilon\} = \sum_{n \geq N} \mathcal{O}(n^{-2}) \qquad (9.1.6)$$

converges. Therefore, by the Borel–Cantelli lemma we get the a.s. convergence result in relation (9.1.3).

The a.s. convergence rate can be evaluated in (9.1.3) in the following way. As

$$\mathbf{P}\{\exists n \geq N : |\xi(t) - \xi_n(t)| \geq g(n)\} \leq \sum_{n \geq N} \frac{\mathcal{E}_n(\xi, t)}{g^2(n)}$$

$$\leq \frac{16B(0)(2\pi + |t|w)^2}{\pi^4(1 - \tilde{w}/w)^2} \sum_{n \geq N} \frac{1}{n^2 g^2(n)}, \qquad (9.1.7)$$

the convergence of the series $\sum_{n \geq N} \frac{1}{n^2 g^2(n)}$ ensures (by the Borel–Cantelli lemma) the existence of a positive integer $N(\omega)$, $\omega \in \Omega$, such that $|\xi(t) - \xi_n(t)| < g(n)$ for all $n \geq N(\omega)$, with probability 1. Therefore we can take, e.g.,

$$|\xi(t) - \xi_n(t)| < \frac{(\ln n)^{(1+\epsilon)/2}}{\sqrt{n}}; \quad \epsilon > 0, \qquad \text{a.s.} \qquad (9.1.8)$$

for all $n \geq N(\omega)$.

Analogously, we can prove for BL weakly stationary Gaussian $\xi(t)$ with continuous spectral density that (9.1.3) follows almost surely at $\tilde{w} = w$.[4]

Following Belyaev's oversampling method, the almost sure sampling restoration of BL and non-BL vectorial processes (with correlated coordinates) were considered by Pogány and Perunič (1991, 1992). Further steps were taken in the restoration of BL and non-BL HRFs, in the a.s. sense, by Pogány and Perunič (1995).

Thus, let $\{\Xi(x); x = (x_1, \ldots, x_n) \in \mathbf{R}^n\}$ be a scalar HRF with the covariance function $\mathcal{K}(x) = \mathbf{E}\Xi(x)\Xi^*(0)$. It is BL with respect to the frequency spectrum support rectangle $\tilde{\mathcal{W}} = \prod_{j=1}^n [-\tilde{w}_j, \tilde{w}_j]$ if $\mathcal{K}(x)$ possesses the spectral representation:

$$\mathcal{K}(x) = \int_{-\tilde{w}_1}^{\tilde{w}_1} \cdots \int_{-\tilde{w}_n}^{\tilde{w}_n} e^{i\langle \lambda, x \rangle} \Phi(d\lambda). \qquad (9.1.9)$$

Then the spectral representation of $\Xi(x)$ is equal to

$$\Xi(x) = \int_{-\tilde{w}_1}^{\tilde{w}_1} \cdots \int_{-\tilde{w}_n}^{\tilde{w}_n} e^{i\langle \lambda, x \rangle} \Upsilon(d\lambda), \qquad (9.1.10)$$

where $\langle \lambda, x \rangle := \sum_{i=1}^n \lambda_i x_i$ is the inner product in $\mathbb{R}^n$. Here $\Upsilon(\cdot)$ is a vectorial orthogonal random measure on $\mathbb{R}^n$, and $\Phi(\cdot)$ denotes the spectral measure of $\Xi(x)$ such that

---

[4]It is interesting to remark here that the results presented above were the main part of the Graduation Thesis by Belyaev, for which V.Ya. Kozlov was the supervisor and A.N. Kolmogorov the research adviser.

$\mathbf{E}\Upsilon(\Delta_1)\Upsilon^*(\Delta_2) = \Phi(\Delta_1 \cap \Delta_2)$, $\Delta_1, \Delta_2 \subset \mathbb{R}^n$. The multiple Kotel'nikov formula states that the BL HRF $\Xi(x)$ can be expressed as a linear combination of quantities $\Xi(x^m)$, where x^m denotes the points of the lattice

$$Lat(\mathcal{W}) := \left\{ x^m = \left(\frac{\pi m_1}{w_1}, \ldots, \frac{\pi m_n}{w_n} \right); m_k \in \mathbf{Z} \right\}, \qquad (9.1.11)$$

where $w_k > \tilde{w}_k, k = \overline{1, n}$. This linear combination is just

$$\Xi(x) = \sum_{m=(m_1,\ldots,m_n)} \Xi(x^m) \prod_{k=1}^{n} \frac{\sin(w_k x_k - \pi m_k)}{w_k x_k - \pi m_k}. \qquad (9.1.12)$$

Precisely, this result holds in the m.s. sense, and also in the a.s sense. Indeed, the truncation error $\mathcal{E}_M(\Xi, x) = \mathbf{E}|\Xi(x) - \Xi_m(x)|^2$, $M = \min_{1 \leq j \leq n}(m_j)$, with respect to the truncated Kotel'nikov series

$$\Xi_m(x) = \sum_{k=1}^{n} \sum_{|j_k| \leq m_k} \Xi(x^j) \prod_{k=1}^{n} \frac{\sin(w_k x_k - \pi m_k)}{w_k x_k - \pi m_k} \qquad (9.1.13)$$

possesses the upper bound

$$\mathcal{E}_M(\Xi, x) \leq A_n \mathcal{K}(0) \left(\frac{16}{\pi^4} \right)^n \prod_{k=1}^{n} \frac{(2\pi + w_k|x_k|)^2}{m_k^2 (1 - \tilde{w}_k/w_k)^2}, \qquad (9.1.14)$$

where A_n is an absolute constant; $A_1 \equiv 1$. It follows immediately from (9.1.14) that $\mathcal{E}_M(\Xi, x) = \mathcal{O}(M^{-2n})$, thus l.i.m.$_{M \to \infty} \Xi_m(x) = \Xi(x)$,[5] respectively

$$\mathbf{P} \left\{ \lim_{M \to \infty} \Xi_m(x) = \Xi(x) \right\} = 1. \qquad (9.1.15)$$

For further details and for the non-BL HRF case one can consult Pogány and Peruničić (1995). The same results for the sampling restoration of BL HRF were derived independently by S. Halikulov, a student of the University of Kiev (as reported by Leonenko 1993).

The almost sure sampling restoration procedure for non-BL WSSP with the help of the BL Kotel'nikov formula is illustrated by the following consideration. Let $\{X(t); t \in \mathbb{R}\}$ be a non-BL scalar WSSP and $C(t)$ its covariance function, i.e.

$$X(t) = \int_{\mathbb{R}} e^{it\lambda} Z(d\lambda); \qquad C(t) = \int_{\mathbb{R}} e^{it\lambda} F(d\lambda). \qquad (9.1.16)$$

Let $\mathcal{L}_w$ be the linear transformation (or filter) with spectral characteristic function $\chi_w(\lambda)$ which coincides with the indicator function of the interval $[-w, w]$. Then we get:

$$\mathcal{L}_w X(t) = \int_{\mathbb{R}} e^{it\lambda} \chi_w(\lambda) Z(d\lambda) = \int_{-w}^{w} e^{it\lambda} Z(d\lambda),$$

---

[5]l.i.m. means the "limes in medio", i.e. the m.s. limit.

and it follows that $\mathcal{L}_w X(t)$ is BL to w. Now, we can expand $\mathcal{L}_w X(t)$ into the Kotel'nikov series $\mathcal{K}\mathcal{L}_w X(t)$. Yet it is not necessary to assume that $F(\{\pm w_k\}) = 0, k \in \mathbf{Z}$ for the a.s. convergence result

$$\mathbf{P}\left\{ \lim_{k \to \infty} \mathcal{K}\mathcal{L}_{w_k} X(t) = X(t) \right\} = 1, \qquad (9.1.17)$$

in which $\{w_k\}_0^\infty$, $w_0 \equiv w$, is a conveniently choosen monotonically increasing divergent positive real sequence. Indeed, since

$$\mathbf{E}(X(t) - \mathcal{L}_{w_k} X(t))(\mathcal{L}_{w_k} X(t) - \mathcal{K}\mathcal{L}_{w_k} X(t))^* = 0,$$

it follows by (9.1.5) that

$$\begin{aligned} \mathcal{E}_{N,k}(X, t) &= \mathbf{E}|X(t) - (\mathcal{K}\mathcal{L}_{w_k} X)_N(t)|^2 \\ &= \int_{\mathbb{R}} (1 - \chi_{w_k}(\lambda)) F(d\lambda) + \mathcal{E}_N(\mathcal{L}_{w_k} X, t). \end{aligned}$$

Here $(\mathcal{K}\mathcal{L}_{w_k} X)_N(t)$ is the truncated part of the Kotel'nikov series $\mathcal{K}\mathcal{L}_{w_k} X(t)$ (in the sense of (9.1.4)). Because

$$\mathbf{P}\left\{ \lim_{N \to \infty} (\mathcal{K}\mathcal{L}_{w_k} X)_N(t) = \mathcal{L}_{w_k} X(t) \right\} = 1,$$

(BL case), the a.s. convergence in (9.1.17) depends on the choice of the sequence $\{w_k\}_0^\infty$. As $\{(C(0))^{-1} F(S); S \in \mathcal{B}_{\mathbb{R}}\}^6$ is a probability measure, it is possible to define a sequence $\{w_k\}_0^\infty$, such that

$$\sum_{k=1}^\infty \int_{\mathbb{R}} (1 - \chi_{w_k}(\lambda)) F(d\lambda) < \infty,$$

therefore (9.1.18) is proved (see Pogány 1989; Pogány and Peruničić 1991). The generalization of this result to non-BL HRF can be found in Pogány (1991; Theorem 3). This restoration procedure is only approximate, since $w_k \uparrow \infty$ as $k \to \infty$.

It is clear that in the foregoing references the principal result is (for BL signals) the convergence of the Kotel'nikov series in the m.s. sense. Then, with the help of the m.s. truncation error upper bound, a.s. convergence is derived. Also, in all the foregoing papers only sufficient conditons for the a.s. restoration of stochastic signals are given.

At this point we have to mention the important results in signal analysis and linear prediction concerning sampling theorems which have been derived by the group of mathematicians from Aachen. Connected to our subject is a very important approach using the so-called *convolution processes* introduced by W. Splettstößer, who translates certain deterministic signal restoration results by Butzer, Stens and himself into the setting of non-BL WSSP with sampling restoration in the mean-square sense. A very detailed overview of these results is given by Butzer et al. (1988) (compare also Splettstößer's

---

$^6\mathcal{B}_{\mathbb{R}}$ denotes the Borel σ-field on $\mathbb{R}$.

papers listed there). However, these considerations differ from the present one, which concentrates on sampling restoration of BL signals in the a.s. sense.

Finally, an HRF with n-dimensional time is *spherically band-limited* when the frequency spectrum support domain is equal to the closed n-dimensional sphere $V_w := \{\lambda; \langle \lambda, \lambda \rangle \le w^2\}$ and $\langle \lambda, \lambda \rangle := \sum_{k=1}^{n} \lambda_k^2$, say. Such HRFs are considered in the well-known book by Yadrenko (1980; pp.167–168). Then there holds

$$\Xi(x) = \int_{V_w} e^{i\langle \lambda, x \rangle} \Upsilon(d\lambda) = \int_{\mathbb{R}^n} \chi_w(\lambda) e^{i\langle \lambda, x \rangle} \Upsilon(d\lambda),$$

where $\chi_w(\lambda)$ denotes the characteristic function of V_w. The multiple Fourier-series of $\chi_w(\lambda)e^{i\langle \lambda, x \rangle}$ on the n-dimensional cube $[-w, w]^n$ will be

$$\chi_w(\lambda)e^{i\langle \lambda, x \rangle} = \frac{1}{(2w)^n} \sum_{k=1}^{n} \sum_{j_k \in \mathbf{Z}} e^{i\langle j, x \rangle \pi/w} \int_{V_w} e^{i\langle \lambda, x - (\pi/w)j \rangle} d\lambda.$$

Using generalized spherical coordinates in the foregoing relation for the integration procedure, we get

$$\chi_w(\lambda)e^{i\langle \lambda, x \rangle} = \left(\frac{\pi}{2}\right)^{n/2} \sum_{k=1}^{n} \sum_{j_k \in \mathbf{Z}} e^{i\langle j, x \rangle \pi/w} \frac{J_{\frac{n}{2}}(|wx - \pi j|)}{|wx - \pi j|^{n/2}}.$$

Here, $J_m(|u|)$ denotes the Bessel function of the first kind of degree m, and $|u|^2 := \langle u, u \rangle$. Then it follows that

$$\Xi(x) = \left(\frac{\pi}{2}\right)^{n/2} \sum_{k=1}^{n} \sum_{j_k \in \mathbf{Z}} \frac{J_{\frac{n}{2}}(|wx - \pi j|)}{|wx - \pi j|^{n/2}} \Xi\left(\frac{\pi}{w} j_1, \ldots, \frac{\pi}{w} j_n\right) \qquad (9.1.18)$$

in the mean-square sense (Yadrenko (1980; Theorem 3, p. 167)). This book (p. 168) contains the following short remark:

"...the a.s. sense sampling restoration formula (9.1.18) of the spherically BL HRF's could be also proved...",

but without any specific continuity conditions upon the observed field or upon its spectral measure.

9.2   The Piranashvili–Lee theory

For the sampling reconstruction of random signals from certain more general classes of stochastic processes the classical paper of Zakai (1965) is of great importance. Zakai introduced the definiton of the so-called (w, δ)-BL signals. Namely, if the covariance $B(t, s)$ of the L_2-process $\xi(t)$ satisfies

$$\int_{\mathbb{R}} \frac{B(t, t)}{1 + t^2} dt < \infty, \qquad (9.2.1)$$

then the (w, δ)-BL process $\xi(t)$ also satisfies the Kotel'nikov-formula (9.1.3) in the a.s. sense. It is obvious that a weakly stationary stochastic process satisfies (9.2.1), since its covariance is bounded above by $B(0)$. More about (w, δ)-BL processes will be found in that part of the present chapter where Lee's work is referenced.[7]

A new point of view was introduced into a.s. sense signal restoration by Piranashvili (1967). He considered an L_2-process $\{\xi(t); t \in \mathbb{R}\}$ with entire, exponentially bounded sample functions and derived new results for analytical, non-stationary processes with exponentially bounded kernel function $f(t, \lambda)$ (the exponential type of the sample function is equal to the exponential type of $f(t, \lambda)$; see Belyaev (1959; Theorem 4) and Piranashvili (1967; Theorem 3)). This kernel appears in the covariance function:

$$B(t, s) = \int_\Lambda \int_\Lambda f(t, \lambda) f^*(s, \mu) F(d\lambda, d\mu), \qquad (9.2.2)$$

where Λ is a parameter set and $F(\Delta', \Delta'') = \mathbb{E}\xi(\Delta')\xi^*(\Delta'')$, $\Delta', \Delta'' \subset \Lambda$ is a positive definite measure of bounded variation on $\mathbb{R}^2$. Because $f(t, \lambda)$ is an entire function, we deduce that $f(t, \lambda) = \sum_{n=0}^\infty f^{(n)}(0, \lambda) t^n / n!$. Put

$$c(\lambda) = \limsup_{n \to \infty} \sqrt[n]{|f^{(n)}(0, \lambda)|}, \qquad (9.2.3)$$

and let $\sup_{\lambda \in \Lambda} c(\lambda) = \gamma < \infty$. Thus, as the exponential type of $f(t, \lambda)$ is equal to γ, then for all $w > \gamma$ the relation (9.1.3) holds in the almost sure sense (see Piranashvili (1967; §11, Theorem 1)).

There are certain straightforward consequences. The first is that if $B(t, s)$ possesses the Karhunen representation (one specifies F in (9.2.2))

$$B(t, s) = \int_\Lambda f(t, \lambda) f^*(s, \lambda) F(d\lambda), \qquad (9.2.4)$$

then for all $w > \gamma = \sup_{\lambda \in \Lambda} c(\lambda)$ the Kotel'nikov-formula (9.1.3) is valid in the a.s. sense. The second consequence is the fact that a.s. restoration is valid for so-called harmonizable processes, when the spectrum support set Λ is bounded. In fact, for harmonizable processes we get the Loève representation (one specifies $f(t, \lambda)$ in (9.2.2))

$$B(t, s) = \int_\Lambda \int_\Lambda e^{i(t\lambda - s\mu)} F(d\lambda, d\mu). \qquad (9.2.5)$$

Here, $c(\lambda) = |\lambda|$ since $f(t, \lambda) = e^{it\lambda}$. Therefore (9.1.3) holds in the a.s. sense for all $w > \gamma = \sup |\Lambda|$. By specifying F and f in the spectral representation (9.1.1), we get the same result for BL processes as in the harmonizable case.

Also, Piranashvili modified the original Kotel'nikov-series with the weight function $\mathrm{sinc}^q \beta(t - n\pi/w)$, where β depends on w, q and γ. Indeed, let $q \in \mathbb{N}, 0 < \beta < (w - \gamma)/q$, where γ is the exponential type of $\xi(t)$. Then we have

$$\xi(t) = \sum_{n \in \mathbb{Z}} \xi(n\pi/w) \frac{\sin(wt - n\pi)}{wt - n\pi} \frac{\sin^q \beta(t - n\pi/w)}{[\beta(t - n\pi/w)]^q}, \qquad (9.2.6)$$

in the a.s. sense. Clearly $\mathcal{E}_N(\xi, t) = \mathcal{O}(N^{-2q-2})$ for the m.s. truncation error in (9.2.6) (taking $q = 0$ in (9.2.6) we come back to the Belyaev results). So, with this modified series one achieves a greater convergence rate.

---

[7] It seems that Zakai was the first Western author to report on the work of Belyaev (1959).

Finally, Piranashvili also considered stochastic processes having covariance (9.2.2), where the analytical continuation of $f(t, \lambda)$ with respect to t is entire, and where

$$|f(t, \lambda)| \leq \tilde{L}_f(\lambda)(1 + |t|^m)e^{c(\lambda)|\Im(t)|}, \tag{9.2.7}$$

for some $m \in \mathbf{N}$, where $\sup_{\lambda \in \Lambda} \tilde{L}_f(\lambda) < \infty$, $\sup_{\lambda \in \Lambda} c(\lambda) = \gamma < \infty$. Then (9.2.6) is valid in the a.s. sense for all $w > \gamma$ and all $0 < \beta < (w - \gamma)/q$ if $q \geq m$ (Piranashvili (1967; Theorem 6)).

Remark 2 *It should be pointed out that the last sentence in the Piranashvili article (above, just after Theorem 9.6),*

"In this case $\mathbf{E}|\xi(t)|^2$ increases on the real axis polynomially as $|t| \to \infty$."

turned out to be one of the fundamentals in Lee's approach nine years later. Indeed, since $\mathbf{E}|\xi(t)|^2 = B(t, t)$, the quotation announces Lee's concept of processes with polynomially bounded covariance.

The very simple truncation error upper bound $\mathcal{E}_N(\xi, t) = \mathcal{O}(N^{-2})$ was derived, following Belyaev and Piranashvili, in Gulyás (1970). With the aid of this bound a non-stationary, exponentially bounded process was shown to be approximated by its Kotel'nikov-series almost surely.

Zakai's and Piranashvili's ideas were generalized by Lee (1976a, 1976b, 1977, 1978). We give next a short presentation of some of Lee's results.

A BL function on $\mathbb{R}$ is of the form

$$g(t) = \int_{-w}^{w} e^{it\lambda} \nu(d\lambda) \tag{9.2.8}$$

for a measure $\nu(\cdot)$ of finite variation on $[-w, w]$. If $\nu(\cdot)$ is absolutely continuous, then $g(t)$ is *conventionally band-limited*, or BL, to $w > 0$. Such functions can be described by the fact that they pass through the linear transformation $\mathcal{L}$ without distortion, i.e. the convolution[8] $g * \mathcal{L}(t) = \mathcal{L}g(t) \equiv g(t)$ for all t. (Here $\mathcal{L}$ is the inverse Fourier transform of $\chi_w(\lambda)$, the indicator function of the interval $[-w, w]$.)

Let us choose a function ψ that is equal to 1 on $[-w, w]$, and is the inverse Fourier transform of an even C^∞-function with support $[-w - \delta, w + \delta]$, $\delta > 0$. Such a function belongs to the Schwartz space $\mathcal{S}$, and its analytical continuation satisfies

$$|\psi(z)| \leq C_n(1 + |z|)^{-n} e^{(w+\delta)|\Im(z)|}, \tag{9.2.9}$$

for some $n \in \mathbf{N}$. We define a class of BL processes, whose covariance functions $B(t, s)$ satisfy

$$\int_{\mathbb{R}} \frac{B(t, t)dt}{(1 + t^2)^q} < \infty, \tag{9.2.10}$$

for some $q \in \mathbf{N}$.

Let μ_q be the measure with density function $(1 + t^2)^{-q}$, and let $L_2(\mu_q)$ be the Hilbert space $L_2(\mu_q) = \{h; \int_{\mathbb{R}} |h|^2 d\mu_q < \infty\}$.

---

[8]Here, the convolution is defined by $x * y(t) = \int_{\mathbb{R}} x(u)y(t - u)du$.

A function $f \in L_2(\mu_q)$ is (w, δ)-BL if $f * \psi(t) = f(t)$. Denote this function class by $B_q(w, \delta)$. Cambanis and Masry characterized $B_1(w, \delta)$ and showed that the space $B_1(w, \delta)$ is practically independent of δ. Lee did the same for $B_q(w, \delta)$ (Cambanis and Masry 1976; Lee 1976b). Therefore instead of $B_q(w, \delta)$ we write $B_q(w)$ in the sequel. According to this notation we shall replace "(w, δ)-BL" with "(w)-BL".

Let $\{\xi(t); t \in \mathbb{R}\}$ be the L_2-process whose covariance $B(t, s)$ satisfies $B(t, t) \leq \mathcal{P}(t)$ for some polynomial $\mathcal{P}(t)$. It follows immediately that the integral (9.2.10) converges for $\deg(\mathcal{P}) \leq 2q - 2$; see Remark 2. Then we define $\xi(t)$ to be (w)-BL if almost all of its sample functions are (w)-BL. So, we define $BP_q(w)$ to be a class of L_2-processes whose covariance satisfies (9.2.10) (or is polynomially bounded) and has almost all sample functions in $B_q(w)$.

A stochastic process will be called a *Piranashvili process* if its covariance function can be represented in the form (9.2.2).

A Piranashvili process is (w)-BL and satisfies (9.2.10) with $m < q$ in the sense of (9.2.7). Conversely, when $B(t, s)$ satisfies (9.2.10) and $\xi(t)$ is (γ)-BL, then $\xi(t)$ is a Piranashvili process of exponential type $w > \gamma$ and $m \geq q$ (Lee 1976a; Theorem 4.1). Thus, any process $\xi(t) \in BP_{q'}(\gamma)$ has a modified Kotel'nikov series (9.2.6) with $w > \gamma, \beta < (w - \gamma)/q, q \geq q'$, and it converges almost surely (Lee 1976a; (12).

Lee (1976b; Theorem 5) gives a complete characterization of the space $BP_q(w)$, as follows: let $\xi(t)$ have polynomially bounded covariance. Then the following assertions are equivalent:

(a) $\xi(t) \in BP_q(w)$,

(b) $B(t, \cdot) \in B_q(w), \ t \in \mathbb{R}$,

(c) $\xi(t) = \sum_{j=0}^{q-1} \xi^{(j)}(0)t^j/j! + \eta(t)t^q/q!$, where $\eta(t)$ is harmonizable with the support $[-w, w] \times [-w, w]$.

From (c) we clearly see why the definition of $BP_q(w)$ does not depend on δ, and also it is obvious that for the class of weakly stationary stochastic processes the (w)-BL definition coincides with the ordinary BL one.

Let $\xi(t)$ have polynomially bounded covariance $B(t, s)$. Then the modified Kotel'nikov-series (9.2.6), or its generalized variant

$$\xi(t) = \sum_{n \in \mathbf{Z}} \xi(n\pi/w) \frac{\sin(wt - n\pi)}{wt - n\pi} \varphi^\wedge(\beta(t - n\pi/w)),$$

interpolates exactly only (w)-BL processes, where φ is some test function with support $[-w, w]$; $\varphi^\wedge = \mathcal{F}\varphi$ denotes the Fourier transform. Approximate interpolation with the modified Kotel'nikov series needs sharper assumptions on $B(t, s)$ (see, e.g., Lee 1977; Theorem 4).

It has already been mentioned that Lloyd was one of the pioneer workers in a.s. sense sampling restoration. Let us return briefly to the famous article of Lloyd (1959). Here a necessary and sufficient condition on the random spectral measure $\zeta(\cdot)$ for a WSSP is given, which allows a process (not necessarily BL) to be exactly restored from its samples $\{\xi(n\pi/w)\}_{n \in \mathbf{Z}}$. Precisely, if $H(\xi)$ is the Hilbert space of the process

$\{\xi(t); t \in \mathbb{R}\}$, and $\mathcal{M}$ denotes the L_2-closed linear span of the set $\{\xi(n\pi/w)\}_{n\in\mathbb{Z}}$, then $\xi(t)$ can be exactly restored from $\{\xi(n\pi/w)\}_{n\in\mathbb{Z}}$ if $H(\xi) = \mathcal{M}$.[9] That will be the case if and only if ζ (accordingly F) has support Λ such that its translates by $n\pi/w$ are disjoint for all integer n (Lloyd 1959, Theorem 3).[10]

For convenience Lloyd chose a coefficient sequence $\{K_n(t)\}_{n\in\mathbb{Z}}$, such that

$$\xi(t) = \sum_{n\in\mathbb{Z}} \xi(n\pi/w)K_n(t), \qquad (9.2.11)$$

in a certain sense. Here we are interested in his a.s. restoration result. Thus, define

$$K(u) := \frac{1}{2w} \int_\Lambda e^{iu\lambda} d\lambda; \quad K_n(t) := K(t - n\pi/w). \qquad (9.2.12)$$

Theorem 9.1 *Suppose that the spectral measure F of the weakly stationary stochastic process $\xi(t)$ with covariance $B(t)$ has an open support Λ whose translates $\{\Lambda - n\pi/w;\ n \in \mathbf{Z}\}$ are disjoint. Suppose further that there are real numbers a, b; $a > 1/2$, $b > 0$, $a + b/2 > 1$; such that*

$$\max\left\{\sup_{t\in\mathbb{R}} |t^a K(t)|, \sup_{t\in\mathbb{R}} |t^b B(t)|\right\} < \infty. \qquad (9.2.13)$$

Then the sampling series (9.2.11) with coefficients (9.2.12) converges a.s., i.e.

$$\mathbf{P}\left\{\lim_{N\to\infty} \sum_{|n|\le N} \xi(n\pi/w)K(t - n\pi/w) = \xi(t)\right\} = 1. \qquad (9.2.14)$$

This result is found in Lloyd (1959; Theorem 4).

Following Lloyd's idea and his own earlier considerations, Lee generalized the last result in the following way. Let $\{\xi(t); t \in \mathbb{R}\}$ be an L_2-process whose covariance satisfies (9.2.10). Let Λ be an open set that contains the support of the spectral measure F of $\xi(t)$. Suppose that $\{\Lambda + n\pi/w\}$[11] are disjoint for all $n \in \mathbb{Z}$. Then we have

$$\mathbf{P}\left\{\lim_{N\to\infty} \sum_{|n|\le N} \xi(n\pi/w)\Theta(t - n\pi/w) = \xi(t)\right\} = 1, \qquad (9.2.15)$$

for all $t \in \mathbb{R}$, where $\Theta(u)$ is the modified Lloyd kernel:

$$\Theta(u) = \frac{1}{2w} \int_{Cl(A)} e^{iu\lambda} \varrho(\lambda) d\lambda, \qquad (9.2.16)$$

and ϱ is a C^∞-function equal to 1 on $Cl(A)$ and 0 outside Λ_0; Λ_0 is an open set with disjoint translates by $n\pi/w$ and $Cl(A)$ denotes the closure of A; $\mathrm{supp}(F) \subseteq A \subseteq Cl(A) \subseteq \Lambda_0$, (Lee 1978; Theorem 4).

---

[9] $H(\xi)$ and $\mathcal{M}$ are subspaces of $L_2(\Omega)$.

[10] In the terminology of ergodic theory, ζ has wandering support with respect to the translates $T\Lambda : \Lambda \to \Lambda + n\pi/w$.

[11] $\Lambda + n\pi/w$ denotes the set $\{\lambda : \lambda - n\pi/w \in \Lambda\}$.

It is also important to note here that (9.2.14) holds for any harmonizable process, see Rao (1967). Finally, a sequence of papers by Rao (1989a; 1989b; 1991) and Chang and Rao (1983) is concerned with the m.s. sampling restoration of harmonizable processes and harmonizable random fields including the important isotropic case, using the bimeasure technique.[12]

9.3   The Gaposhkin–Klesov theory

The development of the almost sure sampling restoration of BL WSSP from uniformly spaced samples began with Belyaev's and Lloyd's papers in 1959; these led to the general Klesov theorem for BL HRF (Klesov 1984).

At about the time that Lee generalized the BL property to the (w)-BL property for functions and processes, and characterized the band-limited Piranashvili processes, Gaposhkin completely solved the problem of uniformly sampled almost sure Kotel'nikov-series restoration for BL stationary and for harmonizable processes. That is, he gave necessary and sufficient conditions on the random measure $\zeta(\cdot)$ in (9.1.2) so that (9.1.3) holds uniformly on compact subsets of $\mathbb{R}$. These are the only necessary and sufficient conditions for a.s. sampling restoration that have been given subsequent to the work of Lloyd. Therefore we now present these results.

Investigations aim to establish the convergence of a sequence of stochastic intergals

$$\mathcal{I}_n(\omega) := \int_{|\lambda| \le w} Q_n(\lambda)\zeta(d\lambda); \quad \omega \in \Omega, \tag{9.3.1}$$

on a probability space $(\Omega, \mathcal{A}, \mathbf{P})$, where $\zeta(\cdot) \in L_2(\Omega)$ is a random spectral measure with orthogonal increments, and $Q_n(\lambda)$ denotes a sequence of complex-valued functions. Gaposhkin derived a general theorem for the convergence, in the a.s. sense, of the sequence (9.3.1) (among other results), and applied this theorem to the Kotel'nikov-formula for BL WSSP and for harmonizable processes.

Theorem 9.2 *Let $\zeta(d\lambda)$ be a random spectral measure with orthogonal increments $\{\mathcal{I}_n(\omega)\}_{n=1}^\infty$, supported on $[-w, w]$. Let $\{Q_n(\lambda)\}_{n=1}^\infty$ be a sequence of complex-valued functions satisfying the following conditions for a certain $\lambda_0 \in \mathbb{R}$:*

(a) *for $b > 0$, $n \in \mathbf{N}$, $|Q_n(\lambda)| \le C_1 \min\{(n|\lambda - \lambda_0|)^b, (n|\lambda - \lambda_0|)^{-b}\}$;*
(b) *when $2^m \le k < n < 2^{m+1}$, $m \in \mathbf{N} \cup \{0\}$ then*

$$|Q_n(\lambda) - Q_k(\lambda)| \le C_2 \min_{i=1,2}\{(n - k)^{a_i}|\lambda - \lambda_0|^{b_i}n^{-c_i}, (n|\lambda - \lambda_0|)^{-a}\};$$

(c) *$a, a_i, b_i, c_i \ i = \overline{1,2}$ satisfy:*

$$a, a_i, c_2 > 1/2; \quad c_1 + b_1 \ge a_1 > c_1; \quad c_2 + b_2 \ge a_2.$$

Then

$$\mathbf{P}\left\{\lim_{n\to\infty}\mathcal{I}_n(\omega) = 0\right\} = 1. \tag{9.3.2}$$

---

[12]The author of this chapter takes pleasure in thanking Professor M.M. Rao, who enabled him to gain insight into the papers listed here.

For other details, compare Gaposhkin (1977; Theorem 2).

Now we shall apply the result (9.3.2) to the Kotel'nikov-series (9.1.3) and formulate the principal Gaposhkin results.

Theorem 9.3 *The Kotel'nikov formula* (9.1.3) *for the BL WSSP* $\xi(t)$ *holds in the a.s. sense if and only if*

$$\mathbf{P}\left\{\lim_{m\to\infty}\left(\int_{|\lambda+w|\leq2^{-m}}\zeta(d\lambda) - \int_{|\lambda-w|\leq2^{-m}}\zeta(d\lambda)\right) = 0\right\} = 1. \qquad (9.3.3)$$

Theorem 9.4 *When* (9.3.3) *is satisfied, then* (9.1.3) *converges uniformly for all* $t \in [T_1, T_2]$ *for almost all sample functions of the process* $\xi(t)$.

For the proof see Gaposhkin (1977; Theorem 3).

Remark 3 *Let* $\{\xi(t); t \in \mathbb{R}\}$ *be a BL WSSP with the covariance* (9.1.1). *It can be easily shown that*

$$\mathbf{E}\left|\xi(t) - \sum_{n\in\mathbf{Z}}\xi(n\pi/w)\frac{\sin(wt - n\pi)}{wt - n\pi}\right|^2 = \sin^2(wt)[F(\{-w\}) + F(\{w\})], \qquad (9.3.4)$$

see, e.g. Wong (1971; p. 107). *Since* $\mathbf{E}|\zeta(\Delta)|^2 = F(\Delta)$, $\Delta \subset \mathbb{R}$ *and* ζ *has orthogonal increments, taking* $t \neq n\pi/w$ *it is necessary and sufficient that*

$$\zeta(\{-w\}) = \zeta(\{w\}) = 0 \quad a.s. \qquad (9.3.5)$$

for the validity of (9.1.3) *in the m.s. sense. Therefore if* (9.1.3) *converges a.s. to* $\xi(t)$, *then* (9.3.5) *holds. On the other hand* (9.3.5) *is an obvious consequence of* (9.3.3). *So* (9.3.5) *is valid and it is enough to consider the open support* $(-w, w)$ *instead of* $[-w, w]$.

The condition (9.3.3) could be carried over to the class of harmonizable processes.

Theorem 9.5 *Let* $\{\xi(t), t \in \Lambda \subseteq \mathbb{R}\}$ *be a harmonizable stochastic process with covariance* (9.2.5). *If* $\sup|\Lambda| = w < \infty$ *then* (9.1.3) *is valid for almost all sample functions of* $\xi(t)$ *if and only if*

$$\mathbf{P}\left\{\lim_{m\to\infty}\left(\int_{\Lambda\cap[-w,-w+2^{-m}]}\zeta(d\lambda) - \int_{\Lambda\cap[w,w-2^{-m}]}\zeta(d\lambda)\right) = 0\right\} = 1. \qquad (9.3.6)$$

If $\sup|\Lambda| = \tilde{w} < w$, then (9.3.6) is automatically satisfied, therefore the Piranashvili result for harmonizable processes holds.

It should be remarked that the Belyaev oversampling result is an immediate consequence of (9.3.3) (Gaposhkin 1977; Consequence 1).

The characterization of (9.3.3) in terms of the spectral measure F appearing in (9.1.1) is more familiar. Indeed, the assertion of Theorem 9.4 is valid if[13]

$$F(\{\pm w\}) = 0, \quad \int_{-w}^{w} \left(\lg\lg \left(2 + \frac{1}{w^2 - \lambda^2} \right) \right)^2 F(d\lambda) < \infty. \tag{9.3.7}$$

The uniform a.s. convergence of (9.1.3) with respect to t on compact subsets of $\mathbb{R}$ (Theorem 9.4) is also achieved in following cases:

(a) $\xi(t)$ is Gaussian and (9.3.5) is satisfied;

(b) $F(d\lambda)$ is absolutely continuous with respect to the ordinary Lebesgue measure $d\lambda$ and possesses a continuous or bounded derivative $f(\lambda)$—the spectral density of the process $\xi(t)$;

(c) for $\lambda \to \pm w$, $\varepsilon > 0$, we have $f(\lambda) = \mathcal{O}((|\lambda^2 - w^2| \lg^2 |\lambda^2 - w^2| \lg^{3+\varepsilon} |\lg|\lambda^2 - w^2|)^{-1})$;

(d) $\{B(n\pi/w)\}_{n \in \mathbb{N}}$ decreases monotonically.

Gaposhkin (1977; Consequences 2a, 3, 4, 6).

With the help of (9.3.3) the Belyaev a.s. convergence rate could be sharpened in (9.1.3). Indeed, consider a function $\kappa(u) \uparrow \infty$, as $u \to \infty$ with the property: $(\exists C < 2)(\kappa(2u) \leq C\kappa(u))$; $u > 0$, and assume that

$$F(\{\pm w\}) = 0; \quad \int_{-w}^{w} \kappa \left(\frac{1}{w^2 - \lambda^2} \right) F(d\lambda) < \infty. \tag{9.3.8}$$

Then we have

$$\lim_{n \to \infty} \left\{ \max_{t \in T} |\xi(t) - \xi_N(t)| \sqrt{\kappa(n)} \right\} = 0, \tag{9.3.9}$$

in the a.s. sense, if and only if

$$\lim_{m \to \infty} \left[\int_{[-w, -w+2^{-m}]} \zeta(d\lambda) - \int_{[w, w-2^{-m}]} \zeta(d\lambda) \right] \sqrt{\kappa(2^m)} = 0 \tag{9.3.10}$$

in the a.s. sense, for a compact $T \subset \mathbb{R}$.

About six years later Klesov (1984) made a generalization of (9.3.7) to the BL HRF case. The approach follows that of Klesov (1983), however the mathematical tools used in the proofs are special to this paper. Therefore we present here only the final result, the general Klesov theorem, and some related results.

Consider a HRF $\{\Xi(x); x \in \mathbb{R}^n\}$, BL to $\tilde{\mathcal{W}}$, with covariance $\mathcal{K}(x)$ spectrally represented in (9.1.8) and (9.1.7) respectively. From (9.1.10) and (9.1.11) it follows that the sampling rate with $\tilde{w}_j = w_j$, $j = \overline{1, n}$, is minimal. So, we consider this limit case in the oversampling restoration procedure. However, (9.1.10) does not hold without certain additional conditions upon the HRF $\Xi(x)$. We take $\tilde{\mathcal{W}} \equiv \mathcal{W}$, and denote by $\partial\mathcal{W}$ the boundary of $\mathcal{W}$.

---

[13] $\lg$ denotes $\log_2$.

Theorem 9.6 *The multiple Kotel'nikov series* (9.1.12) *for* $x \notin Lat(\mathcal{W})$ *converges almost surely to the initial* BL HRF $\Xi(x)$ *if*

$$\int_{\mathcal{W}} \prod_{j=1}^{n} \left(\lg\lg \left(2 + \frac{1}{w_j^2 - \lambda_j^2} \right) \right)^2 \Phi(d\lambda) < \infty, \qquad (9.3.11)$$

and $\Phi(\{\partial\mathcal{W}\}) = 0$.

Consequently, if $\Xi(x)$ is Gaussian, then (9.1.12) is valid for $\Phi(\{\partial\mathcal{W}\}) = 0$ and for all $x \notin Lat(\mathcal{W})$. This interesting fact and the following results, which in particular concern the question of how to avoid the restriction $\Phi(\{\partial\mathcal{W}\}) = 0$ in the a.s. convergence in (9.1.12), are given in Klesov (1983, 1984).

Theorem 9.7 *Assume that*

$$\sum_{j=1}^{n} \sum_{m_j \geq 0} |\mathcal{K}(x^m)| \prod_{j=1}^{n} \frac{\lg\lg(4(m_j + 1))}{(m_j + 1)\lg(2(m_j + 1))} < \infty. \qquad (9.3.12)$$

Then (9.1.12) *follows in the a.s. sense.*

We can also mention Klesov's generalizations of consequences of Theorem 9.2. The relation (9.1.12) holds in the a.s. and the m.s. sense if:

(a) the frequency spectrum support domain $\mathcal{V}$ is contained in $\mathcal{W}$ and $d(\mathcal{V}, \partial\mathcal{W}) = \min_{x \in \mathcal{V}; \, y \in \partial\mathcal{W}} |x - y| > 0$; or,

(b) $\Xi(x)$ has a bounded spectral density.

Further results on the a.s. sampling restoration for Gaussian band-limited homogeneous random fields can be found in Klesov (1985).

9.4   Irregular-derivative Kotel'nikov series

We consider the sampling procedure in which the sample points are perturbed from uniform spacing; this is often called *jittered* or *irregular sampling*. Take a set of points from the real line $\{t_n\}_{n \in \mathbf{Z}}$ having no accumulation point, and replace the weight function $\sin(\cdot)/(\cdot)$ in (9.1.3) with some other convenient function $g(t, t_n)$. The exact sampling restoration of $\xi(t)$ by the series

$$\xi(t) = \sum_{n \in \mathbf{Z}} \xi(t_n) g(t; t_n) \qquad (9.4.1)$$

can in some sense be thought of as an *irregular Kotel'nikov series*; but (9.4.1) will only hold when certain additional assumptions are made about the sequence (t_n) and upon the sequence of the weight functions $(g(t; t_n))$.

This kind of problem was studied in detail by Beutler (1961) (for m.s. convergence only); stronger results were obtained by Seip (1990a, 1990b) and Pogány (1994, 1995).

Let $D := \sup_{n \in \mathbf{Z}} |t_n - n|$. Then the famous Kadec 1/4 -theorem allows the sampling reconstruction

$$f(t) = \sum_{n \in \mathbf{Z}} f(\pi t_n / w) \frac{\sin(wt - t_n \pi)}{wt - t_n \pi}, \qquad (9.4.2)$$

with the restriction $D < 1/4$. Seip (1990a) transferred this deterministic signal result to the case of BL WSSP $\{\xi(t); t \in \mathbb{R}\}$, and confirmed the validity of the same Kadec condition for the uniform m.s. convergence in

$$\xi(t) = \pi \sum_{n \in \mathbb{Z}} \xi(\pi t_n / w) \frac{G(tw/\pi)}{G'(t_n)(wt - t_n \pi)}, \tag{9.4.3}$$

if there is a so-called *guard-band* $\varrho \in (0, w)$, that is, if $\xi(t)$ is BL to $w - \varrho$. Here $g(t, t_n)$ is the weight function

$$G(z) := (z - t_0) \prod_{n=1}^{\infty} \left(1 - \frac{z}{t_{-n}}\right)\left(1 - \frac{z}{t_n}\right). \tag{9.4.4}$$

In Seip (1990a) it is also proved that if $D < 1/8$, almost sure convergence holds in the relation (9.4.3) on compact subsets of the real t-axis.

The restoration of a BL WSSP $\xi(t)$ by the Kotel'nikov formula which uses samples of its $r - 1$ derivatives, $r \in \mathbb{N}$, as well as those $\xi(t)$ itself is called r^{th}-*order derivative sampling restoration* (rDSR in the sequel). For the WSSP $\xi(t)$ BL to $w > 0$ one defines its pth order mean-square derivative $\xi^{(p)}(t)$, $p \in \mathbb{N} \cup \{0\}$. (Since $\xi(t)$ is BL, it is infinitely smooth, thus, the sampling procedure is relevant for any integer r). Here we mention that some results by Higgins (1991) on deterministic signal restoration give the principal tool in deriving the generalization of (9.4.3) to rDSR.

Theorem 9.8 *Let* $\{\xi(t); t \in \mathbb{R}\}$ *be a weakly stationary stochastic process, BL to* $r(w - \varrho)$ *and with covariance function*

$$B(t) = \int_{r(\varrho - w)}^{r(w - \varrho)} e^{it\lambda} F(d\lambda). \tag{9.4.5}$$

Then we have

$$\xi(t) = G^r(tw/\pi) \sum_{n \in \mathbb{Z}} \sum_{j=0}^{r-1} \sum_{k=0}^{r-j-1} \binom{r - k - 1}{j} \frac{\gamma_{rnj}}{(r - k - 1)!}$$

$$\times \frac{\xi^{(r-j-k-1)}(t_n \pi / w)}{(t - t_n \pi / w)^{k+1}}, \tag{9.4.6}$$

(i) *uniformly in m.s. if* $D < 1/4r$;
(ii) *in the a.s. sense on compact subsets of the real t-axis if* $D < 1/8r$.
Here,

$$\gamma_{rnj} := \lim_{z \to t_n \pi / w} \frac{d^j}{dz^j} \left(\frac{z - t_n \pi / w}{G(zw/\pi)}\right)^r. \tag{9.4.7}$$

Compare Pogány (1994; Theorem 4).

With some specifications on the parameter $D := \sup_{n \in \mathbf{Z}} |t_n - n|$ we generate some new results from (9.4.6). So, uniformly spaced sampling means that $D = 0$, i.e. $t_n = n$. Then it follows easily that $G(z) = \sin(\pi t)$. Also, we have

$$\gamma_{rnj} = (-1)^{nr} w^{j-r} \lim_{z \to 0} \frac{d^j}{dz^j} \left(\frac{z}{\sin z} \right)^r. \tag{9.4.8}$$

Let us define $\Gamma_{r,j} := (-1)^{nr} w^{r-j} \gamma_{rnj}$. In the classical paper by Linden and Abramson (1960) the $\Gamma_{r,j}$ are expressed in terms of generalized Bernoulli numbers; e.g. $\Gamma_{r,0} = 1$, $\Gamma_{r,2} = r/3$, $\Gamma_{r,4} = r(5r+2)/15$, $\Gamma_{r,6} = 5(35r^2 + 42r + 16)/63$, etc. and $\Gamma_{r,j} = 0$ for j odd. So, the uniformly sampled r^{th} derivative Kotel'nikov formula becomes

$$\xi(t) = \frac{\sin^r (wt)}{w^r} \sum_{n \in \mathbf{Z}} (-1)^{nr} \sum_{j=0}^{\lceil \frac{r-1}{2} \rceil} \sum_{k=0}^{r-2j-1} \binom{r-k-1}{2j}$$

$$\times \frac{\Gamma_{r,2j}}{(r-k-1)!} w^{2j+k+1} \frac{\xi^{(r-2j-k-1)}(n\pi/w)}{(wt - n\pi)^{k+1}}. \tag{9.4.9}$$

Here (9.4.9) is valid in the m.s. and in the a.s. sense as well, since $D = 0$ (see Theorem 9.8).

In avoiding the Kadec 1/4-condition, Seip constructed certain sampling series with the help of the sample point set $\{t_n\}_{n \in \mathbf{Z}} \subset \mathbb{R}$ having *uniform density* 1, i.e., such that

$$D := \sup_{n \in \mathbf{Z}} |t_n - n| < \infty; \quad \inf_{m \neq n} |t_m - t_n| > 0. \tag{9.4.10}$$

Let S be the Schwartz class of infinitely smooth and rapidly decaying functions. The class of BL functions $B_w = \bigcup_{k \geq 0} B_k(w)$ consists of all functions that are the Fourier transform of a distribution supported on $[-w, w]$. B_w consists precisely of all entire functions $f(z)$ satisfying (9.2.7) (Lee 1976b).

Theorem 9.9 *Let $\{t_n\}_{n \in \mathbf{Z}}$ be any real sequence having uniform density 1, and let $0 < \epsilon < \varrho < w$. Let $h \in S \cap B_\epsilon$ with $h(0) = 1$. If $\xi(t)$ is WSSP BL to $w - \varrho$, i.e., its covariance is of the form*

$$B(t) = \int_{\varrho-w}^{w-\varrho} e^{it\lambda} F(d\lambda), \tag{9.4.11}$$

then we have

$$\xi(t) = \pi G(tw/\pi) \sum_{n \in \mathbf{Z}} \xi(\pi t_n/w) \frac{h(t - t_n \pi/w)}{G'(t_n)(wt - t_n \pi)}, \tag{9.4.12}$$

where the convergence is a.s., and uniform in the m.s. on compact subsets of the t-axis.

Compare Seip (1990b; Theorem 5).

Now, one can formulate a generalization of (9.4.12) to the irregular derivative a.s. Kotel'nikov formula.

Theorem 9.10 *Let* $\{t_n\}_{n\in\mathbb{Z}}$ *be any real sequence of uniform density 1 and let* $0 < \epsilon/r < \varrho < w, r \in \mathbf{N}$. *Let* $h \in \mathcal{S} \cap B_\epsilon$ *with* $h(0) = h_0 \neq 0$. *If* $\xi(t)$ *is a weakly stationary stochastic process BL to* $r(w - \varrho)$, *i.e., its covariance is of the form* (9.4.5), *then we have*

$$\xi(t) = h_0^{-1} G^r(tw/\pi) \sum_{n\in\mathbb{Z}} \sum_{j=0}^{r-1} \sum_{k=0}^{r-j-1} \sum_{m=0}^{r-j-k-1} \xi^{(r-j-k-m-1)}(t_n\pi/w)$$

$$\times \frac{h^{(k)}(t - t_n\pi/w)}{j!k!(r - j - k - m - 1)!} \frac{\gamma_{rnj}}{(t - t_n\pi/w)^{m+1}}, \tag{9.4.13}$$

where the convergence is uniform in the m.s. sense on compact subsets of the real t-axis, and almost sure as well.

Specifying r, h, G, t_n in the Kotel'nikov formula (9.4.13), we get the sequence of results already presented (Pogány 1995).

In the next section we study the generalization of Gaposhkin's necessary and sufficient condition (9.3.3) to the case of uniformly spaced a.s. derivative sampling restoration.

9.5 Derivative sampling and the Gaposhkin theory

We now extend Gaposhkin's necessary and sufficient condition (9.3.3) to the uniformly spaced a.s. rDSR procedure which is contained in (9.4.9).

Consider now the class $PW_w := B_w \cap L_\infty$, the so-called Paley–Wiener functions; for convenience we now take the band $[-w, w]$ to be $[-\pi, \pi]$. The class PW_π forms a Banach space endowed with the usual L_∞-norm: $\|f\| = \sup_{t\in\mathbb{R}} |f(t)|$. Higgins (1991) proved that for a deterministic function $f \in PW_{r\pi}$ there holds

$$f(t) = \sin^r(\pi t) \sum_{n\in\mathbf{Z}} (-1)^{nr} \sum_{j=0}^{r-1} \sum_{k=0}^{r-j-1} \frac{\Gamma_{r,j}\, \pi^{j-r}}{j!(r - j - k - 1)!} \frac{f^{(r-j-k-1)}(n)}{(t - n)^{k+1}} \tag{9.5.1}$$

on compact subsets of $\mathbf{C}$, but only if the band $(-r\pi, r\pi)$ is open. Therefore, for any real λ the function $f_\lambda(t) = e^{it\lambda}$, $\|f_\lambda\| = 1$ possesses the representation (9.5.1) on the band $(-r\pi, r\pi)$. We are interested in the exact value of (9.5.1) at the end-points $\pm r\pi$ of the observed band. Thus, denote by $A_r(f, t)$ the series in (9.5.1). Since $\Gamma_{r,(j \text{ odd})} = 0$, it follows that

$$A_r(f_{r\pi}, t) = \sin^r(\pi t) \sum_{n\in\mathbf{Z}} \sum_{j=0}^{\lceil (r-1)/2 \rceil} \sum_{k=0}^{r-2j-1} \frac{(-1)^j\, r^{r-2j-k-1}}{(2j)!\,(r - 2j - k - 1)!} \frac{\Gamma_{r,2j}}{[\pi(t - n)]^{k+1}}$$

$$\times \left[(-1)^{(r-k-1)/2}|_{r-k-1\equiv 0(\text{mod } 2)} + i(-1)^{(r-k)/2}|_{r-k\equiv 0(\text{mod } 2)} \right], \tag{9.5.2}$$

where $a|_m$ denotes an a which depends only on m.[14] Similarly one obtains the value $A_r(f_{-r\pi}, t) = A_r^*(f_{r\pi}, t)$. Thus, for the WSSP $\xi(t)$ BL to $r\pi$ we have

$$E|\xi(t) - A_r(\xi, t)|^2 = \int_{|\lambda| < r\pi} |e^{it\lambda} - A_r(f_\lambda, t)|^2 F(d\lambda)$$

$$+ |e^{itr\pi} - A_r(f_{r\pi}, t)|^2 \{F(\{r\pi\}) + F(\{-r\pi\})\}. \quad (9.5.3)$$

It is easy to see that $A_1(f_{\pm\pi}, t) = \cos(\pi t)$ from the Fourier series expansion of $f_\lambda(t)$ on $[-\pi, \pi]$, however for $r > 1$, $\Re(A_r(f_{r\pi}, t)) \neq \cos(r\pi t)$ and $\Im(A_r(f_{r\pi}, t)) \neq \sin(r\pi t)$ (Higgins 1991).

The random spectral measure $\zeta(\cdot)$ possesses orthogonal increments and $F(\Delta) = E|\zeta(\Delta)|^2$, so the following result is proved.

Theorem 9.11 *The uniformly sampled r^{th}-order derivative Kotel'nikov formula*

$$\xi(t) = \sin^r(\pi t) \sum_{n \in \mathbb{Z}} (-1)^{nr} \sum_{j=0}^{\lceil (r-1)/2 \rceil} \sum_{k=0}^{r-2j-1} \frac{\pi^{2j-r} \Gamma_{r, 2j}}{(2j)!(r-2j-k-1)!}$$

$$\times \frac{\xi^{(r-2j-k-1)}(n)}{(t-n)^{k+1}} \quad (9.5.4)$$

holds in m.s. if and only if

$$\zeta(\{-r\pi\}) = \zeta(\{r\pi\}) = 0 \quad a.s. \quad (9.5.5)$$

Now, we are ready to establish the derivative variant of the Gaposhkin's necessary and sufficient condition (9.3.3).

Theorem 9.12 *Let $\{\xi(t); t \in \mathbb{R}\}$ be a BL to $r\pi$ WSSP. Then (9.5.4) is valid in the a.s. sense iff*

$$\mathbf{P}\left\{ \lim_{m \to \infty} \left(\int_{[-r\pi, -r\pi + 2^{-m}]} \mathcal{P}_{r-1}^+(\lambda) \zeta(d\lambda) \right.\right.$$

$$\left.\left. - \int_{[r\pi, r\pi - 2^{-m}]} \mathcal{P}_{r-1}^-(\lambda) \zeta(d\lambda) \right) = 0 \right\} = 1, \quad (9.5.6)$$

where

$$\mathcal{P}_{r-1}^\pm(\lambda) = \sum_{j=0}^{\lceil (r-1)/2 \rceil} \sum_{k=0}^{r-2j-1} \frac{4^{\lceil k/2 \rceil} \pi^{2(j+\lceil k/2 \rceil)} \Gamma_{r, 2j}(-1)^j i^{r-1}}{(2j)!(r-2j-k-1)!} \lambda^{r-2j-k-1}$$

$$\times \left[2\pi(1 - (-1)^k) \beta_{2\lceil k/2 \rceil + 2}(\lambda \pm r\pi) + i(1 + (-1)^k) \beta_{2\lceil k/2 \rceil + 1}(\lambda \pm r\pi) \right],$$

---

[14] $\lceil x \rceil$ denotes the integer part of x.

and $\beta_s(\cdot)$ denotes the periodic continuation of the Bernoulli polynomial of degree s from its basic interval to $[-r\pi, r\pi]$.

Proof Arguing in the same manner as in Remark 3, we clearly see with the aid of Theorem 9.11 that it is enough to consider the restoration on $(-r\pi, r\pi)$. Denote by $A_{N,r}(\xi, t)$ the truncated variant of the rth-order derivative Kotel'nikov series $A_r(\xi, t)$, i.e.,

$$A_{N,r}(\xi, t) = \sin^r(\pi t) \sum_{|n| \leq N} (-1)^{nr} \sum_{j=0}^{\lceil (r-1)/2 \rceil} \sum_{k=0}^{r-2j-1} \frac{\pi^{2j-r} \, \Gamma_{r,2j}}{(2j)!(r-2j-k-1)!}$$

$$\times \frac{\xi^{(r-2j-k-1)}(n)}{(t-n)^{k+1}}. \tag{9.5.7}$$

The remainder term $\mathcal{R}_N(\xi, t) := \xi(t) - A_{N,r}(\xi, t)$ is split into two terms $\mathcal{R}_N^{(1)}(\xi, t)$, $\mathcal{R}_N^{(2)}(\xi, t)$ in the following way:

$$\mathcal{R}_N(\xi, t) = -\sin^r(\pi t) \sum_{|n| > N} (-1)^{nr} \sum_{j=0}^{\lceil (r-1)/2 \rceil} \sum_{k=0}^{r-2j-1} \frac{\pi^{2j-r} \, \Gamma_{r,2j} \, \xi^{(r-2j-k-1)}(n)}{(2j)! \, (r-2j-k-1)! \, n^{k+1}}$$

$$+ \sin^r(\pi t) \sum_{|n| > N} (-1)^{nr} \sum_{j=0}^{\lceil (r-1)/2 \rceil} \sum_{k=0}^{r-2j-1} \frac{\pi^{2j-r} \, \Gamma_{r,2j} \, \xi^{(r-2j-k-1)}(n)}{(2j)!(r-2j-k-1)!}$$

$$\times \left(\frac{1}{(t-n)^{k+1}} + \frac{1}{n^{k+1}} \right). \tag{9.5.8}$$

Let $\mathcal{T} \subset \mathbb{R}$ be compact. For $n > 2 \sup |\mathcal{T}|$ we have

$$\max_{t \in \mathcal{T}} |\mathcal{R}_N^{(2)}(\xi, t)|$$

$$\leq \max_{0 \leq j \leq \lceil \frac{r-1}{2} \rceil} \Gamma_{r,2j} \sum_{|n| > N} \sum_{j=0}^{\lceil (r-1)/2 \rceil} \sum_{k=0}^{r-2j-1} \left| \xi^{(r-2j-k-1)}(n) \left(\frac{1}{(t-n)^{k+1}} + \frac{1}{n^{k+1}} \right) \right|$$

$$\leq 4 \max_j \Gamma_{r,2j} \max\{\sup |\mathcal{T}|, 2^r\} \sum_{n=N+1}^{\infty} \sum_{j=0}^{\lceil (r-1)/2 \rceil} \sum_{k=0}^{r-2j-1} \frac{|\xi^{(r-2j-k-1)}(n)|}{n^2}. \tag{9.5.9}$$

Obviously we have

$$\sum_{n \in \mathbb{N}} \sum_{j=0}^{\lceil (r-1)/2 \rceil} \sum_{k=0}^{r-2j-1} \mathbf{E} \left| \xi^{(r-2j-k-1)}(n) \right| n^{-2}$$

$$\leq \frac{\pi^2}{6} \sum_{j=0}^{\lceil (r-1)/2 \rceil} \sum_{k=0}^{r-2j-1} \sqrt{|B^{(2(r-2j-k-1))}(0)|;}$$

consequently, by the Loève theorem, $\lim_{N \to \infty} \mathcal{R}_N^{(2)}(\xi, t) = 0$ in the a.s. sense.

Let $t \neq n$ and let $\mathcal{R}^{(3)} := -\pi^r \sin^{-r}(\pi t)\mathcal{R}^{(1)}$. It remains to prove that

$$\mathcal{R}_N^{(3)}(\xi, t) = \sum_{|n|>N} (-1)^{nr} \sum_{j=0}^{\lceil(r-1)/2\rceil} \sum_{k=0}^{r-2j-1} \frac{\pi^{2j} \Gamma_{r,\,2j} \, \xi^{(r-2j-k-1)}(n)}{(2j)! \, (r-2j-k-1)! \, n^{k+1}} \qquad (9.5.10)$$

vanishes a.s. as $N \to \infty$.

With the help of the spectral representation of the mth m.s. derivative of the process $\xi(t)$, that is, assuming a representation of the form

$$\xi^{(m)}(t) = \int_{-r\pi}^{r\pi} (i\lambda)^m e^{it\lambda} \zeta(d\lambda),$$

we obtain

$$\mathcal{R}_N^{(3)}(\xi, t) = \sum_{j=0}^{\lceil(r-1)/2\rceil} \sum_{k=0}^{r-2j-1} \frac{\pi^{2j} \Gamma_{r,\,2j} \, i^{r-2j-k-1}}{(2j)! \, (r-2j-k-1)!}$$

$$\times \int_{|\lambda|<r\pi} \lambda^{r-2j-k-1} \left\{ \frac{1-(-1)^k}{2} \sum_{n=N+1}^{\infty} \frac{\cos n(\lambda + r\pi)}{n^{2\lceil k/2\rceil+2}} \right.$$

$$\left. + i \frac{1+(-1)^k}{2} \sum_{n=N+1}^{\infty} \frac{\sin n(\lambda + r\pi)}{n^{2\lceil k/2\rceil+1}} \right\} \zeta(d\lambda). \qquad (9.5.11)$$

Therefore, the problem is to estimate the remainders in the curly brackets in (9.5.11). Let $s \in \mathbf{N}$ and introduce

$$\alpha_{2s-1}(\mu) = \sum_{n=1}^{\infty} \frac{\sin n\mu}{n^{2s-1}};$$

$$\bar{\alpha}_{2s}(\mu) = \sum_{n=1}^{\infty} \frac{\cos n\mu}{n^{2s}}. \qquad (9.5.12)$$

It is well known that $\alpha, \bar{\alpha}$ are the Fourier series of absolutely integrable functions (see Fikhtengol'c 1969; 463-9), and that these coincide (except for a numerical factor) with the so-called *Bernoulli polynomials* $\beta_{2s-1}(\mu)$, $\beta_{2s}(\mu)$ of degree $2s-1$, $2s$ respectively, for $\mu \in (0, 2\pi)$ (Karamata 1949). More precisely

$$\alpha_{2s-1}(\mu) = (-1)^{s-1} \frac{(2\pi)^{2s-1}}{2} \beta_{2s-1}(\mu),$$

$$\bar{\alpha}_{2s}(\mu) = (-1)^{s-1} \frac{(2\pi)^{2s}}{2} \beta_{2s}(\mu). \qquad (9.5.13)$$

Applying the familiar Dirichlet formula for the Nth partial sum of a Fourier series to $\bar{\alpha}_{2s}(\mu)$ for a fixed $\mu \in (0, \pi)$, we have by the first Riemann lemma, that

$$\sum_{n=1}^{N} \frac{\cos(n\mu)}{n^{2s}} = \frac{1}{\pi} \int_{-\mu}^{\mu} \overline{\alpha}_{2s}(\mu + x) \frac{\sin(Nx)}{x} dx + o(1)$$

$$= (-1)^{s-1} \frac{(2\pi)^{2s}}{\pi} \beta_{2s}(\mu) \int_{0}^{N\mu} \frac{\sin x}{x} dx$$

$$+ \frac{2}{\pi} \sum_{j=0}^{2s} \beta_j \sum_{l=1}^{\lceil j/2 \rceil} \binom{2\lceil j/2 \rceil}{2l} \mu^{2\lceil j/2 \rceil - 2l} \int_{0}^{\mu} x^{2l-1} \sin(Nx) dx + o(1)$$

$$= (-1)^{s-1} \frac{(2\pi)^{2s}}{\pi} \beta_{2s}(\mu) \int_{0}^{N\mu} \frac{\sin x}{x} dx + \mathcal{O}(N^{-1}) + o(1), \qquad (9.5.14)$$

uniformly in N and μ. Here $\beta_{2s}(\mu) = \sum_{j=0}^{2s} \beta_j \mu^j$ and $\beta'_{2s}(\mu) = \beta_{2s-1}(\mu)$ orders the coefficients β_j.

Finally, we have

$$\sum_{n=N+1}^{\infty} \frac{\cos(n\mu)}{n^{2s}} = (-1)^{s-1} \frac{(2\pi)^{2s}}{\pi} \beta_{2s}(\mu) \left(\frac{\pi}{2} - \int_{0}^{N\mu} \frac{\sin x}{x} dx \right) + \mathcal{O}(N^{-1}) \quad (9.5.15)$$

$$= (-1)^{s-1} \frac{(2\pi)^{2s}}{\pi} \beta_{2s}(\mu) \int_{N\mu}^{\infty} \frac{\sin x}{x} dx + \mathcal{O}(N^{-1}), \qquad (9.5.16)$$

and similarly we have

$$\sum_{n=N+1}^{\infty} \frac{\sin(n\mu)}{n^{2s-1}} = (-1)^{s-1} \frac{(2\pi)^{2s}}{2\pi^2} \beta_{2s-1}(\mu) \left(\frac{\pi}{2} - \int_{0}^{N\mu} \frac{\sin x}{x} dx \right) + \mathcal{O}(N^{-1}) \quad (9.5.17)$$

$$= (-1)^{s-1} \frac{(2\pi)^{2s}}{2\pi^2} \beta_{2s-1}(\mu) \int_{N\mu}^{\infty} \frac{\sin x}{x} dx + \mathcal{O}(N^{-1}). \qquad (9.5.18)$$

The case $s = 1$ gives the Kotel'nikov formula (9.1.3), discussed in detail in §9.1; moreover Klesov (1983, 1984) considers the case $s = 1$ for BL HRF.

Now, since μ plays the role of $\lambda + r\pi$, $\lambda \in (-r\pi, r\pi)$, if $(2m - 1)\pi \leq \mu \leq 2m\pi$, we take $\overline{\alpha}_{2s}(\mu) = \overline{\alpha}_{2s}(2m\pi - \mu)$ and for $(2m - 2)\pi \leq \mu \leq (2m - 1)\pi$, one defines $\overline{\alpha}_{2s}(\mu) = -\overline{\alpha}_{2s}((2m - 1)\pi - \mu)$, $m = \overline{1, r}$. Accordingly, we set $\alpha_{2s-1}(\mu) = -\alpha_{2s-1}(2m\pi - \mu)$ and $\alpha_{2s-1}(\mu) = \alpha_{2s-1}((2m - 1)\pi - \mu)$, for the ranges of μ as above. In this way a complete estimation of the remainder in the curly brackets in the relation (9.5.11) is achieved. So, it is clear that the neighbourhood of $\pm r\pi$ is critical, and we now concentrate our attention on the behaviour of $\mathcal{R}_N^{(3)}(\xi, t)$ at these points.

Taking $m = \lceil \lg N \rceil$, we define the Gaposhkin kernel $Q_N(\lambda)$ in (9.3.1) to satisfy the conditions of Theorem 9.2. From the representation (9.5.11) of $\mathcal{R}_N^{(3)}(\xi, t)$ using (9.5.15)–(9.5.18), we have

$$\mathcal{R}_N^{(3)}(\xi, t) = \frac{\pi}{2} \left[\int_{[-r\pi, -r\pi+2^{-m}]} \mathcal{P}_{r-1}^{+}(\lambda) \zeta(d\lambda) - \int_{[r\pi, r\pi-2^{-m}]} \mathcal{P}_{r-1}^{-}(\lambda) \zeta(d\lambda) \right]$$

$$+ \left[\int_{(-r\pi, 0)} Q_N^{(1,r)}(\lambda) \zeta(d\lambda) + \int_{(0, r\pi)} Q_N^{(2,r)}(\lambda) \zeta(d\lambda) \right] + \mathcal{R}_N^{(4)}(\xi, t)$$

$$(9.5.19)$$

where

$$
\mathcal{P}_{r-1}^{\pm}(\lambda) = \sum_{j=0}^{\lceil (r-1)/2 \rceil} \sum_{k=0}^{r-2j-1} \frac{4^{\lceil k/2 \rceil} \pi^{2(j+\lceil k/2 \rceil)} \Gamma_{r,2j}(-1)^j i^{r-1}}{(2j)!(r-2j-k-1)!} \lambda^{r-2j-k-1}
$$

$$
\times \left[2\pi (1-(-1)^k)\beta_{2\lceil k/2 \rceil +2}(\lambda \pm r\pi) + i(1+(-1)^k)\beta_{2\lceil k/2 \rceil +1}(\lambda \pm r\pi) \right];
$$

$$
Q_N^{(1,r)}(\lambda) = \begin{cases} \mathcal{P}_{r-1}^{(1)}(\lambda) \displaystyle\int_0^{N(\lambda+r\pi)} \frac{\sin x}{x} \, dx & \lambda \in (-r\pi, -r\pi + 2^{-m}] := I_1 \\[3ex] \mathcal{P}_{r-1}^{(2)}(\lambda) \displaystyle\int_{N(\lambda+r\pi)}^{\infty} \frac{\sin x}{x} \, dx & \lambda \in (-r\pi + 2^{-m}, 0] := I_2, \end{cases}
$$

$$
Q_N^{(2,r)}(\lambda) = \begin{cases} \mathcal{P}_{r-1}^{(3)}(\lambda) \displaystyle\int_{N(\lambda-r\pi)}^{\infty} \frac{\sin x}{x} \, dx & \lambda \in (0, r\pi - 2^{-m}] := I_3 \\[3ex] \mathcal{P}_{r-1}^{(4)}(\lambda) \displaystyle\int_0^{N(\lambda-r\pi)} \frac{\sin x}{x} \, dx & \lambda \in (r\pi - 2^{-m}, r\pi) := I_4, \end{cases}
$$

and $\mathcal{P}_{r-1}^{(j)}(\lambda)$, $j = \overline{1,4}$, are obtained from $\mathcal{P}_{r-1}^{\pm}(\lambda)$ using the remark on the periodic extension of α_{2s-1} and $\overline{\alpha}_{2s}$ from $[0, \pi]$ to $[-r\pi, r\pi]$. Finally, (9.5.14) and the first Riemann lemma give us

$$
\mathbf{E}|\mathcal{R}_N^{(4)}(\xi, t)|^2 = \mathcal{O}(N^{-2}), \tag{9.5.20}
$$

therefore $\mathbf{P}\{\lim_{N\to\infty} \mathcal{R}_N^{(4)}(\xi, t) = 0\} = 1$.

It remains to examine the conditions (a) and (b) in Theorem 9.2 with respect to the Gaposhkin kernels $Q_N^{(1,r)}(\lambda)$, $Q_N^{(2,r)}(\lambda)$. For instance, we have

$$
\left| Q_N^{(1,r)}(\lambda) \right| \le N|\lambda + r\pi| \sup_{I_1} \left| \mathcal{P}_{r-1}^{(1)}(\lambda) \right| \qquad \lambda \in I_1,
$$

$$
\left| Q_N^{(1,r)}(\lambda) \right| \le \frac{\sup_{I_2} \left| \mathcal{P}_{r-1}^{(2)}(\lambda) \right|}{N|\lambda + r\pi|} \qquad \lambda \in I_2.
$$

So, the condition (a) is satisfied with $b = 1$. Accordingly, if $2^m \le M < N < 2^{m+1}$, then

$$
\left| Q_M^{(1,r)}(\lambda) - Q_N^{(1,r)}(\lambda) \right| \le \max \left\{ \sup_{I_1} \left| \mathcal{P}_{r-1}^{(1)}(\lambda) \right|, \sup_{I_2} \left| \mathcal{P}_{r-1}^{(2)}(\lambda) \right| \right\}
$$

$$
\times \left| \int_{M(\lambda+r\pi)}^{N(\lambda+r\pi)} \frac{\sin x}{x} \, dx \right|
$$

$$
\le C \min\{(N-M)|\lambda + r\pi|; \; 1 - M/N; \; (N|\lambda + r\pi|)^{-1}\}.
$$

This shows that $a = b_1 = c_2 = 1$; $b_2 = c_1 = 0$ in condition (b). Consequently, since we can prove the validity of these same conditions for $Q_N^{(2,r)}(\lambda)$, we obtain

$$\mathbf{P}\left\{\lim_{N\to\infty} \mathcal{R}_N^{(3)}(\xi,t) = 0\right\} = 1. \tag{9.5.21}$$

Finally, from (9.5.8), (9.5.11), (9.5.19) and Theorem 9.2, the equivalence between the a.s. convergence in (9.5.4) and that in (9.5.6) follows. This ends the proof of the theorem.

Remark 4 *In general the condition* (9.5.6) *is not equivalent to the pair of conditions:*

$$\mathbf{P}\left\{\lim_{m\to\infty} \int_{[-r\pi,-r\pi+2^{-m}]} \mathcal{P}_{r-1}^+(\lambda)\zeta(d\lambda) = 0\right\} = 1, \tag{9.5.22}$$

$$\mathbf{P}\left\{\lim_{m\to\infty} \int_{[r\pi,r\pi-2^{-m}]} \mathcal{P}_{r-1}^-(\lambda)\zeta(d\lambda) = 0\right\} = 1. \tag{9.5.23}$$

Indeed, (9.5.6) *implies* (9.5.22) *and* (9.5.23), *but the converse is not true. For example, one can define a discrete random measure in the following way:*

$$\zeta(\mp r\pi \pm 2^{-m}) = \frac{a_m(U_m + V_m)}{\mathcal{P}_{r-1}^\pm(\mp r\pi \pm 2^{-m})}; \qquad \sum_{m=1}^\infty a_m^2 < \infty,$$

where $\{U_m\}_{m\in\mathbf{N}}$ *is a convergent sequence of orthonormal random variables,* $\{V_m\}_{m\in\mathbf{N}}$ *is a divergent sequence of orthonormal random variables and* U_m, V_m *are orthogonal to each other. Moreover, let us assume that* $\sum_{m=1}^\infty a_m V_m$ *diverges in the almost sure sense. Thus it is possible that* (9.5.6) *is valid, but that* (9.5.22) *and* (9.5.23) *do not hold.*

9.6 Gaposhkin theory and irregular sampling

It is interesting to note an additional aspect of the irregular sampling restoration approach which is closely connected to the so-called *missing data* problem (see, e.g., Parzen 1984). This comes from work of Houdré (1995, 1994), who extended the Gaposhkin condition (9.3.3) to irregular sampling in the case of almost sure sampling restoration of BL weakly stationary stochastic processes. We sketch these results in the last section of this chapter.

Theorem 9.13 *Let* $\xi(t)$ *be a weakly stationary stochastic process, BL to* $(-w, w)$, $0 < w < \pi$, *and let* $\{t_n\} = \{n, s_1, \cdots, s_l\}_{n\in\mathbf{Z}\setminus\{n_1,\cdots,n_r\}}$, *where the real sequence* $\{t_n\}$ *satisfies the Kadec* $\frac{1}{4}$-*condition. Then we have*

$$\xi(t) = \sum_{n\in\mathbf{Z}} \Psi_n(t)\xi(t_n), \tag{9.6.1}$$

in the m.s. sense uniformly on compact subsets of $\mathbb{R}$.

The interpolating series (9.6.1) converges almost surely if and only if

$$\lim_{p \to \infty} \left\{ e^{-iwt} \int_{[-w, -w+2^{-p}]} \zeta(d\lambda) + e^{iwt} \int_{[w-2^{-p}, w]} \zeta(d\lambda) \right\} = 0, \qquad (9.6.2)$$

in the a.s. sense. The weight-function system $\{\Psi_n(t)\}$ is given by

$$\Psi_n(t) = \frac{w}{\pi} \left\{ \frac{\sin w(t_n - t)}{w(t_n - t)} + \frac{w}{\pi} \sum_{p=1}^{r} \sum_{q=1}^{r} \frac{\sin w(n_p - t)}{w(n_p - t)} \frac{\sin w(t_n - n_q)}{w(t_n - n_q)} V_{q,p} \right.$$

$$\left. - \frac{w}{\pi} \sum_{p=1}^{l} \sum_{q=1}^{l} \frac{\sin w(s_p - t)}{w(s_p - t)} \frac{\sin w(t_n - s_q)}{w(t_n - s_q)} V_{q,p} \right\},$$

where $V_{q,p}$ is the (q, p)-entry of the inverse of the matrix

$$V = \left(I - \frac{w}{\pi} \frac{\sin w(f_p - f_q)}{w(f_p - f_q)} \right)_{p,q=1,\dots,r+l},$$

where $t \in \mathbb{R}$ and $f_p = n_p$ (s_p) when $p = \overline{1, r}$ ($p = \overline{r+1, l}$).
The proof is in Houdré (1994; Theorem 9.18).

Under the same conditions on $\{t_n\}$, the almost sure convergence in the sampling series (9.6.1) holds for Gaussian BL weakly stationary stochastic processes, when the random spectral measure has *independent increments*; see Houdré (1994; Corollary 9.22).

If the spectral density $f_\xi(\lambda)$ of the BL weakly stationary stochastic process $\xi(t)$ belongs to $L_{1+\epsilon}(-w, w)$ $\epsilon > 0$ and if the sequence of samples $\{t_n\}$ satisfies the assumptions of Theorem 9.13, then (9.6.1) holds in the almost sure sense (Houdré 1994, Corrolary 9.23). More information and further references can be found in the Houdré articles.

Finally, a generalization of the Gaposhkin condition (9.3.3) to the case of irregular derivative sampling, i.e., the synthesis of the conditions (9.5.6) and (9.6.2), remains an open problem.

10

ABSTRACT HARMONIC ANALYSIS AND
THE SAMPLING THEOREM*

10.1 Introduction

The step from Fourier analysis to abstract harmonic analysis is a natural one which unites a number of apparently disparate results into one general framework with a concise and elegant notation. The classical sampling theorem is a good example of this, since a number of different versions can be obtained from the one abstract result due to I. Kluvánek (1965) simply by choosing the appropriate underlying locally compact abelian group. J.R. Higgins (1985) gives several examples where the sampling theorem is obtained for the real line $\mathbb{R}$, the circle $\mathbb{T}^1$, the torus $\mathbb{T}^2$, the integers $\mathbb{Z}$, the dyadic group and Euclidean space $\mathbb{R}^n$. Haar measure, a translation-invariant measure associated with the locally compact abelian group, plays a prominent role in the abstract theory; reciprocity relations and measure relations between the Haar measures of certain groups and sets emerge as natural and useful. The coset decomposition formula (10.2.3) below plays a particularly important part as it allows the analysis to be carried out in the space of square-integrable functions on a compact abelian group.

As well as providing a unifying framework, abstract harmonic analysis also brings out the group structure underlying a great deal of Fourier analysis and can clarify the significance of particular features of a result. For example in the classical sampling theorem a function f in the Paley–Wiener space $PW_{[-\pi w, \pi w]}$ can be expressed as

$$f(t) = \sum_{k \in \mathbb{Z}} f\left(\frac{k}{w}\right) \frac{\sin \pi(wt - k)}{\pi(wt - k)} \qquad (10.1.1)$$

and so is determined by its values (samples) at the points k/w, $k = 0, \pm 1, \pm 2, \ldots$. The abstract approach reveals the relationship between the sampling interval $1/w$ and the band-width $2\pi w$ (the Lebesgue measure $|[-\pi w, \pi w]|$ of the support $[-\pi w, \pi w]$ of the Fourier transform of f). The abstract approach also makes it clear that the regular sampling theorems under consideration can be viewed as expansions of a function with respect to an orthogonal basis.

In this chapter the real line $\mathbb{R}$ is replaced by a locally compact abelian group G and its dual Γ. Thus instead of a function being reconstructed from values at regularly

---

*The authors of this chapter are grateful to Frank Bonsall, Jim Clunie, Nick Dudley Ward, Simon Eveson, Walter Hayman, Rowland Higgins, Jonathan Partington and A.B. Raha for patiently listening to them and for their many helpful comments, corrections and criticisms. M.M. Dodson owes a great deal to J.H. Williamson who introduced him to Haar measure many years ago.

spaced points in $\mathbb{R}$, a function on a local compact abelian group G is reconstructed from samples taken at points in a discrete subgroup H of G. The classical band-limited condition corresponds to the dual of H being compact.

Those parts of signal theory based on Fourier analysis can often be placed in this abstract framework; for example, an abstract treatment of conventional digital-to-analogue conversion which turns out to be equivalent to windowing in the sense that the output signal is a local average. Also we include analyses of aliasing (§10.5.3) and a multichannel version of the abstract sampling theorem (§10.5.4). The simplicity and scope of these results will we hope convince the reader of the merits of the abstract approach, which it must be said has a price. The absence of familiar properties of and relationships between the real line $\mathbb{R}$ and the integers $\mathbb{Z}$ occasionally gives rise to some technical difficulties which require a specialized knowledge for an adequate discussion. Some of these points we will pass over, contenting ourselves with the appropriate references. The notation used is derived largely from Rudin's monograph *Fourier analysis on groups* (1962) which is the basic reference book.

10.2   Locally compact abelian groups

A *locally compact abelian group* G is an abelian group which is also a locally compact Hausdorff topological space in which the group operations are continuous. The group operation will usually be written additively. Thus the addition map

$$G \times G \to G : (x, y) \mapsto x + y$$

and the inverse map

$$G \to G : x \mapsto -x$$

are required to be continuous; these can be combined into the single requirement that the map $(x, y) \mapsto x - y$ is continuous. For each $a \in G$, the translation $G \to G$ given by

$$x \to x + a$$

is a homeomorphism of G to itself under which the identity 0 goes to a. Groups which are isomorphic and homeomorphic will not be distinguished.

Compactness is a generalization of finiteness. In Euclidean space $\mathbb{R}^n$, the compactness of a set is, by the Heine–Borel theorem, equivalent to the set being closed and bounded. Being a closed and bounded subset of the plane, the multiplicative circle group

$$\mathbb{T}^1 = \{z \in \mathbb{C} : |z| = 1\}$$

or its additive (isomorphic and homeomorphic) counterpart $\mathbb{R}/\mathbb{Z} = [0, 1]$ mod 1, where $\mathbb{Z}$ is the additive group of integers, are compact.

In a *locally compact* space, each point has a compact neighbourhood. The group $\mathbb{R}$ is locally compact since any point α say has a closed and bounded (and hence compact) neighbourhood consisting of the closed bounded interval $[\alpha - 1, \alpha + 1]$. The group $\mathbb{Z}$ is discrete and so locally compact.

The periodicity 2π in the classical theory of Fourier series is associated with the additive compact group of numbers $[0, 2\pi)$ mod 2π, corresponding to the repetitions of the intervals for periodic functions. This group is isomorphic to the additive quotient group $\mathbb{R}/\mathbb{Z} = [0, 1)$ mod 1 and to the circle group $\mathbb{T}^1$. In the classical theory of Fourier series, the group G could be taken to be any of these groups, depending on the convention adopted for the Fourier transform, while in Fourier analysis, G is the additive locally compact group $\mathbb{R}$ of real numbers.

Note that the relatively compact set $[0, 1)$ is a complete set of coset representatives or *transversal* of $\mathbb{R}/\mathbb{Z}$.

10.2.1 *Haar measure and integral*

A locally compact abelian group G has a positive regular Borel measure called the *Haar measure* and denoted by m_G. It is translation-invariant, finite on compact subsets of G and unique up to a multiplicative constant. Thus for any Borel set $E \subseteq G$ and $x \in G$, $m_G(E + x) = m_G(E)$; and if E is compact, then $m_G(E) < \infty$. The existence of Haar measure can be proved using the Riesz representation theorem for positive invariant linear functionals. For more details see (Rudin 1962) and the references cited there; Nachbin's book (1965) contains a comprehensive and self-contained account of Haar measure and integral. Rudin (1977, Theorem 5.14) gives a short proof of the existence of Haar measure for compact abelian groups.

The regularity of Haar measure implies that the measure of any non-empty open set U in G is positive. Thus the Haar measure of a compact group is positive and finite. The Haar measure of a compact group is often taken to be 1 but this is restrictive and will not be done here. Another measure m_G' on G is given by $m_G'(E) = m_G(-E)$ for each measurable set E in G ($-E = \{-x : x \in E\}$). This is readily seen to be another Haar measure, so that by uniqueness $m_G'(E) = c\, m_G(E)$. But $m_G'(E) = m_G(-E) = c\, m_G(E)$ and it follows by considering a symmetric set $S = -S$ (take $S = E \cup -E$) with positive and finite measure that $c = 1$.

The *Haar integral* denoted by

$$\int_G f(x) \, dm_G(x)$$

can be constructed from the Haar measure m_G for suitable functions $f : G \to \mathbb{C}$ as a limit of expressions of the form $\sum_{j=1}^n c_j \, m_G(E_j)$ where $c_j \in \mathbb{C}$ and $E_j = \{x \in G : f(x) = c_j\} = f^{-1}(c_j)$. Since m_G is unique up to a multiplicative constant and is translation-invariant, it follows that the Haar integral is also unique up to a multiplicative constant, and that for each $y \in G$, the integral is translation-invariant in the sense that $\int_G f(x - y) \, dm_G(x) = \int_G f(x) \, dm_G(x)$. In addition, since $m_G(-E) = m_G(E)$, it follows that $dm_G(-x) = dm_G(x)$, or more precisely

$$\int_G f(-x) \, dm_G(x) = \int_{-G} f(y) \, dm_G(-y) = \int_G f(y) \, dm_G(y). \qquad (10.2.1)$$

For any subset E of G, $m_G(E) = \int_G \chi_E(x) \, dm_G(x)$, where χ_E is the characteristic function of E ($\chi_E(x) = 1$ or 0 according as $x \in E$ or $x \notin E$).

The normalization of the Haar measure is fixed by one (non-trivial) integrand function. For example when $G = \mathbb{R}$, the Haar measure can be taken to be the usual Lebesgue measure (so that the Haar measure of the unit interval is 1), corresponding to taking $\int_{\mathbb{R}} \chi_{[0,1]}(x)\, dx = 1$.

The set of functions f for which the integral $\int_G |f(x)|^p dm_G(x)$ exists for some positive exponent p is denoted by $L^p(G)$. When $p \geq 1$, $L^p(G)$ is a Banach space, i.e., $L^p(G)$ is a linear space which is complete with respect to the norm

$$\|f\|_{G,p} = \left(\int_G |f(x)|^p dm_G(x) \right)^{1/p}$$

(Rudin 1962, Appendix E7). When $G = \mathbb{R}$, $L^2(G)$ is the Hilbert space of square integrable functions defined on $\mathbb{R}$; $L^1(\mathbb{T}^1)$ is the Banach space of absolutely integrable functions defined on $\mathbb{T}^1$. Here only the cases $p = 1, 2, \infty$ arise.

The norm will be written $\|f\|$ when the group and exponent are clear from the context. Two functions f, g in $L^p(G)$ which differ on a set of Haar measure 0 (or which are the same "almost everywhere" in Haar measure) are considered to be identical.

10.2.2 Convolution

The convolution $f * g = f *_G g$ of the functions f, g over the group G is analogous to that in classical Fourier analysis and is defined by

$$f * g(x) = \int_G f(x - y)g(y)\, dm_G(y) = \int_G f(y)g(x - y)\, dm_G(y)$$

whenever the integral $\int_G |f(y)g(x - y)|\, dm_G(y)$ exists (see Loomis 1953, §31A; or Rudin 1986, §1.1.6). The two forms are equal by translation-invariance of the Haar measure. The dependence of the convolution on the group G will only be expressed when there is ambiguity, as in the abstract sampling theorem discussed below.

The convolution $f * g$ of two functions f, g in the space

$$L^1(G) = \left\{ f : G \to \mathbb{C} : \int_G |f(x)|\, dm_G(x) < \infty \right\}$$

of integrable functions is defined almost everywhere in $L^1(G)$, making $L^1(G)$ a Banach algebra with multiplication given by convolution. The space $L^2(G)$ of square integrable functions is a Hilbert space with inner product

$$\langle f, g \rangle = \int_G f(x)\overline{g(x)}\, dm_G(x)$$

and provides the setting for the sampling theorem.

10.2.3 Discrete groups

When the abelian group G is discrete, the cardinality

$$\sum_{x \in E} 1 = \#E$$

of any subset E of G is clearly translation-invariant and is finite for compact subsets. The cardinality or *counting* measure of a set is thus a Haar measure for discrete abelian groups. By uniqueness and translation invariance, a Haar measure m_G on a discrete abelian group G must be a multiple of counting measure and so is given by

$$m_G(E) = m_G(\{0\}) \sum_{x \in E} 1 = m_G(\{0\}) \#E,$$

where $\#E$ is the cardinality of E and $m_G(\{0\})$ is the measure of the point 0 (indeed of any point $x \in G$). It follows that the Haar integral of a function $f : G \to \mathbb{C}$ over a discrete abelian group G must be of the form

$$\int_G f(x) dm_G(x) = m_G(\{0\}) \sum_{x \in G} f(x). \tag{10.2.2}$$

When $G = \{g_j : j = 1, 2, \ldots\}$ is infinite, the infinite sum (if it exists) is defined to be the limit of a finite partial sum:

$$m_G(\{0\}) \sum_{x \in G} f(x) = m_G(\{0\}) \lim_{n \to \infty} \sum_{j=1}^{n} f(x_j).$$

Thus when G is discrete, $L^1(G) = \ell^1(G)$, the set of summable sequences in G. The particular normalization of $m_G(\{0\})$ in the following will depend on the circumstances. Note that it is often taken to be 1 (Rudin 1962, §1.1.3) but as with Haar measure on a compact group, this is unnecessarily restrictive.

10.2.4 Haar measure on quotient groups

Let H be a closed subgroup of G and let $[x] = x + H$ denote a coset of H, i.e., an element in the quotient group G/H. Then G can be decomposed into a disjoint union of cosets:

$$G = \bigcup_{[x] \in G/H} [x].$$

The quotient group G/H is given the topology which makes the natural projection from G to G/H given by each point $x \in G$ being sent to its coset $[x]$ continuous. When G is locally compact, the subgroup H and the quotient group G/H are also locally compact and abelian with Haar measures m_H and $m_{G/H}$.

Let $f \in L^1(G)$ and let $F : G \to \mathbb{C}$ be defined by

$$F(x) = \int_H f(x + h) \, dm_H(h).$$

Then for any x' in the coset $x + H$ (i.e., $x' = x + h'$ for some $h' \in H$),

$$F(x') = \int_H f(x + h' + h) \, dm_H(h) = \int_H f(x + h) \, dm_H(h) = F(x)$$

by the translation invariance of m_H. Thus F is constant on cosets of H and so a function $F_1 : G/H \to \mathbb{C}$ can be defined by

$$F_1([x]) = F(x).$$

Hewitt and Ross (1970, 28.54) show that $F_1 \in L^1(G/H)$ and that every function in $L^1(G/H)$ is of this form. Moreover given the Haar measure of G, the Haar measures on H or G/H can be normalized so that the fundamental *coset decomposition formula*

$$\int_G f(x)\, dm_G(x) = \int_{G/H} F_1([x])\, dm_{G/H}([x])$$

$$= \int_{G/H} \int_H f(x+h)\, dm_H(h)\, dm_{G/H}([x]) \qquad (10.2.3)$$

holds (Hewitt and Ross 1970, 28.54 (iii)); Rudin (1962, §2.7.3) gives a simpler proof when f has compact support. This formula and the normalizations involved are used repeatedly in the discussion of the abstract sampling theorem below. The decomposition (10.2.3) occurs in classical sampling theory and has often been used in the derivation of aliasing estimates (see, for example, Butzer et al. 1988, §3.4; Beaty and Higgins 1994; and Faridani 1994, Theorem 3.6. For a suitable function $f : \mathbb{R} \to \mathbb{C}$,

$$\int_{\mathbb{R}} f(x)\, dx = \sum_{k \in \mathbb{Z}} \int_k^{k+1} f(x)\, dx = \sum_{k \in \mathbb{Z}} \int_0^1 f(y+k)\, dy$$

$$= \int_0^1 \sum_{k \in \mathbb{Z}} f(x+k)\, dx = \int_{\mathbb{R}/\mathbb{Z}} \sum_{k \in \mathbb{Z}} f(x+k)\, d[x],$$

since $\mathbb{R}/\mathbb{Z}$ is isomorphic to $[0, 1]$ mod 1. This argument relies on special properties of the real line and the abstract result requires a much deeper approach. For finite groups the decomposition formula implies Lagrange's theorem:

$$\#G = \#H\, \#G/H.$$

10.2.5 *Characters and the dual group*

A *character* of G is a homomorphism $\gamma: G \to \mathbb{T}^1$, the multiplicative circle group $\{z \in \mathbb{C} : |z| = 1\}$. That is for each $x, x' \in G$,

$$\gamma(x + x') = \gamma(x)\, \gamma(x') \qquad (10.2.4)$$

and $|\gamma(x)| = 1 = \gamma(0)$. The characters form an additive group under the operation

$$(\gamma + \gamma')(x) := \gamma(x)\, \gamma'(x),$$

with identity the constant homomorphism γ_0, given by $\gamma_0(x) = 1$, and inverse $-\gamma$ given by

$$(-\gamma)(x) = \gamma(x)^{-1} = \overline{\gamma(x)}.$$

Alternatively, instead of $\mathbb{T}^1$, one could consider the isomorphic additive group $\mathbb{R}/\mathbb{Z}$ of real numbers modulo 1 with the character given by $\gamma(x) = e^{-2\pi i \langle x, \gamma \rangle}$ (Faridani 1994).

The *continuous characters* form a group, called the *dual group* and will be written $G^\wedge$ or Γ. The character γ evaluated at the point $x \in G$ is often written $\gamma(x) = (x, \gamma)$ in order to bring out the duality of G and Γ (to be discussed below in §10.2.7). With this notation, for each $x, x' \in G$ and $\gamma \in \Gamma$,

$$(x + x', \gamma) = \gamma(x + x') = \gamma(x)\gamma(x') = (x, \gamma)(x', \gamma),$$
$$(x, \gamma + \gamma') = \gamma(x)\gamma'(x) = (x, \gamma)(x, \gamma'),$$

from which it follows that $(x, 0) = 1 = (0, \gamma)$ and

$$(-x, \gamma) = (x, -\gamma) = (x, \gamma)^{-1} = \overline{(x, \gamma)}.$$

A natural way of giving the group Γ a topology is via the Gelfand theory of commutative Banach algebras (Loomis 1953, Appendix D). The characters γ in the group Γ can be identified with the (locally compact) maximal ideal space Δ of the Banach algebra $L^1(G)$, consisting of the kernels $\ker h$ of the non-zero complex homomorphisms $h : L^1(G) \to \mathbb{C}$ (Loomis 1953, §34C; Rudin 1962, §1.2.3). However this needs the abstract Fourier transform, to be introduced in the next section. More directly, Γ can be given the topology of uniform convergence on compact subsets of G (i.e., for each compact set $K \subseteq G$, each positive ε and each character $\gamma_0 \in \Gamma$, the sets

$$U(K, \varepsilon, \gamma_0) = \{\gamma : |(x, \gamma) - (x, \gamma_0)| < \varepsilon \text{ for all } x \in K\}$$

are open and form a base for the topology for Γ (Loomis 1953, §34C)). It can be verified that the group operations in Γ are continuous in this topology and that this topology coincides with the Gelfand topology which thus makes Γ a locally compact abelian group.

The form of the characters for a particular group G can be determined from the functional equation (10.2.4). For instance, when G is the finite group $\mathbb{Z}_6 = \{0, 1, 2, \ldots, 5\}$ of residues modulo 6, the character $\gamma : \mathbb{Z}_6 \to \mathbb{T}^1$ is of the form

$$\gamma(x) = e^{2\pi\alpha_\gamma(x)}$$

where $\alpha_\gamma(x) \in \mathbb{R}$. Since $\gamma(x) = \gamma(1)^x$, $\alpha_\gamma(x) = x\alpha_\gamma(1)$. Moreover, since $6 \times x \equiv 0 \bmod 6$, $\gamma(6 \times 1) = 1$, whence $\alpha_\gamma(6 \times 1) = 6\alpha_\gamma(1) \in \mathbb{Z}$. Choose without loss of generality $\alpha_\gamma(1) = k_\gamma/6$, where $k_\gamma \in \{0, 1, \ldots, 5\}$. Then

$$\gamma(x) = e^{\pi i x k_\gamma/3} = e^{\pi i x \gamma/3} = (x, \gamma)$$

where we have identified k_γ and γ. It follows that the dual group $\mathbb{Z}_6^\wedge$ is isomorphic to $\mathbb{Z}_6$.

In a similar way, when G is the real line $\mathbb{R}$, solving (10.2.4) gives the simple and familiar expression

$$(x, \gamma) = e^{ix\gamma},$$

where the homomorphism γ is identified with the real number γ in the exponent. The symmetry of the character here makes $\mathbb{R}$ its own dual. The circle group $\mathbb{T}^1$, which as has been mentioned is isomorphic to the additive group $\mathbb{R}/\mathbb{Z}$, has as its dual group the additive group of integers $\mathbb{Z}$, i.e., $(\mathbb{T}^1)^{\wedge} = \mathbb{Z}$. The characters here are of the form e^{ikx}, where $k \in 2\pi\mathbb{Z} = \{2\pi k' : k' \in \mathbb{Z}\}$ which is isomorphic to $\mathbb{Z}$ and x is in the additive group $\mathbb{R}/\mathbb{Z} = [0, 1) \bmod 1$. Further discussion of these examples can be found in Rudin (1962, §1.2.7), Loomis (1953, §36C) or Hewitt and Ross (1963, Chapter 4, §15).

Although we are not distinguishing between isomorphic groups, their characters can differ in interesting ways. For example the multiplicative group of positive reals $(0, \infty) = \mathbb{R}^+$ is isomorphic (under the logarithm) to the additive group of reals $\mathbb{R}$. The characters of $\mathbb{R}^+$ are of the form x^{iy} where $x = e^r$ and y are positive real numbers; this is associated with the Mellin transform (Titchmarsh 1962, p. 7) and is discussed further in §10.5.1.

10.2.6 Characters on compact abelian groups

It is a simple consequence of the translation-invariance that the characters $\{(\cdot, \lambda) : \lambda \in \Gamma\}$ on a compact abelian group G are an orthogonal family. We can suppose that G and hence Γ each contain more than one element since otherwise there is nothing to prove. For convenience, given x, x' in the group G, define the Kronecker symbol $\delta_{x,x'}$ by

$$\delta_{x,x'} = \begin{cases} 1 & \text{if } x = x' \\ 0 & \text{if } x \neq x' \end{cases}$$

and define $\delta_{\gamma,\gamma'}$ for $\gamma, \gamma' \in \Gamma$ similarly. Now since $(x, 0) = 1$, it follows that

$$\int_G \chi_G(x)(x, 0)\, dm_G(x) = \int_G dm_G(x) = m_G(G)$$

and when $\gamma \neq 0$ there exists (since $\Gamma \neq \{0\}$) an element $x' \in G$ such that $(x', \gamma) \neq 1$. Hence by translation-invariance of the measure m_G,

$$\int_G (x, \gamma)\, dm_G(x) = (x', \gamma) \int_G (x - x', \gamma)\, dm_G(x) = (x', \gamma) \int_G (x, \gamma)\, dm_G(x)$$

and so $(1 - (x', \gamma)) \int_G (x, \gamma)\, dm_G(x) = 0$. Since $(x', \gamma) \neq 1$, it follows that when $\gamma \neq 0$,

$$\int_G (x, \gamma)\, dm_G(x) = 0.$$

Thus

$$\int_G (x, \gamma)\overline{(x, \gamma')}\, dm_G(x) = \int_G (x, \gamma - \gamma')\, dm_G(x) = m_G(G)\, \delta_{\gamma,\gamma'}, \qquad (10.2.5)$$

and the characters on a compact abelian group are orthogonal (and orthonormal when $m_G(G) = 1$). These results are readily verified for $\mathbb{Z}_6$ and they include the Fourier series case:

$$\int_0^{2\pi} e^{itu}\, dt = \begin{cases} 2\pi & \text{if } u = 0 \\ 0 & \text{if } u \neq 0. \end{cases}$$

If $m_G(G) = 1$ the characters are orthonormal.

10.2.7 *Pontryagin duality*

While finite abelian groups and the real line are isomorphic to their respective duals, this is not so for every locally compact abelian group. It is an important result of Pontryagin that the groups G and $\Gamma^\wedge = (G^\wedge)^\wedge$ are isomorphic, in other words, the dual of the dual of G is isomorphic to G or

$$(G^\wedge)^\wedge = \Gamma^\wedge \cong G$$

(Rudin 1962, §1.7). The isomorphism $\Phi : G \to \Gamma^\wedge$ say takes each $x \in G$ to a homomorphism $\Phi_x : \Gamma \to S^1$, so that for any $\gamma, \gamma' \in \Gamma$, $\Phi_x(\gamma) \in \mathbb{T}^1$, $\Phi(\gamma + \gamma') = \Phi_x(\gamma)\,\Phi_x(\gamma')$ and so on. The characters introduced above behave in the same way as the complex numbers $\Phi_x(\gamma)$ and so could be written (γ, Φ_x). However for convenience they are written (x, γ) in the inverse Fourier transform in (10.3.5) below since the integrating variable γ should prevent confusion with the $\gamma(x)$, the value of the character γ at x.

From §10.2.5, the dual group of the circle group $\mathbb{T}^1$ is the additive group $\mathbb{Z}$ of integers, whence the dual group of $\mathbb{Z}$ is $\mathbb{T}^1$. Similar results hold in higher dimensions; for example the dual group of $\mathbb{R}^n$ is also $\mathbb{R}^n$ and the dual group of the n-dimensional torus $\mathbb{T}^n = \mathbb{T}^1 \times \cdots \times \mathbb{T}^1$ is $\mathbb{Z}^n$. These examples illustrate the general result, which is fundamental to the sampling theorem, that the dual of a compact abelian group is discrete (Rudin 1962, §1.2.5). In turn, Pontryagin duality implies that the dual of a discrete locally compact abelian group is compact. A finite abelian topological group G, being compact and discrete, has a dual $G^\wedge$ which is also compact and discrete (and so finite) and isomorphic to G.

10.2.8 *Annihilators*

Annihilators link the dual groups of subgroups with quotient groups. In view of applications later on, it is convenient to explain annihilators in terms of the subgroup Λ of the dual group Γ (Λ has no connection with irregular sampling, discussed in (Higgins 1996, Chapter 10)).

Given a closed subgroup Λ of Γ (taking Λ to be closed allows the continuous characters of Λ to be extended to Γ (Rudin 1962, Theorem 2.1.4)), the *annihilator* of Λ denoted by $\Lambda^\perp$ is the subset of G defined by

$$\Lambda^\perp = \{x \in G : (x, \lambda) = 1 \text{ for all } \lambda \in \Lambda\}.$$

The annihilator is a group (since $(x, \lambda + \lambda') = (x, \lambda)\,(x, \lambda') = 1$) and is closed by the continuity of the characters. In particular $\Gamma^\perp = \{0\}$, the identity of G. The annihilator of $\Lambda^\perp$ is (isomorphic to) Λ, i.e., $(\Lambda^\perp)^\perp = \Lambda$ (Rudin 1962, §2.1). As examples, the annihilator $\{0, 2, 4\}^\perp$ of the subgroup $\{0, 2, 4\}$ of $\mathbb{Z}_6$ is $\{0, 3\}$ since for each $j \in \{0, 2, 4\}$ and each $k \in \{0, 3\}$

$$(j, k) = e^{\pi i j k / 3} = 1.$$

Again for each real $w \neq 0$, the annihilator $(2\pi w\mathbb{Z})^\perp$ of the subgroup $2\pi w\mathbb{Z}$ of $\mathbb{R}$ is $\mathbb{Z}/w$ and for each non-zero integer k, the annihilator $(k\mathbb{Z})^\perp$ of the subgroup $k\mathbb{Z}$ of $\mathbb{Z}$ is $\{e^{2\pi i r/k} : r = 0, 1, \ldots, k - 1\}$, the k-th roots of unity on the circle.

10.2.9 *Quotient groups and characters*

Applications later on make it convenient to explain characters for the quotient group Γ/Λ of the dual group Γ and a subgroup Λ. The characters of the quotient group Γ/Λ are derived from the original characters on Γ and the annihilator $\Lambda^\perp$ of Λ.

Let $\Lambda^\perp = H$. Then the characters of Γ/Λ are of the form $(h, [\gamma])$ for some $h \in H$, $\gamma \in \Gamma$ since when $h \in H$, (h, γ) is constant on cosets of $\Gamma/H^\perp$. For suppose γ' also lies in the coset $[\gamma] = \gamma + \Lambda$. Then $\gamma' = \gamma + \lambda$ for some $\lambda \in \Lambda = H^\perp$ and so

$$\frac{(h, \gamma)}{(h, \gamma')} = (h, \gamma)(h, -\gamma') = (h, \gamma - \gamma') = (h, -\lambda) = 1$$

and $(h, \gamma) = (h, \gamma')$. Thus for each $h \in H = \Lambda^\perp$, we can define the character

$$(h, [\gamma]) = (h, \gamma + \Lambda) := (h, \gamma) \tag{10.2.6}$$

for Γ/Λ, for any $\gamma \in \Gamma$. It can be verified that $H = \Lambda^\perp$ is the dual of Γ/Λ and that $\Lambda = H^\perp$ is the dual of G/H, i.e.,

$$H = (\Gamma/\Lambda)^\wedge = (\Gamma/H^\perp)^\wedge \quad \text{and} \quad \Lambda = H^\perp = (G/H)^\wedge = (G/\Lambda^\perp)^\wedge, \tag{10.2.7}$$

so that $\Gamma/\Lambda = H^\wedge$ (Rudin 1962, Theorem 2.1.2).

10.3 Fourier theory

The abstract development is very similar to the classical theory (Rudin 1962, §1.5; Bachman 1964, p. 227; and Loomis 1953, §36B).

10.3.1 *The Fourier transform*

The Fourier transform of the function $f : G \to \mathbb{C}$ in $L^1(G)$ is denoted either by $f^\wedge : \Gamma \to \mathbb{C}$ or by $\mathcal{F}f : \Gamma \to \mathbb{C}$; and defined by

$$f^\wedge(\gamma) = (\mathcal{F}f)(\gamma) = \int_G f(x)(x, -\gamma)dm_G(x). \tag{10.3.1}$$

There should be no confusion with the notation for duality between groups. The classical case (see (Higgins 1996, §2.2)) corresponds to $G = \mathbb{R}$, when the character is of the form e^{itu} and

$$f^\wedge(u) = \frac{1}{\sqrt{2\pi}} \int_\mathbb{R} f(t)e^{-itu}dt = \int_\mathbb{R} f(t)(t, -u)dm_\mathbb{R}(t), \tag{10.3.2}$$

the Haar measure $m_\mathbb{R}$ of $E \subset \mathbb{R}$ being given by $m_\mathbb{R}(E) = 1/\sqrt{2\pi}|E|$, where $|E|$ is the Lebesgue measure of E. The dependence of the symbols $\wedge$ and $\mathcal{F}$ on G will not usually be expressed. The Fourier transform $\mathcal{F}$ is a linear map or operator (since $\mathcal{F}(f + g) = (f + g)^\wedge = f^\wedge + g^\wedge = \mathcal{F}(f) + \mathcal{F}(g)$, $\mathcal{F}(cf) = (cf)^\wedge = cf^\wedge = c\mathcal{F}(f)$) which takes an integrable function (in $L^1(G)$) to a continuous function.

As an important example, suppose G is a compact abelian group (so that $m_G(G)$ is finite and $L^2(G) \subseteq L^1(G)$) and consider the Fourier transform $\chi_G^\wedge$ of the characteristic function χ_G for G ($\chi_G(x) = 1$ for each $x \in G$). Then χ_G lies in $L^2(G) \subseteq L^1(G)$ and by the definition of the Fourier transform (10.3.1) and the fact that the characters on a compact abelian group are orthogonal (10.2.5),

$$\chi_G^\wedge(\gamma) = \int_G \chi_G(x)\,(x, -\gamma)\,dm_G(x) = \delta_{\gamma,0}\,m_G(G), \qquad (10.3.3)$$

where for each $\gamma, \gamma' \in \Gamma, \delta_{\gamma,\gamma'}$ is 0 or 1 according as γ and γ' are equal or distinct. This simply expresses the orthogonality of the characters on G (see §10.2.6) in terms of the Fourier transform.

When the locally compact group G is not compact, the linear Fourier transform (10.3.1) can, as in the classical case, be extended from $L^1(G)$ to $L^2(G)$ (and the notation $f^\wedge$ retained), using the fact that $L^1(G) \cap L^2(G)$ is a dense subset of $L^2(G)$ (Rudin 1962, §1.6). This is essentially Plancherel's theorem, to be discussed below.

Convolution again

The convolution $f * g$ of two integrable functions f, g ($\in L^1(G)$) is well-defined by Fubini's theorem. The Fourier transform $(f * g)^\wedge$ of the convolution of the integrable functions f, g is given by

$$(f * g)^\wedge(\gamma) = f^\wedge(\gamma)\,g^\wedge(\gamma), \qquad (10.3.4)$$

(Rudin 1962, Theorem 1.6.3) since by translation invariance and Fubini's theorem,

$$\begin{aligned}
(f * g)^\wedge(\gamma) &= \int_G (f * g)(x)(x, -\gamma)dm_G(x) \\
&= \int_G \int_G f(x - y)g(y)dm_G(y)(x, -\gamma)dm_G(x) \\
&= \int_G \left(\int_G f(x - y)(x, -\gamma)dm_G(x) \right) g(y)dm_G(y) \\
&= \int_G \left(\int_G f(z)(z, -\gamma)dm_G(z) \right) (y, -\gamma)g(y)dm_G(y) \\
&= \left(\int_G f(z)(z, -\gamma)dm_G(dz) \right) \left(\int_G f(y)(y, -\gamma)dm_G(dy) \right) \\
&= f^\wedge(\gamma)\,g^\wedge(\gamma),
\end{aligned}$$

much as in the classical case.

The inverse Fourier Transform

The inverse Fourier transform $\psi^\vee : \Gamma \to \mathbb{C}$ of each ψ in $L^1(\Gamma)$ is defined by

$$\psi^\vee(x) = \int_\Gamma \psi(\gamma)(x, \gamma)dm_\Gamma(\gamma) \qquad (10.3.5)$$

and is continuous. Strictly speaking, the character (x, γ) should be written (γ, Φ_x) where $\Phi : G \to \Gamma^\wedge$ is the Pontryagin isomorphism and $\Phi_x : \Gamma \to \mathbb{T}^1$ is an element (homomorphism) of the dual group of Γ—see §10.2.7. Again, the dependence of the symbol $\vee$ on Γ will not be expressed. Thus the repeated Fourier transform

$$\int_\Gamma f^\wedge(\gamma)(y, -\gamma)\, dm_\Gamma(\gamma) = \int_\Gamma \left(\int_G f(x)(x, -\gamma)\, dm_G(x) \right) (y, -\gamma)\, dm_\Gamma(\gamma)$$

of f will be written $f^{\wedge\wedge}$ and the Fourier transform followed by the inverse Fourier transform will be written $f^{\wedge\vee}$.

The Inversion Theorem

The Haar measure on the dual group Γ of G can be normalized so that for f lying in $B(G) \cap L^1(G)$, the inversion theorem holds (Rudin 1962, §1.5), i.e.,

$$f(x) = \int_\Gamma f^\wedge(\gamma)(x, \gamma)\, dm_\Gamma(\gamma) = f^{\wedge\vee}(x) \qquad (10.3.6)$$

for (Haar) almost all points $x \in G$. Here $B(G)$ is the set of all finite linear combinations of positive definite functions and is a dense subset of $L^1(G)$ and of $L^2(G)$ (Rudin 1962, p. 26). Thus the map $\psi \mapsto \psi^\vee$ is the inverse of $\mathcal{F}$, i.e., $f^{\wedge\vee} = (\mathcal{F}^{-1}\mathcal{F})f = f$. When f is continuous, (10.3.6) holds pointwise; and $f^{\wedge\wedge}(x) = f(-x)$ for each x in G (pointwise). If f is not continuous, the representation is interpreted as holding everywhere except on a set of Haar measure 0.

When G is compact (so that Γ is discrete), the inversion theorem (10.3.6) implies that for each $x \in G$,

$$1 = \chi_G(x) = \int_\Gamma \chi_G^\wedge(\gamma)\, (x, \gamma)\, dm_\Gamma(\gamma).$$

But Γ is discrete and by 10.3.3, $\chi_G^\wedge$ is a delta function. Hence the reciprocity relation

$$1 = m_\Gamma(\{0\}) \sum_{\gamma \in \Gamma} \chi_G^\wedge(\gamma)\, (x, \gamma) = m_\Gamma(\{0\})\, m_G(G) \qquad (10.3.7)$$

holds between the dual Haar measures. The inversion theorem (10.3.6) holds if the Haar measure for compact groups is chosen to be unity but this makes the Haar measure on the discrete dual group counting measure. However (10.3.7) also ensures that the inversion theorem holds and avoids the inconsistencies that arise from the usual normalizations for finite groups. Thus for a finite group G of order n, choosing $m_G(G) = \#G$, the cardinality of G, forces the point (Haar) measure of the identity $m_\Gamma(\{0\})$ of the dual group Γ to be given by $m_\Gamma(\{0\}) = 1/(\#G)$ and $m_\Gamma(\Gamma) = 1$.

The inverse Fourier transform can be regarded as an operator $\mathcal{F}^{-1}$ which sends the function ψ in $L^1(\Gamma) \cap L^2(\Gamma)$ to $\psi^\vee$ in $L^2(G)$ and which by the inversion theorem is inverse to $\mathcal{F}$. This operator can be extended to the whole of $L^2(G)$ in the same way as for the Fourier transform, so that the representation holds in norm. However as with the Fourier transform, this extension is usually done in the context of Plancherel's theorem.

Plancherel's Theorem

First we use convolution to show that $\mathcal{F}$ is an isometry on $L^1(G) \cap L^2(G)$. Consider $g = f * \tilde{f}$, where $f \in L^1(G) \cap L^2(G)$ and $\tilde{f}(x) = \overline{f}(-x)$. Then g is continuous, integrable, positive definite and so in $L^1(G) \cap B(G)$. By the Fourier transform of a convolution (10.3.4),

$$g^\wedge = (f * \tilde{f})^\wedge = f^\wedge \, \tilde{f}^\wedge = f^\wedge \, \overline{f^\wedge} = |f^\wedge|^2$$

and the inversion theorem then gives

$$\int_G |f(x)|^2 \, dm_G(x) = \int_G f(x) \, \tilde{f}(-x) \, dm_G(x) = g(0)$$

$$= \int_\Gamma g^\wedge(\gamma) \, dm_\Gamma(\gamma) = \int_\Gamma |f^\wedge(\gamma)|^2 \, dm_\Gamma(\gamma)$$

or in terms of the inner product and the $L^2(G)$-norm $\| \cdot \|_G$

$$\langle f, f \rangle = \|f\|_G^2 = \|f^\wedge\|_\Gamma^2 = \langle f^\wedge, f^\wedge \rangle = \langle \mathcal{F}f, \mathcal{F}f \rangle \qquad (10.3.8)$$

(Rudin 1962, Theorem 1.6.1). Thus the Fourier transform regarded as the operator $\mathcal{F}$ restricted to the dense subset $L^1(G) \cap L^2(G)$ of $L^2(G)$ is an isometry. More is true and $\mathcal{F}(L^1(G) \cap L^2(G))$ is a dense linear subspace of $L^2(\Gamma)$. It follows that the operator $\mathcal{F}$ can be extended isometrically to $L^2(G)$ from $L^1(G) \cap L^2(G)$; whence if $f \in L^2(G)$, $\mathcal{F}f \in L^2(\Gamma)$, then (10.3.8) holds, i.e.,

$$\|f\|_G = \|\mathcal{F}f\|_\Gamma = \|f^\wedge\|_\Gamma;$$

and $\mathcal{F}(L^2(G)) = L^2(\Gamma)$. Since it is an isometry, $\mathcal{F}$ is one to one, onto, bounded (by definition (Hewitt and Ross 1963, p. 454), $\|\mathcal{F}\| = \sup_{\|f\|=1} \|\mathcal{F}f\|_\Gamma = 1$) and so a continuous invertible linear operator on $L^2(G)$ to $L^2(\Gamma)$. Indeed

$$\|f\|_G = \langle f, f \rangle = \|\mathcal{F}f\|_\Gamma = \langle \mathcal{F}f, \mathcal{F}f \rangle = \langle f, \mathcal{F}^*\mathcal{F}f \rangle,$$

whence $\mathcal{F}^* \mathcal{F} = \mathcal{I}_{\mathcal{L}\in(G)}$ and $\mathcal{F}\mathcal{F}^* = \mathcal{I}_{\mathcal{L}\in(\Gamma)}$, i.e., $\mathcal{F}$ is a unitary operator, with inverse $\mathcal{F}^{-1} = \mathcal{F}^*$, the adjoint of $\mathcal{F}$; see (Rudin 1962, §1.6) and (Bachman 1964).

Again just as in the classical case, Plancherel's theorem together with the polarization identity

$$4f\bar{g} = |f + g|^2 - |f - g|^2 + i|f + ig|^2 - i|f - ig|^2$$

yield the Parseval formula that for each f, g in $L^1(G) \cap L^2(G)$,

$$\int_G f(x)\overline{g(x)} \, dm_G(x) = \int_\Gamma f^\wedge(\gamma)\overline{g^\wedge(\gamma)} \, dm_\Gamma(\gamma) \qquad (10.3.9)$$

which likewise extends to $L^2(G)$, giving for each f, $g \in L^2(G)$,

$$\langle f, g \rangle = \langle f^\wedge, g^\wedge \rangle = \langle \mathcal{F}f, \mathcal{F}g \rangle.$$

The extended Fourier transform $\mathcal{F}$ can be defined as follows. Since $L^1(G) \cap L^2(G)$ is dense in $L^2(G)$, given any $f \in L^2(G)$, there exists a sequence (f_n) in $L^1(G) \cap L^2(G)$ which tends to $f \in L^2(G)$ as $n \to \infty$ in the $L^2(G)$-norm, i.e.,

$$\lim_{n \to \infty} \int_G |f(x) - f_n(x)|^2 \, dm_G(x) = \lim_{n \to \infty} \|f - f_n\|_G^2 = 0.$$

Then $(\mathcal{F}f_n) = (f_n^\wedge)$ is a Cauchy sequence in the complete space $L^2(\Gamma)$ and therefore converges to a limit denoted by $\mathcal{F}f$ or $f^\wedge$ in $L^2(\Gamma)$, i.e., the Fourier transform $\mathcal{F}f = f^\wedge$ of a function $f \in L^2(G)$ is defined by

$$(\mathcal{F}f)(\gamma) = f^\wedge(\gamma) := \lim_{n \to \infty} f_n^\wedge(\gamma) = \lim_{n \to \infty} \mathcal{F}f_n = \int_G f_n(x)(x, -\gamma) dm_G(x).$$

This limit $\mathcal{F}f = \lim_{n \to \infty} \mathcal{F}f_n$ does not depend on the choice of the (Cauchy) sequence (f_n) in $L^1(G) \cap L^2(G)$.

The adjoint $\mathcal{F}^*$ has a similar form: given $\varphi \in L^2(\Gamma)$, let (φ_n) be a sequence in $L^1(\Gamma) \cap L^2(\Gamma)$ converging to φ. Then $(\varphi_n^\vee)$ is a Cauchy sequence in $L^2(G)$ (by isometry) and $\mathcal{F}^*\varphi$ is given by its limit in norm, i.e.,

$$(\mathcal{F}^*\varphi)(x) = \varphi^\vee(x) = \lim_{n \to \infty} \varphi_n^\vee(x)$$

where $\|\varphi^\vee - \varphi_n^\vee\|_G^2 = \int_G |\varphi^\vee(x) - \varphi_n^\vee(x)|^2 dm_G(x) = \|\varphi - \varphi_n\|_\Gamma^2 \to 0$ as $n \to \infty$. The inversion theorem holds in $L^2(G)$: for each $f \in L^2(G)$, there exists by density a sequence (f_n) in $L^1(G) \cap B(G)$ converging in $L^2(G)$-norm to f and

$$f(x) = \lim_{n \to \infty} (\mathcal{F}^* f_n^\wedge)(x) = \lim_{n \to \infty} \int_\Gamma f_n^\wedge(\gamma)(x, \gamma) dm_\Gamma(\gamma) = \lim_{n \to \infty} f_n(x),$$

whence $\| f - \mathcal{F}^* f_n^\wedge \| \to 0$ as $n \to \infty$.

An approximating sequence of truncations

There is a particular sequence of functions converging to f which allows us to express the Fourier transform on $L^2(G)$ in a convenient form (with the sequence element in the integrand replaced by a sequence in the range of integration). The support of any $f \in L^2(G)$ is σ-finite, i.e., the support is a countable disjoint union of sets P_j, $j \in \mathbb{N}$, of positive and finite Haar measure (this follows from the construction of the Haar integral in terms of the limit of simple functions). Thus

$$\mathrm{supp} f = \bigcup_{j=1}^\infty P_j$$

where each P_j is of finite positive Haar measure. Write

$$U_n = \bigcup_{j=1}^n P_j,$$

so that $U_n \subseteq U_{n+1}$ and $\mathrm{supp} f = \bigcup_{n=1}^{\infty} U_n = \lim_{n \to \infty} U_n$. For each $n = 1, 2, \ldots$, let

$$f_n = f \, \chi_{U_n}. \tag{10.3.10}$$

Then $f_n \in L^1(G) \cap L^2(G)$ since $|f_n(x)| \le |f(x)|$ for each $x \in G$, and

$$\|f - f_n\|_G^2 = \int_G |f(x) - f_n(x)|^2 dm_G(x) = \int_{\mathrm{supp} f \setminus U_n} |f(x)|^2 dm_G(x)$$

which can be made arbitrarily small for sufficiently large n since

$$\|f\|_G^2 = \sum_{j=1}^{\infty} \int_{P_j} |f(x)|^2 dm_G(x)$$

$$= \int_{U_n} |f(x)|^2 dm_G(x) + \sum_{j>n} \int_{P_j} |f(x)|^2 dm_G(x) < \infty.$$

Hence

$$\mathcal{F}f(\gamma) = \lim_{n \to \infty} \int_G f_n(x)(x, -\gamma) dm_G(x)$$

$$= \lim_{n \to \infty} \int_{U_n} f(x)(x, -\gamma) dm_G(x) \tag{10.3.11}$$

and the limit applies to a sequence of ranges rather than a sequence in the integrand. This has its advantages.

Similarly, as the support of $f^\wedge$ is σ-finite, each $f \in L^2(G)$ has the representation

$$f(x) = \mathcal{F}^* f^\wedge(x) = \lim_{n \to \infty} \int_{V_n} f^\wedge(\gamma)(x, \gamma) dm_\Gamma(\gamma), \tag{10.3.12}$$

where V_n is of finite and positive Haar measure and $\mathrm{supp} f^\wedge = \bigcup_{n=1}^{\infty} V_n = \lim_{n \to \infty} V_n$.

The similarity of these results to the classical Fourier theory is evident. Another result, needed in the next section, is commonly known as the shift theorem in the classical setting and asserts that for any $f \in L^2(G)$, the Fourier transform of the "shift" $f_y(x) = f(x+y)$ is given by

$$f_y^\wedge(\gamma) = \int_G f(x)(x - y, -\gamma) dm_G(x) = (y, \gamma) f^\wedge(\gamma). \tag{10.3.13}$$

10.3.2 The Poisson summation formula

Higgins (1996, §2.3) gives the classical case of this important formula and Loomis (1953, §37E) a proof of the abstract result. In view of applications later, we start with the dual group Γ of G. Suppose that Γ has a closed subgroup Λ and let the Haar measures on Λ and the quotient group Γ/Λ be normalized so that the coset decomposition formula (10.2.3) holds. Then for suitable φ (Loomis chooses φ with $\psi([\gamma]) = \int_\Lambda \varphi(\lambda + \gamma)\, dm_\Lambda(\lambda)$ continuous on Γ/Λ), it can be shown using the Fourier transform of the function ψ that

$$\int_\Lambda \varphi(\lambda) dm_\Lambda(\lambda) = \int_{(\Gamma/\Lambda)^\wedge} \varphi^\wedge(x) dm_{(\Gamma/\Lambda)^\wedge}(x).$$

By using the "shift" formula (10.3.13) and the fact that $(\Gamma/\Lambda)^\wedge = \Lambda^\perp$ (see 10.2.7) it follows that

$$\int_\Lambda \varphi(\gamma + \lambda) dm_\Lambda(\lambda) = \int_{\Lambda^\perp} \varphi^\wedge(x)(x, \gamma) dm_{\Lambda^\perp}(x).$$

Thus by (10.2.2) when Λ and $H = \Lambda^\perp = (\Gamma/\Lambda)^\wedge$ are discrete,

$$m_\Lambda(\{0\}) \sum_{\lambda \in \Lambda} \varphi(\lambda) = m_H(\{0\}) \sum_{h \in H} \varphi^\wedge(h)$$

or

$$m_\Lambda(\{0\}) \sum_{\lambda \in \Lambda} \varphi(\lambda + \gamma) = m_H(\{0\}) \sum_{h \in H} \varphi^\wedge(h)(h, \gamma).$$

10.3.3   Compact abelian groups, bases and normalization

The characters on a compact abelian group G with discrete dual Γ have been shown to be orthogonal in §10.2.6. In fact they are a complete family as well, i.e., for any function f in $L^2(G)$ the set $\{\sum_{j=1}^n a_j(\cdot, \gamma_j) : n = 1, 2, \ldots\}$ of finite linear combinations of the characters on G is dense in $L^2(G)$. This follows from the isometry (10.3.8) established in Plancherel's theorem, which for discrete Γ reduces to Parseval's completeness relation

$$\|f\|_G^2 = \|\mathcal{F}f\|_\Gamma^2 = m_\Gamma(\{0\}) \sum_{\gamma \in \Gamma} |f^\wedge(\gamma)|^2 = \sum_{j=1}^\infty |a_j|^2$$

for each $f \in L^2(G)$.

Completeness can be demonstrated directly (Loomis, §38C) and it also follows simply from the dual Γ of the compact group G being discrete and using the representation (10.3.12) for f to construct a sequence $(\sum_{j=1}^n a_j(\cdot, \gamma_j))$ of sums which converges to f in $L^2(G)$-norm. Indeed since Γ is discrete, $\operatorname{supp} f^\wedge \subseteq \Gamma$ is countable and is of the form $\operatorname{supp} f^\wedge = \{\gamma_j \in \Gamma : j \in \mathbb{N}\}$. Let $T_n = \{\gamma_j \in \operatorname{supp} f : j \le n\}$. Then by (10.3.12),

$$f(x) = \mathcal{F}^* f^\wedge(x) = m_\Gamma(\{0\}) \lim_{n \to \infty} \sum_{j=1}^n f^\wedge(\gamma_j)(x, \gamma_j) = m_\Gamma(\{0\}) \sum_{j=1}^\infty f^\wedge(\gamma_j)(x, \gamma_j)$$

in $L^2(G)$-norm. Moreover since the characters $(\cdot, \gamma_j)$ are orthogonal, the coefficients in the representation are unique and the set of characters forms a basis for $L^2(G)$. This is a standard result for orthonormal sets in Hilbert spaces (Rudin 1986, §4.9) but we give a proof because the choice of Haar measure precludes orthonormality.

Suppose $f(x) = \lim_{n \to \infty} \sum_{j=1}^{n} a_j(x, \gamma_j)$ and for each $n = 1, 2, \ldots$, let $f(x) - \sum_{j=1}^{n} a_j(x, \gamma_j) = \varepsilon_n(x)$. Then

$$\int_G \left| f(x) - \sum_{j=1}^{n} a_j(x, \gamma_j) \right|^2 dm_G(x) = \|\varepsilon_n\|_G^2$$

so that $\varepsilon_n \to 0$ in $L^2(G)$-norm as $n \to \infty$. Since the characters $(\cdot, \gamma_j)$ are orthogonal (10.3.3),

$$\int_G f(x)(x, -\gamma_k) dm_G(x) = a_k m_G(G) + \int_G \varepsilon_n(x)(x, -\gamma_k) dm_G(x).$$

Also

$$\left| \int_G \varepsilon_n(x)(x, -\gamma_k) dm_G(x) \right| \le m_G(G)^{1/2} \|\varepsilon_n\|,$$

so that for each $k = 1, 2, \ldots$,

$$a_k = \frac{f^\wedge(\gamma_k)}{m_G(G)} + O(\|\varepsilon_n\|) = m_\Gamma(\{0\}) f^\wedge(\gamma_k) + O(\|\varepsilon_n\|)$$

by the reciprocity relation (10.3.7). Hence $|a_k - m_\Gamma(\{0\}) f^\wedge(\gamma_k)|$ can be made arbitrarily small and $a_k = m_\Gamma(\{0\}) f^\wedge(\gamma_k)$. Thus the characters are a basis for $L^2(G)$ and the coefficients a_j in the linear combination are essentially (neglecting the normalizing factor $m_\Gamma(\{0\})$) the Fourier coefficients $f^\wedge(\gamma)$. Note that since G is compact, $L^2(G) \subseteq L^1(G)$ and so the representation for $f \in L^2(G)$ also holds in $L^1(G)$ norm.

It follows from the Stone–Weierstrass theorem that continuous functions on a compact group G can be uniformly approximated by finite linear combinations of characters; this can also be proved directly using the above Fourier series representation (Loomis 1953, §38D).

10.4   An abstract sampling theorem

The sampling theorem can now be put into the framework of abstract harmonic analysis by replacing the real line $\mathbb{R}$ with a locally compact abelian group G and its dual Γ. This was first done by Kluvánek (1965) with the objective of determining a class of functions which can be represented by values taken at points in a given discrete subgroup H of G. An alternative formulation of the sampling problem is to suppose that the band-region (the set of 'frequencies') is known *a priori* and then to proceed to determine a satisfactory sampling set H. This approach, widely used in signal processing, is followed by Dodson et al. (1986) and Higgins (1996) and will be taken here.

The Paley–Wiener space (Higgins 1996, Definition 6.19) is a natural setting for the sampling theorem and the definition extends to abstract harmonic analysis for a subset A of Γ as follows:

$$PW_A = \{ f \in L^2(G) \cap C(G) : \mathrm{supp} f^\wedge(\gamma) \subseteq \mathrm{cl}\, A \},$$

where $C(G)$ is the set of all continuous functions $f : G \to \mathbb{C}$. However the subclass $\widetilde{PW}_A$ of PW_A defined by

$$\widetilde{PW}_A = \{f \in L^2(G) \cap C(G) : f^\wedge(\gamma) = 0 \text{ for all } \gamma \notin A\} \qquad (10.4.1)$$

is more suited to the disjoint translate condition imposed on A (see (10.4.2) below).

A subgroup Λ (assumed to be discrete) of Γ is obtained from A by means of the disjoint translates condition and A is extended to a transversal (i.e., a complete set of coset representatives) B of the compact quotient group Γ/Λ. The assumption that Λ is discrete is not a serious restriction since any connected locally compact group always contains a discrete finitely generated subgroup such that the quotient group is compact (Montgomery and Zippin 1955, §2.21). For example if $\Gamma = \mathbb{R}$, then one can take $\Lambda = \lambda\mathbb{Z}$, where $\lambda > 0$, so that $\Gamma/\Lambda = \mathbb{R}/\lambda\mathbb{Z}$ which is isomorphic to the circle group $\mathbb{T}^1$ under the map $t + \lambda\mathbb{Z} \mapsto e^{2\pi i t/\lambda}$.

The annihilator $H = \Lambda^\perp$ of Λ is the dual of the compact quotient group Γ/Λ (see §10.2.8) and so is also a discrete subgroup of G. In the abstract setting, the sampling theorem asserts that any function $f \in \widetilde{PW}_A$ can be reconstructed from samples $f(h)$ taken at points h in the discrete (sampling) subgroup H. The compactness of the quotient group Γ/Λ is the key; the abstract form of the sampling theorem is established by passing from $L^2(\Gamma)$ to $L^2(\Gamma/\Lambda)$ via the transversal B in Γ of Γ/Λ. The usual course of taking the Haar measures of the compact quotient groups Γ/Λ and G/H to be unity obscures the structure of the Fourier transforms. Transversals of G/H in G (denoted by K) do not appear in the proof of Kluvánek's theorem but are useful in an application considered later (see §10.5.2 below).

In the classical setup, $G = \mathbb{R}$ and in order that the Fourier transform has the form (10.3.2), the Haar measure is given by $1/\sqrt{2\pi}$ of Lebesgue measure (i.e., $m_\mathbb{R}(E) = 1/\sqrt{2\pi}|E|$). The Haar measure of the dual $\Gamma = \mathbb{R}$ is determined by the inversion theorem and is also $1/\sqrt{2\pi}$ of Lebesgue measure. The set $A = B = (-\pi w, \pi w]$ has Haar measure

$$m_\Gamma((-\pi w, \pi w]) = \frac{1}{\sqrt{2\pi}} 2\pi w = \sqrt{2\pi}\,w,$$

$\Lambda = 2w\mathbb{Z}$, $\Gamma/\Lambda = \mathbb{R}/(2w\mathbb{Z})$, $H = \Lambda^\perp = \mathbb{Z}/2w$ (see §10.2.8), $G/H = 2w\,\mathbb{R}/\mathbb{Z}$, $K = (-1/2w, 1/2w]$ and has Haar measure $1/(\sqrt{2\pi}\,w)$; and $\widetilde{PW}_B = \widetilde{PW}_{(-\pi w, \pi w]}$. The relationships between the Haar measures of Λ, Γ/Λ, G/H, B and K will emerge in the next section. Note that $m_\Gamma(B)$ corresponds to the sampling rate and $M_G(K)$ to the sampling interval, the distance between consecutive sampling points. The product of the Haar measures of B and K is unity as required by (10.4.9) below.

10.4.1   The spaces $L^2(\Gamma/\Lambda)$ and $L^2(B)$

By neglecting sets of Haar measure 0, the set of functions $\varphi \in L^2(\Gamma)$ for which $\varphi(\gamma) = 0$ for almost all $\gamma \notin B$, can be written $L^2(\Gamma)\chi_B$, the space of square-integrable functions on Γ which vanish outside B; and similarly for $L^1(\Gamma)$. By an abuse of notation, we shall call these sets $L^2(B)$ and $L^1(B)$ respectively. The related spaces $L^2(K)$ and $L^2(G/H)$ play a less prominent role.

Being a transversal of Γ/Λ, B consists of one and just one point from each distinct coset $[\gamma] = \gamma + \Lambda$, i.e., $B \cap (\Lambda + \gamma)$ consists of a single point in B. Thus translates of B by non-zero elements in Λ are disjoint:

$$B \cap (B + \lambda) = \begin{cases} \emptyset & \text{if } \lambda \neq 0 \\ B & \text{if } \lambda = 0. \end{cases} \tag{10.4.2}$$

The result holds for any subset A of B and is clear in the classical case where the transversal $[-\lambda/2, \lambda/2)$ of Γ/Λ is $\mathbb{R}/\lambda\mathbb{Z}$. An alternative formulation of the disjoint translates condition, which in practice is much easier to verify (Dodson and Silva 1985), is that the difference set $D(B) = \{\omega - \omega' : \omega, \omega' \in B\}$ meets Λ only in $\{0\}$, i.e.,

$$D(B) \cap \Lambda = \{0\}. \tag{10.4.3}$$

The Haar measures of Λ, Γ/Λ, H and G/H will be normalized so that the coset decomposition formula (10.2.3) and the Fourier inversion theorem (10.3.6) hold. Then by (10.3.7) and since $\Lambda^\wedge = G/H$ and $H^\wedge = \Gamma/\Lambda$, it follows that

$$m_\Lambda(\{0\}) \, m_{G/H}(G/H) = m_H(\{0\}) \, m_{\Gamma/\Lambda}(\Gamma/\Lambda) = 1. \tag{10.4.4}$$

Then the Haar measure in Γ of B is given by

$$m_\Gamma(B) = \int_\Gamma \chi_B(\gamma) \, dm_\Gamma(\gamma) = \int_{\Gamma/\Lambda} \int_\Lambda \chi_B(\gamma + \lambda) \, dm_\Lambda(\lambda) \, dm_{\Gamma/\Lambda}([\gamma])$$

$$= m_\Lambda(\{0\}) \int_{\Gamma/\Lambda} dm_{\Gamma/\Lambda}([\gamma]),$$

and so the reciprocity relations (10.3.7) yield the measure relations

$$m_\Gamma(B) = m_\Lambda(\{0\}) \, m_{\Gamma/\Lambda}(\Gamma/\Lambda) = \frac{m_\Lambda(\{0\})}{m_H(\{0\})} = \frac{m_{\Gamma/\Lambda}(\Gamma/\Lambda)}{m_{G/H}(G/H)}, \tag{10.4.5}$$

which imply that $m_\Gamma(B)$ is finite and positive. Thus in the classical case, where $G = \Gamma = \mathbb{R}$, it can be verified that

$$m_{\mathbb{R}/(2w\mathbb{Z})}(\mathbb{R}/(2w\mathbb{Z})) \, m_{2w\mathbb{Z}}(\{0\}) = \sqrt{2\pi} \, w$$

and $m_{2w\mathbb{R}/\mathbb{Z}}(2w\mathbb{R}/\mathbb{Z}) \, m_{\mathbb{Z}/2w}(\{0\}) = 1/(\sqrt{2\pi} \, w)$.

Being a transversal of the compact quotient group, B has finite Haar measure but will not in general be compact or even bounded. Fortunately, the spaces $L^2(B)$ and $L^2(\Gamma/\Lambda)$ are essentially the same. To show this, given any function $\varphi : B \to \mathbb{C}$, define the associated function $\varphi_1 : \Gamma/\Lambda \to \mathbb{C}$ by $\varphi_1([\omega]) = \varphi(\omega)$ for each $\omega \in B$. By the disjoint translates condition (10.4.2), for each $\gamma \in \Gamma$, the sum $\sum_{\lambda \in \Lambda} \chi_B(\gamma + \lambda)\varphi(\gamma + \lambda)$ collapses to just one term and

$$\sum_{\lambda \in \Lambda} \chi_B(\gamma + \lambda)\varphi(\gamma + \lambda) = \varphi(\gamma + \lambda_\gamma) = \varphi_1([\gamma]), \tag{10.4.6}$$

where λ_γ is the unique element in Λ such that $\gamma + \lambda_\gamma \in B$. It can be verified that φ is in $L^1(B)$ or $L^2(B)$ if and only if φ_1 is in $L^1(\Gamma/\Lambda)$ or $L^2(\Gamma/\Lambda)$ respectively. One begins by showing $\varphi \in L^1(B) \cap L^2(B)$ if and only if $\varphi_1 \in L^1(\Gamma/\Lambda) \cap L^2(\Gamma/\Lambda)$. Since Λ is discrete, given any function $\varphi \in L^1(\Gamma) \cap L^2(\Gamma)$ which vanishes outside the transversal B (so that $\varphi = \chi_B \varphi$), by the fundamental coset decomposition formula (10.2.3)

$$m_\Lambda(\{0\}) \int_{\Gamma/\Lambda} \sum_{\lambda \in \Lambda} \varphi(\gamma + \lambda) \, dm_{\Gamma/\Lambda}([\gamma]) = \int_\Gamma \varphi(\gamma) \, dm_\Gamma(\gamma) = \int_B \varphi(\omega) dm_\Gamma(\omega)$$

whence by (10.4.6)

$$m_\Lambda(\{0\}) \int_{\Gamma/\Lambda} \varphi_1([\gamma]) dm_{\Gamma/\Lambda}([\gamma]) = \int_B \varphi(\omega) dm_\Gamma(\omega). \tag{10.4.7}$$

Thus in particular, since $\varphi(\omega) = 0$ for $\omega \notin B$,

$$\|\varphi\|_B^2 = \int_B |\varphi(\omega)|^2 \, dm_\Gamma(\omega) = \int_\Gamma |\varphi(\gamma)|^2 \, dm_\Gamma(\gamma)$$

$$= m_\Lambda(\{0\}) \int_{\Gamma/\Lambda} |\varphi_1([\gamma])|^2 \, dm_{\Gamma/\Lambda}([\gamma]) = m_\Lambda(\{0\}) \|\varphi_1\|_{\Gamma/\Lambda}^2 \tag{10.4.8}$$

and similarly for the integrable case. In the same way, the orthogonality and completeness of the characters $(h, [\gamma])$, where $h \in H$, the annihilator of Λ, in Γ/Λ can be transferred to the corresponding elements (h, ω) in B.

Since Γ/Λ is compact, its characters are orthogonal (see §10.2.6), so that by (10.3.3) and (10.4.7) for each $h \in H$, the annihilator of Λ,

$$\chi_B^\vee(h) = m_\Lambda(\{0\}) \int_{\Gamma/\Lambda} (h, [\gamma]) dm_{\Gamma/\Lambda}([\gamma]) = m_\Gamma \, \delta_{h,0}.$$

Similarly, since K is compact, $m_G(K)$ is finite and

$$m_G(K) = m_H(\{0\}) \, m_{G/H}(G/H).$$

Hence the reciprocity relation

$$m_\Gamma(B) \, m_G(K) = 1 \tag{10.4.9}$$

between the transversals B and K holds (this is 10.3.7 when $K = G$ and $H = \{0\}$). The corresponding orthogonality relation holds for $\lambda \in \Lambda$ (recall $\Lambda^\wedge = G/H$):

$$\chi_K^\vee(\lambda) = m_H(\{0\}) \int_{G/H} ([x], \lambda) \, dm_{G/H}([x]) = m_G(K) \, \delta_{\lambda,0}.$$

The standard band-limited condition that the frequency of a signal is at most πw implies that translates by $2\pi w$ of the spectrum are disjoint. The repetition of the translates of the spectrum along the frequency axis gives a periodic function and is associated with the additive locally compact abelian group $\{2\pi wk : k \in \mathbb{Z}\} = 2\pi w\mathbb{Z}$. The quotient group $\mathbb{R}/2\pi w\mathbb{Z}$ has $(-\pi w, \pi w]$ as a transversal and is isomorphic to $\mathbb{T}^1$ and so is compact.

Note that apart from normalization, the function $\chi_B^\vee : G \to \mathbb{C}$ corresponds to the sinc function in the classical theory and it has the same interpolation property (Higgins 1996, Definition 1.2).

10.4.2 A basis for $L^2(B)$

This basis consists of characters in $L^2(\Gamma)$ restricted to B and relies on the characters $(h, [\gamma])$ in Γ/Λ being a basis. Let $\varphi \in L^2(B)$, i.e., $\varphi \in L^2(\Gamma)$ and $\operatorname{supp} \varphi \subseteq B$. For any $\gamma \in [\omega] = \omega + \Lambda$ where $\omega \in B$, define

$$\varphi_1([\gamma]) := \varphi(\omega).$$

Then for each $\gamma \in \Gamma$, $\varphi(\gamma) = \varphi_1([\gamma]) \chi_B(\gamma)$. Now Γ/Λ is compact with discrete dual H, so that using the decomposition (10.4.7) and $(h, [\omega]) = (h, \omega)$ (see 10.2.6), the Fourier transform $\varphi^\wedge$ of φ is given by

$$\begin{aligned}
\varphi^\wedge(h) &= \int_\Gamma \varphi(\gamma)(h, -\gamma)\, dm_\Gamma(\gamma) \\
&= m_\Lambda(\{0\}) \int_{\Gamma/\Lambda} \sum_{\lambda \in \Lambda} \varphi(\gamma + \lambda)(h, -\gamma - \lambda)\, dm_{\Gamma/\Lambda}([\gamma]) \\
&= m_\Lambda(\{0\}) \int_{\Gamma/\Lambda} \varphi_1([\gamma])(h, -[\gamma])\, dm_{\Gamma/\Lambda}([\gamma]) \\
&= m_\Lambda(\{0\}) \varphi_1^\wedge(h).
\end{aligned}$$

Hence by the inversion theorem (10.3.6), for each $\omega \in B$,

$$\begin{aligned}
\varphi(\omega) &= \varphi_1([\omega]) = m_H(\{0\}) \sum_{h \in H} \varphi_1^\wedge(h)(h, [\omega]) \\
&= \frac{m_H(\{0\})}{m_\Lambda(\{0\})} \sum_{h \in H} \varphi^\wedge(h)(h, \omega) = \frac{1}{m_\Gamma(B)} \sum_{h \in H} \varphi^\wedge(h)(h, \omega) \\
&= m_G(K) \sum_{h \in H} \varphi^\wedge(h)(h, \omega),
\end{aligned}$$

by (10.4.9). Thus completeness is established. It follows that for each $\gamma \in \Gamma$,

$$\varphi(\gamma) = \frac{m_H(\{0\})}{m_\Lambda(\{0\})} \sum_{h \in H} \varphi^\wedge(h)(h, \gamma) \chi_B(\gamma),$$

where the characteristic function χ_B is introduced to define φ throughout Γ. Hence

$$\varphi(\gamma) = \frac{1}{m_\Gamma(B)} \sum_{h \in H} \varphi^\wedge(h)(h, \gamma) \chi_B(\gamma) = m_G(K) \sum_{h \in H} \varphi^\wedge(h)(h, \gamma) \chi_B(\gamma).$$

$$(10.4.10)$$

10.4.3 Kluvánek's sampling theorem

The following version of the sampling theorem is slightly more general than that proved by Kluvánek (1965) and follows (Dodson et al. 1986). The disjoint translates condition (10.4.2) (or the equivalent (10.4.3)) is the key to the proof. The notation, the normalizations and the proof given differ somewhat from Kluvánek's. Recall that $\widetilde{PW}_A$ is given by (10.4.1).

Theorem 10.1 *Let G be a locally compact abelian group with dual group Γ, with Haar measure m_Γ on Γ normalized so that the inversion theorem holds. Let A be a Haar measurable subset of Γ. Suppose that*

(i) *there exists a discrete subgroup Λ of Γ with annihilator H and such that A and its translate $A + \lambda$ are disjoint for $\lambda \neq 0$,*

(ii) *Γ/Λ is compact.*

Then there is a complete set B of coset representatives of Γ/Λ in Γ containing A and such that $m_\Gamma(B)$ is positive and finite. Moreover $\Lambda^\perp = H$ is a discrete subgroup of G and for each $f \in \widetilde{PW}_A$,

$$\|f\|_G^2 = \frac{1}{m_\Gamma(B)} \sum_{h \in H} |f(h)|^2$$

and for each $x \in G$,

$$f(x) = \frac{1}{m_\Gamma(B)} \sum_{h \in H} f(h) \chi_A^\vee(x - h) \tag{10.4.11}$$

absolutely in $L^2(G)$ and uniformly on G.

Proof It follows from the disjoint translates condition satisfied by A that A is a set of coset representatives of Γ/Λ in Γ. Let C be any measurable transversal of Γ/Λ in Γ and let B be the set A together with those points in C which are not in the same coset as any point in A, so that B is a measurable transversal of Γ/Λ containing A. By (10.4.5), $m_\Gamma(B)$ is positive and finite.

Since $H = \Lambda^\perp$, it is the dual of Γ/Λ (see §10.2.8) which is compact and so H is discrete. By Plancherel's theorem (10.3.8) and (10.4.8) with $\varphi = f^\wedge \in L^2(B)$ and $\varphi_1 \in L^2(\Gamma/\Lambda)$ given by (10.4.6),

$$\|f\|_G = \|f^\wedge\|_\Gamma = \|f^\wedge\|_B = \frac{\|\varphi_1\|_{\Gamma/\Lambda}}{\sqrt{m_\Lambda(\{0\})}},$$

and by Plancherel's theorem for Γ/Λ and H,

$$\|\varphi_1\|_{\Gamma/\Lambda} = \|f^{\wedge\vee}\|_H = \|f\|_H,$$

where since H is discrete, $\|f\|_H^2 = m_H(\{0\}) \sum_{h \in H} |f(h)|^2$ by (10.2.2). Hence by the measure relations (10.4.5),

$$\|f\|_G^2 = \frac{m_H(\{0\})}{m_\Lambda(\{0\})} \sum_{h \in H} |f(h)|^2 = \frac{1}{m_\Gamma(B)} \sum_{h \in H} |f(h)|^2,$$

as required.

Under the inverse Fourier transform operator $\mathcal{F}^*$, the basis $\{(h, \cdot) : h \in H\}$ of characters for $L^2(B)$ is sent to the basis

$$\{\chi_B^\vee(\cdot + h) : h \in H\}$$

for $\widetilde{PW}_B$. For convenience write $\widetilde{\psi}_h(\gamma) = \chi_B(\gamma)(h, \gamma)$ and $\psi_h(\gamma) = \chi_A(\gamma)(h, \gamma)$, so that

$$\widetilde{\psi}_h^\vee(x) = \chi_B^\vee(x + h) \quad \text{and} \quad \psi_h^\vee(x) = \chi_A^\vee(x + h).$$

Now $f \in \widetilde{PW}_A$ where $A \subseteq B$, so that $f^\wedge(\gamma) = 0$ for $\gamma \notin B$ and $f^\wedge \in L^2(B)$. Hence by (10.4.10) and since $f^{\wedge\wedge}(x) = f(-x)$,

$$f^\wedge = \frac{1}{m_\Gamma(B)} \sum_{h \in H} f^{\wedge\wedge}(h) \widetilde{\psi}_h$$

$$= \frac{1}{m_\Gamma(B)} \sum_{h \in H} f^{\wedge\wedge}(h) \psi_h$$

$$= \frac{1}{m_\Gamma(B)} \sum_{h \in H} f(h) \psi_{-h}$$

in $L^2(\Gamma)$-norm. That is given $\varepsilon > 0$, there exists an $n_0 = n_0(\varepsilon)$ such that for $n \geq n_0$,

$$\left\| f^\wedge - \frac{1}{m_\Gamma(B)} \sum_{j=1}^{n} f(h_j) \psi_{-h_j} \right\| < \varepsilon.$$

But since $\mathcal{F}$ and $\mathcal{F}^*$ are unitary, the operator norm $\|\mathcal{F}^*\|$ of $\mathcal{F}^*$ is unity, so that

$$\left\| f - \frac{1}{m_\Gamma(B)} \sum_{j=1}^{n} f(h_j) \psi_{-h_j}^\vee \right\| = \left\| \mathcal{F}^* \left(f^\wedge - \frac{1}{m_\Gamma(B)} \sum_{j=1}^{n} f(h_j) \psi_{-h_j} \right) \right\|$$

$$\leq \|\mathcal{F}^*\| \left\| f^\wedge - \frac{1}{m_\Gamma(B)} \sum_{j=1}^{n} f(h_j) \psi_{-h_j} \right\|$$

$$= \left\| f^\wedge - \frac{1}{m_\Gamma(B)} \sum_{j=1}^{n} f(h_j) \psi_{-h_j} \right\| < \varepsilon,$$

whence, as $\psi_{-h}^\vee(x) = \psi_A^\vee(x - h)$ and f is continuous,

$$f(x) = \frac{1}{m_\Gamma(B)} \sum_{h \in H} f(h) \chi_A^\vee(x - h)$$

$$= \frac{1}{m_\Gamma(B) \, m_H(\{0\})} \int_H f(h) \chi_A^\vee(x - h) dm_H(h)$$

and (10.4.11) follows by (10.4.4) and (10.4.5).

The $L^2(G)$-norm convergence of the series is absolute. For let $g_x(y) = \chi_A^\vee(x - y)$, so that $g_x \in \widetilde{PW}_A$. Then by (10.3.13) and since $dm_G(-x) = dm_G(x)$ (see (10.2.1)),

$$
\begin{aligned}
g_x^\wedge(\gamma) &= \int_G g_x(y)(y, -\gamma)\, dm_G(y) = \int_G \chi_A^\vee(x - y)(y, -\gamma)\, dm_G(y) \\
&= \int_G \chi_A^\vee(u)(x - u, -\gamma)\, dm_G(u) = (x, -\gamma) \int_G \chi_A^\vee(u)(u, \gamma)\, dm_G(u) \\
&= (x, -\gamma)\, \chi_A^{\vee\vee}(\gamma) = (x, -\gamma)\, \chi_A(-\gamma).
\end{aligned}
$$

Thus since $|(x, -\gamma)| = 1$, by (10.3.8) and for each $x \in G$,

$$
\begin{aligned}
\sum_{h \in H} |\chi_A^\vee(x - h)|^2 &= \sum_{h \in H} |g_x(h)|^2 \\
&= m_\Gamma(B) \int_G |g_x(y)|^2\, dm_G(y) \\
&= m_\Gamma(B) \int_\Gamma |(g_x)^\wedge(\gamma)|^2\, dm_\Gamma(\gamma) \\
&= m_\Gamma(B) \int_\Gamma |\chi_A(-\gamma)|^2\, dm_\Gamma(\gamma) \\
&= m_\Gamma(B) \int_\Gamma |\chi_A(\gamma)|^2\, dm_\Gamma(\gamma) \\
&= m_\Gamma(B)\, m_\Gamma(A),
\end{aligned}
$$

whence $\sum_{h \in H} |f(h)\chi_A^\vee(x - h)|^2$ is at most

$$
\left(\sum_{h \in H} |f(h)|^2 \sum_{h \in H} |\chi_A^\vee(x - h)|^2 \right)^{1/2} = m_\Gamma(B)\, m_\Gamma(A)^{1/2} \, \|f\|_G.
$$

The convergence of the series to f is also uniform. By (10.4.10) with $\varphi = f^\wedge$, for any $\varepsilon > 0$ there exists an integer $n = n(\varepsilon)$ such that

$$
\int_\Gamma |f^\wedge(\gamma) - \frac{1}{m_\Gamma(B)} \sum_{j=1}^n f(h_j)(h_j, \gamma)\chi_A(\gamma)|^2 dm_\Gamma(\gamma) < \frac{\varepsilon^2}{m_\Gamma(A)}. \qquad (10.4.12)
$$

By the Fourier inversion theorem (10.3.6),

$$
\left| f(x) - m_\Gamma(B)^{-1} \sum_{j=1}^n f(h_j)\chi_A^\vee(x - h_j) \right|
$$

$$
= \left| \int_\Gamma f^\wedge(\gamma)(x, \gamma)\, dm_\Gamma(\gamma) - \frac{1}{m_\Gamma(B)} \sum_{j=1}^n f(h_j) \int_\Gamma \chi_A(\gamma)(x - h_j, \gamma)\, dm_\Gamma(\gamma) \right|
$$

$$
= \left| \int_\Gamma (x, \gamma) \left(f^\wedge(\gamma) - \frac{1}{m_\Gamma(B)} \sum_{j=1}^n f(h_j)\chi_A(\gamma)(-h_j, \gamma) \right) dm_\Gamma(\gamma) \right|
$$

$$\leq \int_A \left| f^\wedge(\gamma) - \frac{1}{m_\Gamma(B)} \sum_{j=1}^n f(h_j)\chi_A(\gamma)(-h_j,\gamma) \right| dm_\Gamma(\gamma)$$

$$\leq \left(\int_A \left| f^\wedge(\gamma) - \frac{1}{m_\Gamma(B)} \sum_{j=1}^n f(h_j)\chi_A(\gamma)(-h_j,\gamma) \right|^2 dm_\Gamma(\gamma) \right)^{1/2} m_\Gamma(A)^{1/2} < \varepsilon$$

by (10.4.12) and uniform convergence is established.

The representation of f can also be proved using the tautology $f^\wedge = f^\wedge \chi_A$ and Parseval's formula (10.3.9) or the convolution theorem (proofs in the classical setting are to be found in Dodson and Silva (1985)). From the reciprocity relations (10.4.9), the norm $\|f\|_G$ also satisfies

$$\|f\|_G^2 = m_\Lambda(\{0\})\|f\|_H^2 = m_G(K) \sum_{h \in H} |f(h)|^2$$

and f is given by

$$f(x) = m_G(K) \sum_{h \in H} f(h)\chi_A^\vee(x - h) = \frac{1}{m_\Lambda(\{0\})} (f *_H \chi_A^\vee)(x)$$
$$= m_{G/H}(G/H)(f *_H \chi_A^\vee)(x),$$

i.e., $f = m_{G/H}(G/H)(f *_H \chi_A^\vee)$. The classical Whittaker–Kotel'nikov–Shannon sampling theorem follows by specializing G to $\mathbb{R}$.

Corollary 10.2 Let $f \in L^2(\mathbb{R})$ be continuous with $\mathrm{supp} f^\wedge \subset [-\pi w, \pi w]$, i.e., let $f \in PW_{[-\pi w, \pi w]}$. Then

$$f(x) = \sum_{k \in \mathbb{Z}} f\left(\frac{k}{w}\right) \frac{\sin \pi(wx - k)}{\pi(wx - k)}.$$

Proof Since $f(x) = (1/\sqrt{2\pi}) \int_\mathbb{R} f^\wedge(u)e^{iux} du$, there is no loss in generality in supposing that $f^\wedge$ vanishes outside $(-\pi w, \pi w]$. Put $G = \mathbb{R}$ and let the Haar measure be $1/\sqrt{2\pi}$ times Lebesgue measure. Then $\Gamma = \mathbb{R}$ and has the same Haar measure and the characters $(x, \gamma) = e^{ix\gamma}$. Also $A = B = (-\pi w, \pi w]$ so that $\Lambda = 2\pi w\mathbb{Z}$, $\Gamma/\Lambda = \mathbb{R}/2\pi w\mathbb{Z}$ and the annihilator H of Λ is $\mathbb{Z}/w$. Hence $m_\Gamma(B) = (1/\sqrt{2\pi})2\pi w = \sqrt{2\pi}\, w$ and

$$\chi_A^\vee(x) = \frac{1}{\sqrt{2\pi}} \int_{(-\pi w, \pi w]} e^{ix\gamma} d\gamma = \frac{1}{\sqrt{2\pi}} \frac{2\sin \pi wx}{x}.$$

Substituting in (10.4.11) gives

$$f(x) = \frac{1}{\sqrt{2\pi}\, w} \sum_{k \in \mathbb{Z}} f\left(\frac{k}{w}\right) \frac{1}{\sqrt{2\pi}} \frac{2\sin \pi w(x - k/w)}{x - k/w},$$

and the corollary follows after a little rearranging.

The flexibility in allowing A to be a subset of a transversal is useful in applications to signal processing involving band-pass or multiband signals (Higgins 1996, Chapter 13). Determining sampling rates for these signals is tantamount to finding the discrete groups Λ. The optimal regular rate corresponds to A being a complete set of coset representatives for Γ/Λ; this can be a stringent condition, amounting to a tessellation of the group Γ by translates of A by elements of Λ.

In physical applications, and particularly in X-ray crystallography, the dual space is often referred to as *reciprocal* space. This reciprocity arises from the nature of the characters and is exemplified by the relation $m_\Gamma(B)\, m_G(K) = 1$ given in (10.4.9).

10.5   Some applications

By the appropriate choice of the group G, Kluvánek's extension includes many sampling theorems; we now give a few more examples.

10.5.1   *Some other groups*

First the case of a compact abelian group is considered. The function $f : G \to \mathbb{C}$ in $\widetilde{PW}_A$ does not depend on the Haar measure on G. However the terms $m_\Gamma(B)$ and $\chi_B^\vee$ evidently depend on the Haar measure of Γ and hence on that of G. These two dependencies cancel in the representation (10.4.11) as is seen explicitly when G is a compact abelian group. In this case the sampling theorem can be expressed in a form independent of the normalizations of the Haar measures.

Theorem 10.3 *Let G be a compact abelian group and let $f \in \widetilde{PW}_A$. Then A and B are finite and*

$$f(x) = \frac{1}{\#B} \sum_{h \in H} f(h) \sum_{\gamma \in A} (x - h, \gamma) = \frac{1}{\#B} \sum_{\gamma \in A} (x, \gamma) \sum_{h \in H} f(h)\overline{(h, \gamma)},$$

where $\#B$ is the number of elements in B.

Proof   Recall that A satisfies the disjoint translates condition (10.4.2) for the subgroup Λ of Γ and that $B \supseteq A$ is a transversal of Γ/Λ in Γ. Since G is compact, its dual Γ is a discrete group. Thus the compact quotient group Γ/Λ is also discrete and so finite. Hence the complete set of coset representatives B of Γ/Λ is also finite, as is A. In addition, the annihilator H of Λ is discrete and by virtue of being a subgroup of G, is compact as well. Hence H is also finite. By definition,

$$\chi_A^\vee(x) = \int_\Gamma \chi_A(\gamma)(x, \gamma)\, dm_\Gamma(\gamma) = m_\Gamma(\{0\}) \sum_{\gamma \in \Gamma} (x, \gamma)\chi_A(\gamma).$$

But $m_\Gamma(B) = m_\Gamma(\{0\})\#B$ and the result follows from (10.4.11).

The additive group $\mathbb{Z}_6$

Sampling theorems for finite abelian groups are discussed by Stanković and Stanković (1984) but we will now consider the simple example of $\mathbb{Z}_6$. Suppose $A = B = \{0, 1\} \subset \mathbb{Z}_6^\wedge\ (\cong \mathbb{Z}_6)$, and suppose that the Fourier transform $f^\wedge$ of $f : \mathbb{Z}_6 \to \mathbb{C}$ vanishes outside the set $\{0, 1\}$. Then $\#B = 2$ and $B \subset \Gamma$ is a transversal of $\mathbb{Z}_6^\wedge/\Lambda$ where $\Lambda = \{0, 2, 4\} \subset \mathbb{Z}_6^\wedge$ and has annihilator $H = \Lambda^\perp = \{0, 3\} \subset \mathbb{Z}_6$ (see §10.2.8). Also since $\mathbb{Z}_6^\wedge$ is discrete and the character $(x, \gamma) = e^{\pi x \gamma/3}$,

$$\chi_{\{0,1\}}^\vee(x) = \int_{\mathbb{Z}_6^\wedge} \chi_{\{0,1\}}(\gamma)(x, \gamma)\, dm_{\mathbb{Z}_6^\wedge}(\gamma) = m_{\mathbb{Z}_6^\wedge}(\{0\}) \sum_{\gamma=0}^{1} e^{\pi i x \gamma/3}$$

$$= m_{\mathbb{Z}_6^\wedge}(\{0\})(1 + e^{\pi i \gamma/3}).$$

Then by Theorem 10.3,

$$f(x) = \frac{1}{\#B} \sum_{h=0}^{3} f(h) \sum_{\gamma \in B} \chi_{\{0,1\}}(x - h, \gamma)$$

$$= \frac{1}{2}\left[f(0)(1 + e^{\pi i x/3}) + f(3)(1 - e^{\pi i x/3}) \right]$$

$$= \frac{1}{2}\left[f(0) + f(3) + e^{\pi i x/3}(f(0) - f(3)) \right].$$

If $f(0) = f(3)$, the function f is constant. Of course f could be determined by solving a system of simultaneous linear equations to find $f^\wedge$ but the sampling theorem is simpler and more direct.

The multiplicative group $\mathbb{R}^+$

As another example we shall give a sampling theorem associated with the case $G = \mathbb{R}^+$, the (strictly) positive reals, related to the Mellin transform (Titchmarsh 1962, p. 7). The group $\mathbb{R}^+$ has not been distinguished from the isomorphic group $\mathbb{R}$ and the resulting sampling theorem can be obtained by a simple direct substitution (the exponential function). Nevertheless we include it as it gives an irregular sampling result of some self-contained interest.

Since $\mathbb{R}^+$ is a locally compact multiplicative abelian group, $\mathbb{R}^+$ has a Haar measure $m_{\mathbb{R}^+}$ for which $m_{\mathbb{R}^+}(Ex) = m_{\mathbb{R}^+}(E)$. By uniqueness, $m_{\mathbb{R}^+}(E) = m_{\mathbb{R}}(\log E)$ where $m_{\mathbb{R}}$ is Lebesgue measure on $\mathbb{R}$ and $\log E = \{\log x : x \in E\}$. It follows that

$$dm_{\mathbb{R}^+}(x) = dm_{\mathbb{R}}(\log x).$$

To determine the dual $(\mathbb{R}^+)^\wedge$, let $\gamma \in \mathbb{R}^+$ so that $\gamma : \mathbb{R}^+ \to \mathbb{T}^1$ is a homomorphism satisfying $\gamma(x\,x') = \gamma(x)\gamma(x')$ for $x, x' \in R^+$. Then for each integer n, $\gamma(x^n) = (\gamma(x))^n$. Try $\gamma(x) = \lambda x^\theta$; then $\lambda = 1$ since $\gamma(1) = 1$ and $\theta = i\theta'$, where $\theta' = \theta'(\gamma) \in \mathbb{R}$, since $|e^\theta| = 1$. Identify γ with $\theta'(\gamma)$. Then the dual group of $\mathbb{R}^+$ is $\mathbb{R}$ (and since $\mathbb{R}$ and $\mathbb{R}^+$ are isomorphic, the dual of the dual is isomorphic to $\mathbb{R}^+$). For suitable f, the Fourier transform $f^\wedge$ corresponding to $G = \mathbb{R}^+$ is

$$f^\wedge(\gamma) = F(\gamma) = \int_0^\infty f(x)\, x^{-i\gamma}\, d(\log x) = \int_0^\infty f(x)\, x^{-1-i\gamma}\, dx$$

The inverse "Fourier" transform is

$$F^{\vee}(x) = \int_{-\infty}^{\infty} F(\gamma) x^{i\gamma} d\gamma.$$

Suppose $f^{\wedge}$ vanishes outside $(-w, w] \subset \mathbb{R}$. Then

$$f(x) = \frac{1}{2w} \sum_{k \in \mathbb{Z}} f(e^{k/2w}) \frac{\sin(\log x - k/2w)}{\log x - k/2w}.$$

This is just a consequence of $\mathbb{R}^{+}$ being isomorphic to $\mathbb{R}$ under the log function.

10.5.2 *Filtering sampled signals*

Sampled signals formed from analogue signals are step functions (referred to by engineers as "box car" or "square wave" signals). In many signal processing applications, these functions are smoothed by filtering out the high frequencies and retaining only those in the original analogue signal. This corresponds to truncating a function f (see 10.3.10). It turns out that filtering amounts to replacing the value of the original signal at an instant by an average over a suitable window. The resulting errors for low-pass signals are very small when the width of the window is significantly less than the reciprocal of the bandwidth. More details are in (Beaty et al. 1994); here the abstract treatment of the process will be sketched.

In the abstract setting the function $f \in \widetilde{PW}_A$ is approximated by the step-function

$$\sigma(x) = \sum_{h \in H} f(h) \chi_K(x - h),$$

where K is a transversal of G/H. By Theorem 10.1, the norm $\|\sigma\| = \|f\|$ and the Fourier transform $\sigma^{\wedge}$ is given by

$$\sigma^{\wedge}(\gamma) = \sum_{h \in H} f(h)(-h, \gamma) \chi_K^{\wedge}(\gamma)$$

(the details of the analysis are omitted). The Fourier transform of the step-function σ is supported outside A and a new function r_A with the desired frequency support is obtained by filtering. The Fourier transform $r_A^{\wedge}$ of the filtered reconstructed function and $r_A = \mathcal{F}^* \chi_A \mathcal{F} \sigma$ is given by

$$r_A^{\wedge}(\gamma) = \chi_A(\gamma) \sum_{h \in H} f(h)(-h, \gamma) \chi_K^{\wedge}(\gamma) = m_\Gamma(B) \chi_K^{\wedge}(\gamma) f^{\wedge}(\gamma),$$

since $f^{\wedge}(\gamma) = (1/m_\Gamma(B)) \sum_{h \in H} f(h) (h, \gamma) \chi_A(\gamma)$ in norm. Hence the $L^2(G)$-norm $\|\varepsilon\|$ of the error $\varepsilon = f - r_A$ satisfies

$$\|\varepsilon\| = \|f - r_A\| = \|f^{\wedge} - r_A^{\wedge}\| = \|f^{\wedge} - m_\Gamma(B) \chi_K^{\wedge} f^{\wedge}\| = \|f^{\wedge}(1 - m_\Gamma(B) \chi_K^{\wedge})\|.$$

The reconstruction function r_A is given by the local average or 'windowing' function μ_{-K} where

$$\mu_K(x) = \frac{1}{m_G(K)} \int_{K+x} f(u)\,dm_G(u).$$

For $r_A^\wedge = m_\Gamma(B)\,\chi_K^\wedge f^\wedge$, whence $r_A = m_\Gamma(B)\,(\chi_K^\wedge f^\wedge)^\vee = m_\Gamma(B)\chi_K * f$ and since $m_\Gamma(B)\,m_G(K) = 1$ and $m_G(K) = m_G(-K)$ (see 10.4.9),

$$
\begin{aligned}
r_A(x) &= m_\Gamma \int_G \chi_K(x-u)f(u)\,dm_G(u) = \frac{1}{m_G(K)} \int_G \chi_{-K+x}(u)f(u)\,dm_G(u) \\
&= \frac{1}{m_G(K)} \int_{-K+x} f(u)\,dm_G(u) = \frac{1}{m_G(-K)} \int_{-K+x} f(u)\,dm_G(u) \\
&= \mu_{-K}(x)
\end{aligned}
$$

(see Beaty et al. 1994) but note that the set K there should be replaced by $-K$). Thus if K is symmetric (i.e., $-K = K$) or more generally if $m_G(\mathrm{cl}K) = m_G(K)$ and $\mathrm{cl}K$ is symmetric, then $r_A = \mu_K$. In the classical case, $A = B = (-\pi w, \pi w]$ and $K = (-1/2w, 1/2w]$ whence

$$r_{(-\pi w, \pi w]}(x) = \sqrt{2\pi}\,\frac{1}{\sqrt{2\pi}}w \int_{-1/2w+x}^{1/2w+x} f(u)\,du = \mu_{(-1/2w, 1/2w]}(x).$$

10.5.3 *The aliasing error*

Aliasing occurs when the *assumed band-region* used to generate a sampling series does not wholly contain the *actual band-region* of the function sampled. In the following it is convenient to specify that the assumed band-region is a transversal B of Γ/Λ rather than a set $A \subseteq B$. Then, instead of supposing that $f^\wedge$ vanishes outside B, it is only assumed that $f^\wedge \in L^1(\Gamma)$.

The *alias* of f, denoted $\mathcal{A}_f$, is defined by

$$\mathcal{A}_f(x) = \frac{1}{m_\Gamma(B)} \sum_{h \in H} f(h)\chi_B^\vee(x-h).$$

The *aliasing error* at $x \in G$ is $|f(x) - \mathcal{A}_f(x)|$.

Theorem 10.4 *Let* $f \in C(G) \cap L^2(G)$ *with* $f^\wedge \in L^1(\Gamma)$. *Then for each* $x \in G$ *the aliasing error satisfies the inequality*

$$|(f - \mathcal{A}_f)(x)| \le 2 \int_{\Gamma \backslash B} |f^\wedge(\gamma)|\,dm_\Gamma(\gamma).$$

Proof By the inversion theorem 10.3.6 the error $f - A_f$ can be expressed in the form

$$(f - A_f)(x) = \int_\Gamma f^\wedge(\gamma)(x, \gamma)\, dm_\Gamma(\gamma)$$
$$- \frac{1}{m_\Gamma(B)} \sum_{h \in H} f(h) \int_\Gamma (-h, \gamma) \chi_B(\gamma)(x, \gamma)\, dm_\Gamma(\gamma).$$

The convergence in L^1 of the series in the second term allows the order of summation and integration to be interchanged to obtain

$$(f - A_f)(x) = \int_\Gamma f^\wedge(\gamma)(x, \gamma)\, dm_\Gamma(\gamma) - \int_\Gamma \frac{\chi_B(\gamma)(x, \gamma)}{m_\Gamma(B)} \sum_{h \in H} f(h)(-h, \gamma)\, dm_\Gamma(\gamma).$$

On applying the coset decomposition formula (10.2.3), the Poisson summation formula §(10.3.2) in the form

$$m_H(\{0\}) \sum_{h \in H} f(h)(-h, \gamma) = m_\Lambda(\{0\}) \sum_{\lambda' \in \Lambda} f^\wedge(\gamma + \lambda'),$$

and finally the measure relations (10.4.5) the error simplifies to

$$(f - A_f)(x) = \int_{\Gamma/\Lambda} m_\Lambda(\{0\}) \sum_{\lambda \in \Lambda} f^\wedge(\gamma + \lambda)\Big\{(x, \gamma + \lambda)$$
$$- \sum_{\lambda' \in \Lambda} \chi_B(\gamma + \lambda')(x, \gamma + \lambda')\Big\} dm_{\Gamma/\Lambda}([\gamma]).$$

The summation over λ' has only one non-zero term, given by $\lambda' = \lambda_\gamma$, where λ_γ is the unique element in Λ such that $\gamma + \lambda_\gamma \in B$. Thus

$$f(x) - A_f(x) = \int_{\Gamma/\Lambda} m_\Lambda(\{0\})$$
$$\times \sum_{\lambda \in \Lambda} f^\wedge(\gamma + \lambda)\{(x, \gamma + \lambda) - (x, \gamma + \lambda_\gamma)\} dm_{\Gamma/\Lambda}([\gamma]).$$

Note the integrand vanishes when $\lambda = \lambda_\gamma$, and that

$$|(x, \gamma + \lambda) - (x, \gamma + \lambda_\gamma)| \le 2.$$

The first point is exploited by introducing the factor $\chi_{\Gamma \setminus B}$, whence

$$|f(x) - A_f(x)| \le 2 \int_{\Gamma/\Lambda} m_\Lambda(\{0\}) \sum_{\lambda \in \Lambda} |f^\wedge(\gamma + \lambda)| \chi_{\Gamma \setminus B}(\gamma + \lambda) dm_{\Gamma/\Lambda}([\gamma])$$
$$= 2 \int_\Gamma |f^\wedge(\gamma)| \chi_{\Gamma \setminus B}(\gamma) dm_\Gamma(\gamma) = 2 \int_{\Gamma \setminus B} |f^\wedge(\gamma)| dm_\Gamma(\gamma)$$

where the second to last step follows from (10.2.2). (There is a similar argument in the proof of the aliasing estimate in the case $G = \mathbb{R}^N$ (Higgins 1996, §14.3).)

The aliasing error has been extensively studied over the last three decades, with notable contributions from J. L. Brown Jr. and J. R. Higgins. It is natural to ask whether the aliasing estimate is sharp, i.e., if equality holds for some function, and each of these authors has constructed extremal functions to show the estimate cannot be improved (Brown 1967; Higgins 1991).

An example of an extremal function

Brown's approach to finding an extremal can be followed if one further hypothesis on the group structure is introduced. An important idea is to look for an extremal function which vanishes at each sample point, so that the alias $\mathcal{A}_f$ is the null function. This makes the aliasing error estimate simple to verify. It is also important that the measure of the band-region of the extremal is twice that of the alias. This leads to the extra requirement that the group Λ has a subgroup of index 2, given by $\Lambda' = \{\lambda + \lambda : \lambda \in \Lambda\}$. Then, since $H = \Lambda^\perp$, $(h, \lambda') = 1$ for each $h \in H$, $\lambda' \in \Lambda'$, and hence $H \subset H'$ where $H' = \Lambda'^\perp$.

Choose $h' \in H' \setminus H$ (i.e., $h' \notin H$) and suppose that B' is another transversal for Γ/Λ disjoint from B. Let

$$f(x) = \int_{B \cup B'} (-h', \gamma)(x, \gamma) dm_\Gamma(\gamma) = \chi_{B \cup B'}^\vee(x - h').$$

Since Λ' is discrete, the coset decomposition formula (10.2.3) together with (10.2.2) can be used to write f in the form

$$f(x) = \int_{\Gamma/\Lambda'} m_{\Lambda'}(\{0\}) \sum_{\lambda \in \Lambda'} \chi_{B \cup B'}(\gamma + \lambda)(x - h', \gamma + \lambda) \, dm_{\Gamma/\Lambda'}([\gamma]).$$

Hence, for $h \in H$, noting that $h - h' \neq 0$,

$$\begin{aligned}
f(h) &= \int_{\Gamma/\Lambda'} m_{\Lambda'}(\{0\}) \sum_{\lambda \in \Lambda'} \chi_{B \cup B'}(\gamma + \lambda)(h - h', \gamma + \lambda) \, dm_{\Gamma/\Lambda'}([\gamma]) \\
&= m_{\Lambda'}(\{0\}) \int_{\Gamma/\Lambda'} (h - h', [\gamma]) \, dm_{\Gamma/\Lambda'}([\gamma]) = 0,
\end{aligned}$$

from §10.2.6 (since Γ/Λ' is compact). Therefore the alias $\mathcal{A}_f$ is identically zero and the aliasing error estimate reduces to

$$|f(x)| \leq 2 \int_{\Gamma \setminus B} |f^\wedge(\gamma)| dm_\Gamma(\gamma).$$

From the definition of f, $f^\wedge(\gamma) = (-h, \gamma)\chi_{B \cup B'}(\gamma)$, so the right-hand side is

$$2 \int_{\Gamma \setminus B} |(-h', \gamma)| \chi_{B \cup B'}(\gamma) dm_\Gamma(\gamma) = 2m_\Gamma(B') = 2m_\Gamma(B),$$

and finally, $f(h') = \chi_{B \cup B'}^\vee(0) = 2m_\Gamma(B)$, so that f is extremal at the point $x = h'$.

This construction works in many common cases, in particular when $G = \mathbb{R}^n$. Beaty and Higgins (1994) gave this extremal in the one-dimensional multiband setting.

The finite cyclic abelian group $G = \Gamma = \mathbb{Z}_6 = \{0, 1, 2, 3, 4, 5\}$ is an example in which the construction does not always work. This group is the product of its subgroups $\{0, 3\}, \{0, 2, 4\}$ and therefore if H is chosen to be $\{0, 3\}$, then $H^\perp = \Lambda = \{0, 2, 4\}$ and Λ does not have a subgroup of index 2.

10.5.4 *An abstract multichannel sampling theorem*

A multichannel sampling theorem (Higgins 1996, Chapter 12) in which a certain class of functions can be recovered from samples of several related functions can also be placed in the setting of abstract harmonic analysis. Indeed Kluvánek's theorem is a one-channel sampling theorem, in that the reconstruction of f uses samples of one function, namely f itself.

The general approach followed here was introduced by Papoulis (1977). Each sampled function is linked to the original function via a multiplier transformation on the Fourier transform, i.e., each sampled function has a transform of the form $Mf^\wedge$, where M is the multiplier. Papoulis' theorem allowed a finite number of channels, say N, and required a condition on the N multipliers M_i which in essence allowed the invertibility of a matrix whose entries depend on the M_i (see also Brown 1981; Beaty 1994; and Higgins 1996, Chapter 12).

The following theorem extends multichannel sampling to the setting of abstract harmonic analysis. The main difference from Kluvánek's theorem is that the band-region of the functions considered depends on the number of channels. More precisely, for the N-channel problem, the band-region consists of a union of N transversals for Γ/Λ, say $B_1, \ldots, B_N$.

For any pair $i, j, 1 \le i, j \le N$, let the map $v_{ij} : B_i \to B_j$ be defined by

$$v_{ij}(\gamma) = \gamma + \lambda$$

where $\lambda = \lambda(\gamma, i, j)$ is the unique element in Λ such that $\gamma + \lambda \in B_j$. Then, since $H = \Lambda^\perp$, $(h, \gamma) = (h, v_{ij}(\gamma))$, where $h \in H$ and $\gamma \in B_i$.

Theorem 10.5 *Let $B_1, \ldots, B_N$ be disjoint transversals for Γ/Λ. Suppose that $M_1, \ldots, M_N$ are of bounded variation on $\cup_{i=1}^N B_i$ and vanish outside $\cup_{i=1}^N B_i$. Further assume that there exist functions $R_1, \ldots, R_N \in L^2(\cup_{i=1}^N B_i)$ which again vanish outside $\cup_{i=1}^N B_i$ and satisfy the condition*

$$\sum_{k=1}^N M_k(v_{ij}(\gamma)) R_k(\gamma) = \delta_{ij}$$

for each $\gamma \in B_i$ and each $i = 1, \ldots, N$. Then for $f \in \widetilde{PW}_{\cup_{i=1}^N B_i}$

$$f(x) = \frac{1}{m_\Gamma(B)} \sum_{k=1}^N \sum_{h \in H} (M_k^\vee * f)(h) R_k^\vee(x - h), \qquad (10.5.1)$$

where $m_\Gamma(B)$ is the Haar measure of any transversal for Γ/Λ. The sum converges absolutely and uniformly on G.

Proof Let $s(x)$ be the value of the sampling series on the right-hand side. Express $R_k^\vee$ by the Fourier inversion formula and interchange the order of summation and integration (allowed by the L^1 convergence of the series). Next employ the Poisson summation formula in the form

$$m_H(\{0\}) \sum_{h \in H} (M_k^\vee * f)(h)(-h, \gamma) = m_\Lambda(\{0\}) \sum_{\lambda \in \Lambda} M_k(\gamma + \lambda) f^\wedge(\gamma + \lambda)$$

and simplify, first via (10.4.5), and then by noting that each R_K vanishes outside the union of the transversals $B_1, \ldots, B_N$, to arrive at the equation

$$s(x) = \sum_{i=1}^N \int_{B_i} \sum_{k=1}^N R_k(\gamma)(x, \gamma) \sum_{\lambda \in \Lambda} M_k(\gamma + \lambda) f^\wedge(\gamma + \lambda) dm_\Gamma(\gamma).$$

As each M_k also vanishes outside $\cup_{j=1}^N B_j$ the sum over Λ is finite and can be written

$$\sum_{\lambda \in \Lambda} M_k(\gamma + \lambda) f^\wedge(\gamma + \lambda) = \sum_{j=1}^N M_k\left(v_{ij}(\gamma)\right) f^\wedge\left(v_{ij}(\gamma)\right).$$

Finally, by changing the order of the finite sums, the hypotheses of the theorem can be invoked to obtain

$$\begin{aligned}
s(x) &= \sum_{i=1}^N \int_{B_i} (x, \gamma) \sum_{j=1}^N f^\wedge\left(v_{ij}(\gamma)\right) \sum_{k=1}^N M_k\left(v_{ij}(\gamma)\right) R_k(\gamma) dm_\Gamma(\gamma) \\
&= \sum_{i=1}^N \int_{B_i} (x, \gamma) \sum_{j=1}^N f^\wedge\left(v_{ij}(\gamma)\right) \delta_{ij} dm_\Gamma(\gamma) \\
&= \sum_{i=1}^N \int_{B_i} (x, \gamma) \sum_{j=1}^N f^\wedge\left(v_{ii}(\gamma)\right) = f(x).
\end{aligned}$$

As with Kluvánek's theorem, the sampling series can be shown to converge absolutely and uniformly on G.

10.5.5 *Further developments*

Groups are naturally associated with regular sampling and are consequently not really suitable for studying irregular sampling. However "interlaced" sampling theorems for locally compact abelian groups in which the irregular sampling sets are unions of finitely many cosets of the subgroup H have been obtained recently (Faridani 1994). This approach also allows the analysis of aliasing errors associated with functions which are not band-limited.

Extending the sampling theorem to the non-abelian case is a natural but very difficult question since the dual of a locally compact non-abelian group is not necessarily a group, so that the disjoint translates property would be problematic. Nevertheless, a type of Fourier theory can be developed which suggests at least the possibility of establishing some kind of sampling theorem for some types of non-abelian groups, such as compact groups or Lie groups.

REFERENCES

Chapter 1

Abramowitz, M. and Stegun, I. A. (1965). *Handbook of mathematical functions*. Dover Publications, New York.

Apéry, R. (1979). Irrationalité de $\zeta(2)$ et $\zeta(3)$, Journées arithmétiques de Luminy, *Astérique*, **61**, 11–13.

Askey, R. (1975). *Orthogonal polynomials and special functions*. Regional Conf. Series in Applied Math., SIAM, Philadelphia.

Bernstein, S.N. (1958). On the best approximation of continuous functions by means of polynomials (Kharkov Univ. 1913). In: *Collected works*, vol. I (Engl. Translation: US Atomic Energy Commission Translation Series 3460, Oak Ridge, TN, 109–114).

Butzer, P.L., Flocke, S., and Hauss, M. (1994). Euler functions $E_\alpha(z)$ with complex α and applications. In: *Approximation, probability, and related fields*. Proc. Conference at Santa Barbara, May 1993; G. Anastassiou and S.T. Rachev (eds.). Plenum Press, New York, 127–150.

Butzer, P.L. and Hauss, M. (1991). Stirling functions of first and second kind: Some new applications. In: *Approximation, interpolation, and summability*. Proc. Conf. in honour of Prof. Jakimovski, Tel Aviv, June 4–8, 1990. Research Inst. of Math. Sciences, vol. 4, Weizmann Press, Israel, 89–108.

Butzer, P.L., Hauss, M., and Leclerc, M. (1992). Bernoulli numbers and polynomials of arbitrary complex indices. *Appl. Math. Lett.*, **5**(6), 83–88.

Butzer, P.L., Hauss, M., and Schmidt, M. (1989). Factorial functions and Stirling numbers of arbitrary complex indices. *Resultate Math.*, **16**, 16–48.

Butzer, P.L., Hauss, M., and Stens, R.L. (1991). The sampling theorem and its unique role in various branches of mathematics. In: *Mitteilungen der Mathematischen Gesellschaft in Hamburg*. Festschrift zum 300–jährigen Bestehen der Gesellschaft 3. Teil Band XII, Heft 3, 523–547.

Butzer, P.L. and Nasri-Roudsari, G. (1997). Kramer's sampling theorem in signal analysis and its role in mathematics. In: *Image processing; mathematical methods and applications*, J.M. Blackledge (ed.). The Institute of Mathematics and its Applications, New Series Number 61, Clarendon Press, Oxford, 49–95.

Butzer, P.L. and Nessel, R.J. (1971). *Fourier analysis and approximation*, vol. I. Birkhäuser and Academic Press, Basel and New York.

Butzer, P.L. and Schoettler, G. (1993). Sampling expansions for Jacobi functions and their applications. *Acta Sci. Math.*, **57**, 305–327.

Butzer, P.L., Stens, R.L., and Wehrens, M. (1980). The continuous Legendre transform, its inverse transform, and applications. *Int. J. Math. and Math. Sci.*, **3**, 47–67.

Campbell, L.L. (1964). A comparison of the sampling theorems of Kramer and Whittaker. *J. SIAM*, **12**, 117–130.

Chu, Shih-Chieh (1303). *Ssu Yuan Yü Chien* (= Precious Mirror of the Four Elements; in Chinese).

Davis, H.T. (1962). *The summation of series.* Principia Press, Trinity Univ., San Antonio, TX.

Ferrar, W.L. (1939). Summation formulae and their relation to Dirichlet's series II. *Composito Math.*, **4**, 394–405.

Gould, H.W. (1972). *Combinatorial identities: a standardized set of tables listing 500 binomial coefficient summations.* Published by the author, Morgantown, WV.

Gradstein, I. and Ryshik, I. (1981). *Tables of integrals, series and products.* Academic Press, New York.

Hagen, G. (1891). *Synopsis der höheren Mathematik*, vol 1: *Arithmetische und algebraische Analyse.* Felix L. Dames, Berlin.

Hamburger, H. (1922). Über einige Beziehungen, die mit der Funktionalgleichung der Riemannschen ζ-Funktion äquivalent sind. *Math. Anal.*, 133–144.

Hauss, M. (1989). *Über die Theorie der fraktionierten Stirling Zahlen und deren Anwendungen.* Dipl.-Arbeit, RWTH Aachen.

Hauss, M. (1995). *Verallgemeinerte Stirling, Bernoulli und Euler Zahlen, deren Anwendungen und schnell konvergente Reihen für Zeta Funktionen.* Doctoral Dissertation, RWTH Aachen, Verlag Shaker, Aachen.

Hauss, M. (1997). A Boole-type formula involving conjugate Euler polynomials. In: *Charlemagne and his Heritage: 1200 Years of Civilization and Science in Europe*, vol 2: *Mathematical Arts.* P.L. Butzer, H. Jongen, M. Kerner, and W. Oberschelp (eds.). Brepols Publishers, Turnhout, 361–375.

Hauss, M. (1997). An Euler–Maclaurin-type formula involving conjugate Bernoulli polynomials and an application to $\zeta(2m + 1)$. *Comm. Appl. Analysis*, **1**, 15–32.

Heuser, H. (1989). *Lehrbuch der Analysis*, 6th edn. I. B. G. Teubner, Stuttgart.

Higgins, J.R. (1996). *Sampling theory in Fourier and signal analysis: foundations,* Clarendon Press, Oxford.

Jordan, C. (1950). *Calculus of finite differences.* Chelsea Publishing Company, New York (Original Edition, Budapest, 1939).

Koecher, M. (1987). *Klassische elementare Analysis.* Birkhäuser Verlag, Basel.

Lewin, L. (1981). *Polylogarithms and associated functions.* North-Holland, Amsterdam.

Loeb, D.E. (1992). A generalization of the Stirling numbers. *Discrete Math.*, **103**, 259–269.

Mordell, L.J. (1929). Poisson's summation formula and the Riemann Zeta function. *J. London Math. Soc.*, **4**, 285–291.

Needham, J. (1959). *Science and civilization in China*, vol. 3: *Mathematics and the Sciences of the Heavens and the Earth.* Cambridge University Press, New York.

Nielsen, N. (1965). *Die Gammafunktion.* Chelsea Publishing Company, New York (First Edition 1906).

Remmert, R. (1984). *Funktionentheorie I.* Springer-Verlag, Berlin.

Riordan, J. (1979). *Combinatorial identities.* Wiley, New York, 1968, and Krieger Publ. Co., Huntington, NY.

Rota, G.C. (1975). *Finite operator calculus.* Academic Press, New York.

Titchmarsh, E.C. (1951). *The theory of Riemann's zeta function.* Oxford University Press.

Vandermonde, A. (1772). Mémoire sur des irrationnelles de différens ordre avec une application au cercle, *Mem. Acad. Royale Sci. Paris*, 489–498.

Weil, A. (1976). *Elliptic functions according to Eisenstein and Kronecker.* Springer-Verlag, Berlin.

Weil, A. (1984). *Number theory: an approach through history from Hammurapi to Legendre.* Birkhäuser-Verlag, Boston.

Zayed, A. I. (1993). A proof of new summation formulae by using sampling theorems. *Proc. Amer. Math. Soc.*, **117**, 699–710.

Chapter 2

Briggs, J.P. and Peat, D.F. (1985). *Looking glass universe—The emerging science of wholeness.* Fontana Paperbacks, Glasgow.

Bruns, H. (1903). *Grundlinien des wissenschaftlichen Rechnens*, B.G Teubner, Leipzig.

Davenport, H. (1937). On some infinite series involving arithmetical functions (II), *Quart. Journ. Math.*, **8**, 313–320.

Duffin, R.J. (1957). Representation of Fourier integrals as sums. III. *Proc. Amer. Math. Soc.*, **8**, 272–277.

Euler, Leonhard (1988). *Introduction to analysis of the infinite.* Book I (translation J.D. Blanton). Springer-Verlag, New York.

Hardy, G.H. (1916). Weierstrass's non-differentiable function, *Trans. Amer. Math. Soc.*, **17**, 301–325.

Hardy, G.H. and Wright, E.M. (1979). *An introduction to the theory of numbers* (5th edn). Clarendon Press, Oxford.

Landau, Edmund (1899). *Neuer Beweis der Gleichung $\sum_1^\infty \mu(k)/k = 0$.* Inaugural-Dissertation, Berlin.

Landau, Edmund (1958). *Elementary number theory.* Chelsea, New York.

McCauley, J.L. (1993). *Chaos, dynamics and fractals.* Cambridge University Press.

Pellionisz, A.J. (1989). Neural geometry: Towards a fractal model of neurons; in *Models of brain function*, R.M.J. Cotterill (ed.). Cambridge University Press.

Pellionisz, A.J. (1991). Discovery of neural geometry by neurobiology and its utilization in neurocomputer theory and development. *Proceedings International Conference on Artificial Neural Networks*, T. Kohonen (ed.), Helsinki, 485–493.

Reed, I.S., Tufts, D.W., Truong, T.K., Shih, M.T., Yin, X., and Yu, X. (1990). Fourier analysis and signal processing by use of the Möbius inversion formula. *IEEE Trans. on ASSP*, **38**, 458–470.

Schiff, J.L., Surendonk, T.J., and Walker, W.J. (1992). An algorithm for computing the inverse Z-transform. *IEEE Trans. on Signal Processing*, **40**, 2194–2198.

Schiff, J.L. and Walker, W.J. (1987). A sampling theorem for analytic functions. *Proc. Amer. Math. Soc.*, **99**, 737–740.

Schiff, J.L. and Walker, W.J. (1988). A sampling theorem and Wintner's results on Fourier coefficients. *J. Math. Anal. Appl.*, **133**, 466–471.

Schiff, J.L. and Walker, W.J. (1993). *The arithmetic Fourier transform, analysis, geometry, and groups: A Riemann legacy volume*, H.M. Srivastava and Th.M. Rassias (eds.). Hadronic Press, Florida, 613–625.

Schiff, J.L. and Walker, W.J. (1994). A sampling formula in signal processing and the Prime Number Theorem. *New Zeal. J. Math.*, **23**, 147–155.

Tepedelenliglu, N. (1989). A note on the computational complexity of the Arithmetic Fourier Transform. *IEEE Trans. on ASSP*, **37**, 1146–1147.

Tufts, D.W. (1989). Comments on 'A note on the computational complexity of the Arithmetic Fourier Transform'. *IEEE Trans. on ASSP*, **37**, 1147–1148.

Tufts, D.W. and Sadasiv, G. (1988). The Arithmetic Fourier Transform. *IEEE ASSP Magazine*, **5**, 13–17.

Walker, W.J. (1994). A summability method for the Arithmetic Fourier Transform. *BIT*, **34**, 304–309.

Walker, W.J. (1995). The Arithmetic Fourier Transform and real neural networks—Summability by primes. *J. Math. Anal. Appl.*, **190**, 211–219.

Wigley, N.M. and Jullien, G.A. (1992). On implementing the Arithmetic Fourier Transform. *IEEE Trans. on Signal Processing*, **40**, 2233–2242.

Wintner, A. (1945). *An arithmetical approach to ordinary Fourier Series*. Waverly Press, Baltimore.

Zygmund, A. (1959). *Trigonometric series*, vol. I. Cambridge University Press.

Chapter 3

Beaty, M.G. (1994). Multichannel sampling for multiband signals. *Signal Proc.*, **36**, 133–138.

Beaty, M.G. and Dodson, M.M. (1989). Derivative sampling for multiband signals. *Num. Funct. Analysis Optim.*, **10**, 875–898.

Beaty, M.G. and Higgins, J.R. (1994). Aliasing and Poisson summation in the sampling theory of Paley–Wiener spaces. *J. Fourier Analysis Applic.*, **1**, 67–85.

Boas, R.P. (1954). *Entire functions*. Academic Press, New York.

Butzer, P.L. and Splettstößer, W. (1977). *Approximation und interpolation durch verallgemeinerte Abtastsummen*. Westdeutscher-Verlag, Opladen.

Butzer, P.L. and Schöttler, G. (1993). Sampling expansions for Jacobi functions and their applications. *Acta Sci. Math.* (Szeged), **57**, 305–327.

Cheung, K.W. (1992). A multidimensional extension of Papoulis' generalized sampling expansion with applications in minimum density sampling. In: *Advanced topics in Shannon sampling and interpolation theory*, R.J. Marks II (ed.). Springer-Verlag, New York, 85–119.

Fogel, L. (1955). A note on the sampling theorem. *IRE Tras.–Inform. Theory*, **IT–1**, 47–48.

Gröchenig, K. and Razafinjatovo, H. (1996) On Landau's necessary density conditions for sampling and interpolation of band-limited functions. *J. London Math. Soc.* (2) **54**, 557–565.

Grozev, G.R. and Rahman, Q.I. (1994). Entire functions of exponential type belonging to $L^p(\mathbb{R})$. *J. Lond. Math. Soc.* (2), **50**, 302–317.

Higgins, J.R. (1991). Sampling theorems and the contour integral method. *Applic. Anal.*, **41**, 155–171.

Higgins, J.R. (1996). Sampling theory in Fourier and signal analysis, Oxford University Press.

Hinsen, G. (1993). Irregular sampling of bandlimited L^p functions. *J. Approx. Theory*, **72**, 346–364.

Horng, J.C. (1967). On a new double generalized sampling theorem. *Formosan Science*, **31**, 20–29.

Jagerman, D. and Fogel, L. (1956). Some general aspects of the sampling theorem. *IEEE Trans. Inform. Theory*, **2**, 139–156.

Kempski, B.L. (1995). *Extensions of the Whittaker–Shannon sampling series aided by symbolic computation*. M. Phil. Thesis, Anglia Polytechnic University, Cambridge.

Kress, R. (1972). On the general Hermite cardinal interpolation. *Math. Comp.*, **26**, 925–933.

Linden, D.A. (1959). A discussion of sampling theorems. *Proc. IRE*, **47**, 1219–1226.

Linden, D.A. and Abramson, N.M. (1960). A generalization of the sampling theorem. *Inform. Control*, **3**, 26–31. Errata, *ibid.* (1961), 95–96.

Montgomery, W.D. (1964). The gradient in the sampling of N-dimensional band-limited functions. *J. Electron Contr.*, **17**, 437–447.

Montgomery, W.D. (1965). K-order sampling of N-dimensional band-limited functions. *Internat. J. Control*, **1**, 7–12.

Petersen, D.P. and Middleton, D. (1964). Reconstruction of multi-dimensional stochastic fields from discrete measurements of amplitude and gradient. *Inform. Contr.*, **1**, 445–476.

Rahman, Q.I. (1965). Interpolation of entire functions. *Amer. J. Math.*, **87**, 1029–1076.

Rahman, Q.I. and Schmeisser, G. (1985). Reconstruction and approximation of functions from samples. In *Delay equations, approximation and application*, G. Meinardus and G. Nürnberger (eds.). International Series of Numerical Mathematics, Vol. 74, Birkhäuser-Verlag, Basel, 213–233.

Rahman, Q.I. and Schmeisser, G. (1994). A quadrature formula for entire functions of exponential type. *Math. Comp.*, **63**, 215–227.

Rawn, M.D. (1989). A stable nonuniform sampling expansion involving derivatives. · *IEEE Trans. Inform. Theory*, **35**, 1223–1227.

Razafinjatovo, H.N. (1994). Iterative reconstructions in irregular sampling with derivatives. *J. Fourier Anal. Appl.*, **1**, 281–295.

Shannon, C.E. (1949). Communication in the presence of noise. *Proc. IRE*, **37**, 10–21.

Vaaler, J.D. (1985). Some extremal functions in Fourier analysis. *Bull. Am. Math. Soc.*, **12**, 183–216.

Chapter 4

Atal, B.S. and Hanauer, S. L. (1971). Speech analysis and synthesis by linear prediction of the speech wave. *J. Acoust. Soc. Amer.*, **50**, 637–650.

Brown, J.L., Jr. (1972). Uniform linear prediction of bandlimited processes from past samples. *IEEE Trans. Inform. Theory*, **IT-18**, 662–664.

Brown, J.L., Jr. and Morean, O. (1986). Robust prediction of bandlimited signals from past samples. *IEEE Trans. Inform. Theory*, **IT-32**, 410–412.

Butzer, P.L. and Gessinger, A. (1995). In: *Proceedings of SampTA'95*, Jurmala Latvia, Sept. 1995. Institute of Electronics and Computer Science, Riga, Latvia, 100–107,

Butzer, P.L. and Stens, R.L. (1992). Linear prediction by samples from the past. In: *Advanced topics in Shannon sampling and interpolation theory*, by R.J. Marks, II. Springer-Verlag, Berlin, 157–183.

Chan, R.H. and Strang, G. (1989). Toeplitz equations by conjugate gradients with circulant preconditioner. *SIAM J. Sci. Statist. Comput.*, **10**, 104–119.

Elias, P. (1955). Predictive coding I, II. *IRE Trans. Inform. Th.*, **IT-1**, 16–23, 30–33.

Engl, H.W. (1993). Regularization methods for the stable solution of inverse problems. *Surv. Math. Industry*, **3**, 71–143.

Feichtinger, H.G., Gröchenig, K., and Strohmer, T. (1995). Efficient numerical methods in non-uniform sampling theory. *Num. Mathematik*, **69**, 423–440.

Golub, G.H. and Van Loan, C.F. (1989). *Matrix computations*. Johns Hopkins University Press, Baltimore.

Gruenbacher, D.M. and Hummels, D.R. (1994). A simple algorithm for generating discrete prolate spheroidal sequences. *IEEE Trans. Sig. Proc.*, **42**, 3276–3278.

Kauppinen, J.K., Saarinen, P.E., and Hollberg, M.R. (1994). Linear prediction in spectroscopy. *J. of Molecular Structure*, **324**, 61–74.

Lucky, R.W. (1968). Adaptive redundancy removal in data transmission. *Bell Syst. Tech. J.*, **47**, 549–573.

Makhoul, J. (1975). Linear prediction: A tutorial review. *Proc. IEEE*, **63**, 561–580.

Marks, R.J. II (1991). *Introduction to Shannon sampling and interpolation theory*. Springer-Verlag, New York.

Meyer, Y. (1994). *Wavelets: algorithms and applications*. Society for Industrial and Applied Mathematics, Philadelphia, PA.

Mugler, D.H. (1990). Computationally-efficient linear prediction of a band-limited signal and its derivative, *IEEE Trans. on Information Theory*, **IT-36**, 589–596.

Mugler, D.H. (1992). Computational aspects of an optimal linear prediction formula for band-limited signals. *Computational and applied mathematics, I.* C. Brezinski and U. Kulisch (eds.). Elsevier Science Publishers, Amsterdam, 351–356.

Mugler, D.H. (1995). Linear prediction of a band-limited signal from past samples at arbitary points: An svd-based approach. In: *Proceedings of SampTA'95*, Jurmala, Latvia, Sept. 1995. Institute of Electronics and Computer Science, Riga, Latvia, 113–118.

Mugler, D.H. and Splettstößer, W. (1986). Difference methods for the prediction of band-limited signals. *SIAM J. Appl. Math.*, **46**, 930–941.

Mugler, D.H. and Splettstößer, W. (1987a). Linear prediction from samples of a function and its derivatives. *IEEE Trans. Inform. Theory*, **IT-33**, 360–366.

Mugler, D.H. and Splettstößer, W. (1987b). Some new difference schemes for the prediction of band-limited signals from past samples. *Mathematics in Signal Processing*. T.S. Durrani et al. (ed.). Oxford University Press, 33–43.

Nagy, J.G. (1995). Applications of Toeplitz systems. *SIAM News*, **28** (8), 10–11.

Rabiner, L.R. and Schafer, R.W. (1993). *Digital processing of speech signals*. Macmillan, London.

Slepian, D. (1978). Prolate spheroidal wave functions, Fourier analysis, and uncertainty–V: The discrete case. *Bell Syst. Tech. J.*, **57**, 1371–1430.

Splettstößer, W. (1982). On the prediction of band-limited signals from past samples. *Inform. Sci.*, **28**, 115–130.

Strobach, P. (1991). New forms of Levinson and Schur algorithms. *IEEE Sig. Proc. Magazine*, **8**, 12–36.

Wainstein, L.A. and Zubakov, V.D. (1962). *Extraction of signals from noise.* Prentice-Hall, Englewood Cliffs, NJ.

Wiener, N. (1949). *Extrapolation, interpolation, and smoothing of stationary times series.* MIT Press, Cambridge, MA.

Wingham, D.J. (1992). The reconstruction of a band-limited function and its Fourier transform from a finite mumber of samples at arbitrary locations by singular value decomposition. *IEEE Trans. Sig. Proc.*, **40**, 559–570.

Chapter 5

Annaby, M.H. On Kramer's theorem associated with second order boundary value problems. (To appear.)

Bailey, P.B., Everitt, W.N., and Zettl, A. (1996). Regular and singular Sturm–Liouville problems with coupled boundary conditions. *Proc. Roy. Soc. Edinb.*, **126**, 505–514.

Butzer, P.L. and Nasri-Roudsari, G. (1997). Kramer's sampling theorem in signal analysis and its role in mathematics. In: *Image processing; mathematical methods and applications.* Proc. of IMA Conference, Cranfield University, UK, J.M. Blackledge (ed.). Clarendon Press, Oxford, pp. 49–95.

Butzer, P.L. and Schöttler, G. (1994). Sampling theorems associated with fourth and higher order self-adjoint eigenvalue problems. *J. Comput. Appl. Math.*, **51**, 159–177.

Butzer, P.L., Splettstößer, W. and Stens, R.L. (1988). The sampling theorem and linear prediction in signal analysis. *Jahresber. Deutsch. Math.-Verein.*, **90**, 1–70.

Campbell, L.L. (1964). A comparison of the sampling theorems of Kramer and Whittaker. *J. SIAM*, **12**, 117–130.

Coddington, E.A. and Levinson, N. (1955). *Theory of ordinary differential equations.* McGraw-Hill, New York.

Copson, E.T. (1944). *Theory of functions of a complex variable.* Oxford University Press.

Davis, P.J. (1965). *Interpolation and approximation.* Blaisdell, New York.

Everitt, W.N. (1986). *Linear ordinary quasi-differential expressions.* Lecture notes for the Fourth International Symposium on Differential equations and differential geometry, Beijing, Peoples' Republic of China (Department of Mathematics, University of Beijing, Peoples' Republic of China).

Everitt, W.N. (1995). Some remarks on the Titchmarsch–Weyl m-coefficient and associated differential operators. In: *Differential equations and geometric dynamics; control science and dynamical systems.* Marcel Dekker, Inc., New York.

Everitt, W.N. and Nasri-Roudsari, G. Sturm–Liouville problems with coupled boundary conditions and Lagrange interpolation series. (To appear (a)).

Everitt, W.N. and Nasri-Roudsari, G. Interpolation theorems and Sturm–Liouville boundary value problems. (To appear (b)).

Everitt, W.N., Nasri-Roudsari, G., and Rehberg, J. A note on the Kramer sampling theorem. (To appear.)

Everitt, W.N. and Race, D. (1987). Some remarks on linear ordinary quasi-differential expressions. *Proc. London Math. Soc.*, **54** (3), 300–320.

Everitt, W.N., Schöttler, G., and Butzer, P.L. (1994). Sturm–Liouville boundary value problems and Lagrange interpolation series. *Rend. Math. Appl.*, **14** (7), 87–126.

Everitt, W.N. and Zettl, A. (1991). Differential operators generated by a countable number of quasi-differential expressions on the real line. *Proc. London Math. Soc.*, **64** (3), 524–544.

Genuit, M. and Schöttler, G. (1995). A problem of L.L. Campbell on the equivalence of the Kramer and Shannon sampling theorems. *Computers Math. Applic.*, **30**, 433–443.

Higgins, J.R. (1977). *Completeness and basis properties of sets of special functions.* Cambridge University Press.

Higgins, J.R. (1996). *Sampling theory in Fourier and signal analysis: foundations.* Clarendon Press, Oxford.

Jerri, A.J. (1969). On the equivalence of Kramer's and Shannon's sampling theorem. *IEEE Trans. Inform. Theory*, **IT-15**, 497–499.

Kramer, H.P. (1959). A generalized sampling theorem. *J. Math. Phys.*, **38**, 68–72.

Naimark, M.A. (1968). *Linear differential operators*, vols. I and II (translated from the Russian edition of 1960). Ungar Publishing Company, New York.

Niessen, H.-D. and Zettl, A. (1992). Singular Sturm–Liouville problems: The Friedrichs extension and comparison of eigenvalues. *Proc. London Math. Soc.*, **64** (3), 545–578.

Titchmarsh, E.C. (1939). *The theory of functions* (2nd edn). Oxford University Press.

Titchmarsh, E.C. (1962). *Eigenfunction expansions,* vol. I (2nd edn). Oxford University Press.

Zayed, A.I. (1991). On Kramer's sampling theorem associated with general Sturm–Liouville problems and Lagrange interpolation. *SIAM J. Appl. Math.*, **51**, 575–604.

Zayed, A.I. (1993a). A new role of Green's function in interpolation and sampling theory. *J. Math. Anal. Appl.*, **175**, 222–238.

Zayed, A.I. (1993b). *Advances in Shannon's sampling theory.* CRC Press, Boca Raton.

Zayed, A.I., El-Sayed, M.A., and Annaby, M.H. (1991). On Lagrange interpolation and Kramer's sampling theorem associated with self-adjoint boundary-value problems. *J. Math. Anal. Appl.*, **158**, No. 1, 269–284.

Zayed, A.I., Hinsen, G., and Butzer, P.L. (1990). On Lagrange interpolation and Kramer-type sampling theorems associated with Sturm–Liouville problems. *SIAM J. Appl. Math.*, **50**, 893–909.

Chapter 6

Beaty, M.G., Dodson, M.M., and Higgins, J.R. (1994). Approximating Paley–Wiener functions by smoothed step functions. *J. Approx. Theory*, **78**, 433–445.

Boas, R.P., Jr. (1954). *Entire functions.* Academic Press, New York.

de Boor, C. (1990). *Splinefunktionen.* Birkhäuser, Basel.

de Boor, C. and DeVore, R. (1983). Approximation by smooth multivariate splines. *Trans. Amer. Math. Soc.*, **276**, 775–788.

de Boor, C. and Höllig, K. (1982). Recurrence relations for multivariate B-splines. *Proc. Amer. Math. Soc.*, **85**, 397–400.

de Boor, C. and Höllig, K. (1983). B-splines from parallelepipeds. *J. Anal. Math.*, **42**, 99–115.

Brown, J.L., Jr. (1967/68). On the error in reconstructing a non-bandlimited function by means of the bandpass sampling theorem. *J. Math. Anal. Appl.*, **18**, 74–85; erratum, *ibid.*, **21**, 699.

Buhmann, M.D. and Ron, A. (1994). Radial basis functions: L^p-approximation orders with scattered centers. In: *Wavelets, images, and surface fitting*, Proc. Conf., Chamonix-Mont-Blanc, France, 1993, P.-J. Laurent, A. le Méhauté, and L.L. Schumaker (eds.). A.K. Peters, Wellesley, MA, 93–122.

Butzer, P.L., Engels, W., Ries, S., and Stens, R.L. (1986). The Shannon sampling series and the reconstruction of signals in terms of linear, quadratic and cubic splines. *SIAM J. Appl. Math.*, **46**, 299–323.

Butzer, P.L., Fischer, A., and Stens, R.L. (1990). Generalized sampling approximation of multivariate signals; theory and some applications. *Note Mat.*, **10**, Suppl. No. 1, 173–191.

Butzer, P.L., Fischer, A., and Stens, R.L. (1993). Generalized sampling approximation of multivariate signals; general theory. *Atti Sem. Mat. Fis. Univ. Modena*, **41**, 17–37.

Butzer, P.L. and Hinsen, G. (1989). Two-dimensional nonuniform sampling expansions – an iterative approach. I. Theory of two-dimensional bandlimited signals. II. Reconstruction formulae and applications. *Appl. Anal.*, **32**, 53–67, 69–85.

Butzer, P.L. and Nessel, R.J. (1971). *Fourier analysis and approximation*, Vol. 1: *One-dimensional theory*. Academic Press, New York; Birkhäuser, Basel.

Butzer, P.L., Ries, S., and Stens, R.L. (1987). Approximation of continuous and discontinuous functions by generalized sampling series. *J. Approx. Theory*, **50**, 25–39.

Butzer, P.L. and Splettstößer, W. (1977). A sampling theorem for duration-limited functions with error estimates. *Inform. and Control*, **34**, 55–65.

Butzer, P.L., Splettstößer, W., and Stens, R.L. (1988). The sampling theorem and linear prediction in signal analysis. *Jahresber. Deutsch. Math.-Verein.*, **90**, 1–70.

Butzer, P.L. and Stens, R.L. (1983). The Poisson summation formula: Whittaker's cardinal series and approximate integration. In: *Approximation theory*, Proc. Conf., Edmonton, Canada, 1982. Z. Ditzian et al. (eds.). CMS Conf. Proc., Vol. 3, 19–36. Canadian Math. Soc., Providence, RI.

Butzer, P.L. and Stens, R.L. (1985). A modification of the Whittaker–Kotelnikov–Shannon sampling series. *Aequationes Math.*, **28**, 305–311.

Butzer, P.L. and Stens, R.L. (1992). Sampling theory for not necessarily band-limited functions: A historical overview. *SIAM Rev.*, **34**, 40–53.

Chui, C.K. (1988). *Multivariate splines*, CBMS-NSF Reg. Conf. Ser. Appl. Math., Vol. 54. SIAM, Philadelphia.

Chui, C.K. and Lai, H.-C. (1987). Vandermonde determinant and Lagrange interpolation in $\mathbb{R}^s$. In *Nonlinear and convex analysis*, Proc. Conf., Santa Barbara, California, 1985, B.-L. Lin and S. Simons (eds.). Lecture Notes in Pure and Appl. Math., Vol. 107, 23–35. Dekker, New York.

Dinh-Dung (1992). A modification of the sampling theorem and recovery of functions. In: *Optimal recovery*, Proc. Symp., Varna, Bulgaria, 1989, B. Bojanov and H. Woźniakowski (eds.). Nova Sci. Publ., Commack, NY, 153–177.

Dodson, M.M. and Silva, A.M. (1985). Fourier analysis and the sampling theorem. *Proc. Roy. Irish Acad. Sect. A*, **85**, 81–108.

Engels, W., Stark, E.L., and Vogt, L. (1987). Optimal kernels for a general sampling theorem. *J. Approx. Theory*, **50**, 69–83.

Fischer, A. (1990). *Approximation durch Abtastreihen in mehreren Dimensionen*. Diplomarbeit, RWTH Aachen.

Fischer, A. and Stens, R.L. (1990). Generalized sampling approximation of multivariate signals; inverse approximation theorems. In: *Approximation theory*, Proc. Conf., Kecskemét, Hungary, 1990, J. Szabados and K. Tandori (eds.). Colloq. Math. Soc. János Bolyai, Vol. 58, 275–286. North-Holland Publishing Company, Amsterdam; Jànos Bolyai Math. Soc., Budapest.

Gauss, Carl Friedrich (1900). *Werke*, vol. VIII. Königliche Gesellschaft der Wissenschaften zu Göttingen, Teubner, Leibzig.

Gervais, R., Rahman, Q.I., and Schmeisser, G. (1984). A bandlimited function simulating a duration-limited one. In: *Anniversary volume on approximation theory and functional analysis*, Proc. Conf., Oberwolfach, Germany, 1983, P.L. Butzer, R.L. Stens, and B. Sz.-Nagy (eds.). ISNM Vol. 65, 355–362. Birkhäuser, Basel.

Hettich, U. and Stens, R.L. (to appear). Approximating a bandlimited function in terms of its samples. *Comput. Math. Appl.*

Higgins, J.R. (1985). Five short stories about the cardinal series. *Bull. Amer. Math. Soc.*, **12**, 45–89.

Higgins, J.R. (1996). *Sampling theory in Fourier and signal analysis: Foundations*. Clarendon Press, Oxford.

Hoffman, K. (1975). *Analysis in Euclidean space*. Prentice-Hall, Englewood Cliffs, NJ.

Höllig, K. (1986). Box splines. In: *Approximation theory V*, Proc. Symp., College Station, Texas, 1986, C.K. Chui, L.L. Schumaker, and J.D. Ward (eds.). Academic Press, Boston, 71–95.

Jetter, K. (1987). A short survey on cardinal interpolation by box splines. In: *Topics in multivariate approximation*, Proc. Workshop, Santiago, Chile, 1986 C.K. Chui, L.L. Schumaker, and F.I. Utreras (eds.). Academic Press, Boston, 125–139.

Kotel'nikov, V.K. (1933). On the carrying capacity of the 'ether' and wire in telecommunications (Russian). *Material for the first All-Union conference of questions of communications*. Izd. Red. Upr. Svyazi RKKA, Moscow.

Lloyd, S.P. (1959). A sampling theorem for stationary (wide sense) stochastic processes. *Trans. Amer. Math. Soc.*, **92**, 1–12.

Mersereau, R.M. (1979). The processing of hexagonally sampled two dimensional signals. *Proc. IEEE*, **67**, 930–949.

Mersereau, R.M. and Speake, T.C. (1983). The processing of periodically sampled multidimensional signals. *IEEE Trans. Acount. Speech Signal Process*, **ASSP-31**, 188–194.

Micchelli, C.A. (1979). On a numerically efficient method for computing multivariate B-splines. In *Multivariate approximation theory*, Proc. Conf., Oberwolfach, Germany, 1979, W. Schempp and K. Zeller (eds.). ISNM, Vol. 51, pp. 211–248. Birkhäuser, Basel.

Nessel, R.J. (1967). Contributions to the theory of saturation for singular integrals in several variables. III. Radial kernels. *Nederl. Akad. Wetensch. Proc. Ser. A*, **70**, =*Indag. Math.*, **29**, 65–73.

Nessel, R.J. and Pawelke, A. (1968). Über Favardklassen von Summationsprozessen mehrdimensionaler Fourierreihen. *Compositio Math.*, **19**, 196–212.

Nikol'skiĭ, S.M. (1975). *Approximation of functions of several variables and imbedding theorems.* Springer, Berlin.

Parzen, E. (1956). A simple proof and some extensions of sampling theorems. *Stanford University, Tech. Rep.*, 7.

Peterson, D.P. and Middleton, D. (1962). Sampling and reconstruction of wave number-limited functions in N-dimensional Euclidean space. *Inform. and Control*, **5**, 279–323.

Prosser, R.T. (1966). A multidimensional sampling theorem. *J. Math. Anal. Appl.*, **16**, 574–584.

Ries, S. and Stens, R.L. (1984). Approximation by generalized sampling series. In: *Constructive theory of functions*, Proc. Conf., Varna, Bulgaria, 1984, B. Sendov, P. Petrushev, R. Maleev, and S. Tashev (eds.). Publishing House Bulgarian Acad. Sci., Sofia, 746–756.

Ries, S. and Stens, R.L. (1987). A localization principle for the approximation by sampling series. In: *Theory of the approximation of functions*, Proc. Conf., Kiev, USSR, 1983, N.P. Korneĭcuk, S.B. Stechkin, and S.A. Telyakovskiĭ (eds.). Izdat Nauka, Moscow, 507–510.

Ron, A. (1992). The L_2-approximation orders of principal shift-invariant spaces generated by a radial basis function. In: *Numerical methods in approximation theory*, Vol. 9. Proc. Conf., Oberwolfach, Germany, 1991, D. Braess and L.L. Schumaker (eds.). ISNM Vol. 105, 245–268. Birkhäuser, Basel.

Shannon, C.E. (1949). Communication in the presence of noise. *Proc. IRE*, **37**, 10–21.

Splettstößer, W. (1978). On generalized sampling sums based on convolution integrals. *Arch. Elek. Übertr.*, **32**, 267–275.

Splettstößer, W. (1979). Error estimates for sampling approximation of non-bandlimited functions. *Math. Methods Appl. Sci.*, **1**, 127–137.

Splettstößer, W. (1982). Sampling approximation of continuous functions with multi-dimensional domain. *IEEE Trans. Inform. Theory*, **IT-28**, 809–814.

Splettstößer, W., Stens, R.L., and Wilmes, G. (1981). On approximation by the interpolating series of G. Valiron. *Funct. Approx. Comment. Math.*, **11**, 39–56.

Stark, H. (1979). Sampling theorems in polar coordinates. *J. Opt. Soc. Amer.*, **69**, 1519–1525.

Stein, E.M. and Weiss, G. (1971). *Introduction to Fourier analysis on Euclidean spaces.* Princeton University Press.

Stens, R.L. (1980a). Error estimates for sampling sums based on convolution integrals. *Inform. and Control*, **45**, 37–47.

Stens, R.L. (1980b). Approximation to duration-limited functions by sampling sums. *Signal Process.*, **2**, 173–176.

Stens, R.L. (1984). Approximation of functions by Whittaker's cardinal series. In: *General inequalities 4*, Proc. Conf., Oberwolfach, Germany, 1983, W. Walter (ed.). ISNM, Vol. 71, 137–149. Birkhäuser, Basel.

Theis, M. (1919). Über eine Interpolationsformel von de la Vallée Poussin. *Math. Z.*, **3**, 93–113.

de la Vallée Poussin, Ch. J. (1908). Sur la convergence des formules d'interpolation entre ordonnées équidistantes. *Acad. Roy. Belg. Bull. Cl. Sci.*, **4**, 319–410.

Whittaker, E.T. (1915). On the functions which are represented by the expansions of the interpolation-theory. *Proc. Roy. Soc. Edinburgh Sect. A*, **35**, 181–194.

Chapter 7

Aldroubi, A. and Unser, M. (1992). Families of wavelet transforms in connection with Shannon's sampling theory and the Gabor transform. In: *Wavelets: a tutorial in theory and applications*, Chui, C. K. (ed.). Academic Press, San Diego, 509–528.

Aldroubi, A. and Unser, M. (1993). Families of multiresolution and wavelet spaces with optimal properties. *Numer. Funct. Anal. Optim.*, **14**, 417–446.

Aldroubi, A. and Unser, M. (1994). Sampling procedures in function spaces and asymptotic equivalence with Shannon's sampling theory. *Numer. Funct. Anal. Optim.*, **15**, 1–21.

Antoine, J.P. (1992). Wavelet analysis in image processing. In: *Signal processing VI: theory and applications*, Vandewallen, J., Boite, R., Moonen, M. and Oosterlinck, A. (eds.). Elsevier Science Publishers, Amsterdam, 23–30.

Antoine, J.P., Carette, P., Murenzi, R., and Piette, B. (1993). Image analysis with two-dimensional continuous wavelet transform. *Signal Processing*, **31**, 241–272.

Benedetto, J.J. (1992). Irregular sampling and frames. In: *Wavelets: a tutorial in theory and applications*, Chui, C.K. (ed.), Academic Press, San Diego, 445–507.

Benedetto, J.J. (1994). Frame decompositions, sampling, and uncertainty principle inequalities. In: *Wavelets: mathematics and applications*, Benedetto, J.J. and Frazier, M.W. (eds.). CRC Press, Boca Raton, 247–304.

Benedetto, J.J. and Heller, W. (1990). Irregular sampling and the theory of frames, I. *Note Mat.*, **X**, Suppl. No. 1, 103–125.

Benedetto, J.J. and Teolis, A. (1993). A wavelet auditory model and data compression. *Appl. Comput. Harmon. Anal.*, **1**, 3–28.

Benedetto, J.J. and Walnut, D. (1994). Gabor frames for L^2 and related spaces. In: *Wavelets: mathematics and applications*, Benedetto, J.J. and Frazier, M.W. (eds.). CRC Press, Boca Raton, 97–162.

Beylkin, G., Coifman, R., and Rokhlin, V. (1991). Fast wavelet transforms and numerical algorithms I. *Comm. Pure Appl. Math.*, **XLIV**, 141–183.

de Boor, C. (1987). *A practical guide to splines.* Springer, New York.

de Boor, C. (1990). Quasiinterpolants and approximation power of multivariate splines. In: *Computation of curves and surfaces*, Dahmen, W., Gasca, M. and Micchelli, C.A. (eds.). Kluwer Academic, Dordrecht, 313–345.

de Boor, C., DeVore, R.A., and Ron, A. (1993). On the construction of multivariate (pre)wavelets. *Constr. Approx.*, **9**, 123–166.

de Boor, C., DeVore, R.A., and Ron, A. (1994). Approximation from shift-invariant subspaces of $L^2(\mathbb{R}^d)$. *Trans. Amer. Math. Soc.*, **341**, 787–806.

de Boor, C. and Höllig, K. (1982/83). B-splines from parallelepipeds. *J. Anal. Math.*, **42**, 99–115.

de Boor, C., Höllig, K., and Riemenschneider, S.D. (1986). Convergence of cardinal series. *Proc. Amer. Math. Soc.*, **98**, 457–460.

de Boor, C., Höllig, K. and Riemenschneider, S.D. (1993). *Box splines.* Springer, Berlin.

de Boor. C. and Jia, R.-Q. (1985). Controlled approximation and a characterization of the local approximation order. *Proc. Amer. Math. Soc.*, **95**, 547–553.

Buhmann, M.D. (1989). *Multivariable interpolation using radial basis functions.* Ph.D. Dissertation, University of Cambridge.

Burchard, H.G., Chui, C.K., and Ward, J.D. On polynomial degree and approximation order. (To appear.)

Burchard, H.G. and Lei, J. (1995). Coordinate order of approximation by functional-based approximation operators. *J. Approx. Theory*, **82**, 240–256.

Butzer, P.L., Fischer, A., and Rückforth, K. (1994). Scaling functions and wavelets with vanishing moments. *Comput. Math. Appl.*, **27**, 33–39.

Butzer, P.L., Fischer, A., and Stens, R.L. (1990). Generalized sampling approximation of multivariate signals; theory and some applications. *Note Mat.*, **X**, Suppl. No. 1, 173–191.

Butzer, P.L., Fischer, A., and Stens, R.L. (1992). Generalized sampling approximation of multivariate signals; general theory. In: *Proc. Fourth Meeting on Real Analysis and Measure Theory, Capri 1990 = Atti Sem. Mat. Fis. Univ. Modena,* **XL**, 1–21 and **XLI**, 17–37.

Butzer, P.L. and Nessel, R.J. (1971). *Fourier analysis and approximation,* vol. I: *One-dimensional theory.* Birkhäuser, Basel.

Butzer, P.L., Schmidt, M., and Stark, E.L. (1988). Observations on the history of central B-splines. *Arch. Hist. Exact Sci.*, **39**, 137–156.

Butzer, P.L., Splettstößer, W., and Stens, R.L. (1988). The sampling theorem and linear prediction in signal analysis. *Jahresber. Deutsch. Math.-Verein.*, **90**, 1–70.

Butzer, P.L. and Stens, R.L. (1983). The Poisson summation formula, Whittaker's cardinal series and approximate integration. In: *Proc. Second Edmonton Conf. on Approximation Theory (Edmonton 7–11 June, 1982),* Ditzian, Z. et al. (ed.). Providence, RI: *American Math. Soc. 1983 = Canadian Math. Soc. Proc.*, **3**, 19–36.

Cavaretta, A.S., Dahmen, W., and Micchelli, C.A. (1991). Stationary Subdivision. *Mem. Amer. Math. Soc.*, **453**.

Chui, C.K. (1992). *An introduction to wavelets.* Academic Press, San Diego.

Chui, C.K., Stöckler, J., and Ward, J.D. (1992). Compactly supported box-spline wavelets. *Approx. Theory Appl.*, **8**, 77–100.

Dahlke, S., Dahmen, W., Hochmuth, R., and Schneider, R. (1997). Stable multiscale bases and local error estimation for elliptic problems. *Applied Numerical Mathematics*, **23**, 21–47.

Dahmen, W. (1993). Decomposition of refinable spaces and applications to operator equations. *Numer. Algorithms*, **5**, 229–245.

Dahmen, W. (1995). Multiscale analysis, approximation, and interpolation spaces. In: *Approximation Theory VIII,* Chui, C.K. and Schumaker, L.L. (eds.). World Scientific Publishing Co., Singapore, 47–88.

Dahmen, W. and Kunoth, A. (1992). Multilevel preconditioning. *Numer. Math.*, **63**, 315–344.

Dahmen, W. and Micchelli, C.A. (1984). On the approximation order from certain multivariate spline spaces. *J. Austral. Math. Soc. Ser. B*, **26**, 233–246.

Dahmen, W., Prössdorf, S., and Schneider, R. (1993a). Wavelet approximation methods for pseudodifferential equations II: Matrix compression and fast solution. *Adv. Comput. Math.*, **1**, 259–335.

Dahmen, W., Prössdorf, S., and Schneider, R. (1993b). Multiscale methods for pseudodifferential equations. In: *Recent advances in wavelet analysis*. Schumaker, L.L. and Webb, G. (eds.). Academic Press, Boston, 191–235.

Daubechies, I. (1988). Orthonormal bases of compactly supported wavelets. *Comm. Pure Appl. Math.*, **41**, 909–996.

Daubechies, I. (1990). The wavelet transform, time-frequency localization and signal analysis. *IEEE Trans. Inform. Theory*, **36**, 961–1005.

Daubechies, I. (1992). *Ten lectures on wavelets*. CBMS-NSF Regional Conf. Ser. in Appl. Math., SIAM, Philadelphia.

DeVore, R.A., Jawerth, B., and Popov, V. (1992). Compression of wavelet decompositions. *Amer. J. Math.*, **114**, 737–785.

Dyn, N., Jackson, I.R.H., Levin, D., and Ron, A. (1992). On multivariate approximation by integer translates of a basis function. *Israel J. Math.*, **78**, 95–130.

Feichtinger, H.G. and Gröchenig, K. (1988). A unified approach to atomic decompositions via integrable group representations. In: *Function spaces and applications. Proc. US-Swed. Semin., Lund/Swed.*, Cwikel, M. *et al.* (eds.). *Lecture Notes in Math.*, **1302**, 52–73.

Feichtinger, H.G. and Gröchenig, K. (1994). Theory and practice of irregular sampling. In: *Wavelets: mathematics and applications*, Benedetto, J.J. and Frazier, M.W. (eds.). CRC Press, Boca Raton, FL, 305–363.

Fischer, A. (1995). Multiresolution analysis and multivariate approximation of smooth signals in $C_B(\mathbb{R}^d)$. *J. Fourier Anal. Appl.*, **2**, 162–180.

Fischer, A. (1996). *Multiresolution Analysis und Multivariate Approximation mit Anwendungen auf Abtast- und Quasi-Interpolations-Operatoren*. Doctoral Dissertation RWTH Aachen, Shaker Verlag, Aachen.

Fischer, A. (1997). On wavelets and prewavelets with vanishing moments in higher dimensions. *J. Approx. Theory*, **90**, 46–74.

Fischer, A. and Stens, R.L. (1990). Generalized sampling approximation of multivariate signals; inverse approximation theorems. In: *Proc. Conf. on Approximation Theory, Kecskemét, Hungary, 1990; Colloquia Mathematica Janos Bolyai*, 275–286.

Froment, J. and Mallat, S. (1992). Second generation compact image coding with wavelets. In: *Wavelets: a tutorial in theory and applications*, Chui, C.K. (ed.). Academic Press, San Diego, 655–678.

v. Golitschek, M. (1972). On the convergence of interpolating periodic spline functions of high degree. *Numer. Math.*, **19**, 146–154.

Goodman, T.N.T. and Micchelli, C.A. (1994). Orthonormal cardinal functions. In: *Wavelets: theory, algorithms, and applications*, Chui, C.K., Montefusco, L. and Puccio, L. (eds.). Academic Press, Boston, 53–88.

Gröchenig, K. (1987). Analyse multi-échelle et bases d'ondelettes. *C.R. Acad. Sci. Paris Sér. I Math.*, **305**, 13–15.

Gröchenig, K. (1991). Describing functions: Atomic decompositions versus frames. *Monatsh. Math.*, **112**, 1–42.

Gröchenig, K. (1993). Irregular sampling of wavelet and short-time Fourier transforms. *Constr. Approx.*, **9**, 283–297.

Grossmann, A. and Morlet, J. (1984). Decomposition of Hardy functions into square integrable wavelets of constant shape. *SIAM J. Math. Anal.*, **15**, 723–736.

Grossmann, A., Morlet, J., and Paul, T. (1985). Transforms associated to square integrable group representations, I. General results. *J. Math. Phys.*, **27**, 2473–2479.

Grossmann, A., Morlet, J., and Paul, T. (1986). Transforms associated to square integrable group representations, II. Examples. *Ann. Inst. H. Poincaré*, **45**, 293–309.

Haar, A. (1910). Zur Theorie der orthogonalen Funktionen-Systeme. *Math. Ann.*, **69**, 331–371.

Halton, E.J. and Light, W.A. (1993). On local and controlled approximation order. *J. Approx. Theory*, **72**, 268–277.

Heil, C. and Walnut, D. (1989). Continuous and discrete wavelet transforms. *SIAM Rev.*, **31**, 628–666.

Higgins, J.R. (1996). *Sampling theory in Fourier and signal analysis: foundations.* Clarendon Press, Oxford.

Jackson, I.R.H. (1989). An order of convergence for some radial basis function. *IMA J. Numer. Anal.*, **9**, 567–587.

Jaffard, S. (1992). Wavelet methods for fast resolution of elliptic problems. *SIAM J. Numer. Anal.*, **29**, 965–986.

Jia, R.-Q. and Lei, J. (1993). On approximation by multi-integer translates of functions having global support. *J. Approx. Theory*, **72**, 2–23.

Jia, R.-Q. and Micchelli, C.A. (1991). Using the refinement equations for the construction of pre-wavelets II: Powers of two. In: *Curves and surfaces,* Laurent, P.J., Le Méhauté, A., and Schumacher, L.L. (eds.). Academic Press, Boston, 209–246.

Kölzow, D. (1994). *Wavelets: a tutorial and a bibliography.* Working-out of the material on wavelets, presented at the Second and Third Conference on Measure Theory and Real Analysis, held at Grado, May 10–22, 1992, and September 10 – October 1, 1993, respectively. Erlangen.

Kotel'nikov, V.A. (1933). On the carrying capacity of 'ether' and wire in electro-communications (Russian). Material for the First All-Union conference on Questions of Communications. *Izd. Red. Upr. Svyazi RKKA*, Moscow.

Lei, J. (1994a). L_p-Approximation by certain projection operators. *J. Math. Anal. Appl.*, **185**, 1–14.

Lei, J. (1994b). On approximation by translates of globally supported functions. *J. Approx. Theory*, **77**, 123–138.

Lemarié, P.G. (1990). Deux constructions de bases d'ondelettes: ondelettes à trace et ondelettes interpolantes. *Séminaire d'Analyse Harmonique, Année 1989/90.* Université Paris XI, Orsay, 56–63.

Lemarié, P.G. (1991a). Fonctions à support compact dans les analyses multi-résolutions. *Rev. Mat. Iberoamericana*, **7**, 157–182.

Lemarié, P.G. (1991b). La propriété de support minimal dans les analyses multi-résolution. *C.R. Acad. Sci. Paris Sér. I Math.*, **312**, 773–776.

Lemarié-Rieusset, P.G. and Meyer, Y. (1986). Ondelettes et bases hilbertiennes. *Rev. Mat. Iberoamericana*, **8**, 1–18.

Lewis, R.M. (1994). Cardinal interpolating multiresolutions. *J. Approx. Theory*, **76**, 177–202.

Light, W.A. and Cheney, E.W. (1992). Quasi-interpolation with translates of a function having noncompact support. *Constr. Approx.*, **8**, 35–48.

Mallat, S.G. (1989a). Multiresolution approximation and wavelet orthonormal bases of $L^2(\mathbb{R})$. *Trans. Amer. Math. Soc.*, **315**, 69–87.

Mallat, S.G. (1989b). A theory for multiresolution signal decomposition: the wavelet representation. *IEEE Trans. Patt. Anal. and Mach. Intell.*, **11**, 674–693.

Meyer, Y. (1990). *Ondelettes et opérateurs I*, Hermann, Paris.

Morlet, J. (1983). Sampling Theory and wave propagation. In: *NATO ASI Series, vol. 1: Issues in Acoustic Signal/Image Processing and Recognition*, Chen, C.H. (ed.). Springer, Berlin, 233–261.

Morlet, J., Arens, G., Fourgeau, I., and Giard, D. (1982). Wave propagation and sampling theory. *Geophys.*, **47**, 203–236.

Murenzi, R. (1989). Wavelet transforms associated to the n-dimensional Euclidean group with dilations: signals in more than one dimension. In: *Wavelets, time-frequency methods and phase space (Proc. Marseille, Dec. 1987)*. Combes, J.-M., Grossmann, A., and Tchamitchian, Ph. (eds.). Springer, Berlin, 239–246.

Murenzi, R. (1990). *Ondelettes multidimensionnelles et applications à l'analyse d'images*. Thèse de Doctorat, UCL, Louvain-la-Neuve.

Nachbin, L. (1965). *Haar integral*. Van Nostrand, Princeton.

Reiter, H. (1968). *Classical harmonic analysis and locally compact groups*. Oxford University Press.

Riemenschneider, S.D. (1989). Multivariate cardinal interpolation. In: *Approximation theory VI*, vol. 2, Chui, C.K., Schumacher, L.L. and Ward, J.D. (eds.). Academic Press, Boston, MA, 561–580.

Riemenschneider, S.D. and Shen, Z. (1990). Wavelets and pre-wavelets in low dimensions. *J. Approx. Theory*, **71**, 18–38.

Riemenschneider, S.D. and Shen, Z. (1991). Box splines, cardinal series and wavelets. In: *Approximation theory and functional analysis*. Chui, C.K. (ed.). Academic Press, Boston, MA, 133–149.

Ries, S. and Stens, R.L. (1984). Approximation by generalized sampling series. *Constructive Theory of Functions*, **84**, 746–756.

Rückforth, K. (1994). *Wavelets, Frames, Multiresolutionanalysis und ihre Beziehung zur Approximationstheorie*. Diplomarbeit, RWTH Aachen.

Schoenberg, I.J. (1946). Contributions to the problem of approximation of equidistant data by analytic functions. Part A.—On the problem of smoothing or graduation. A first class of analytic approximation formulae. Part B.—On the second problem of osculatory interpolation. A second class of analytic approximation formulae. *Quart. Appl. Math.*, **IV**, 45–99 and 112–141.

Schoenberg, I.J. (1972). Notes on spline functions I. The limits of the interpolating periodic spline functions as their degree tends to infinity. *Indag. Math.*, **34**, 412–422.

Schoenberg, I.J. (1973). *Cardinal interpolation*. CBMS-NSF Regional Conf. Ser. in Appl. Math., vol. 12, SIAM, Philadelphia.

Shannon, C.E. (1949). Communication in the presence of noise. *Proc. IRE*, **37**, 10–21.

Splettstößer, W. (1978). On generalized sampling sums based on convolution integrals. *Arch. Elek. Übertr.*, **32**, 267–275.

Splettstößer, W. (1979). Error estimates for sampling approximation of non-band-limited functions. *Math. Methods Appl. Sci.*, **1**, 127–137.

Splettstößer, W., Stens, R.L., and Wilmes, G. (1981). On approximation by the inter-polating series of G. Valiron. *Funct. Approx. Comment. Math.*, **11**, 39–56.

Steidl, G. (1995). On multivariate attenuation factors. *Num. Algorithms*, **9**, 245–261.

Stens, R.L. (1980). Error estimates for sampling sums based on convolution integrals. *Inform. and Control*, **45**, 37–47.

Stöckler, J. (1992). Multivariate wavelets. In: *Wavelets: a tutorial in theory and applications*, Chui, C.K. (ed.). Academic Press, New York, 325–355.

Strang, G. (1989). Wavelets and dilation equations: A brief introduction. *SIAM Rev.*, **31**, 614–627.

Strang, G. and Fix, G. (1973). A Fourier analysis of the finite-element variational method. In: *Constructive aspects of functional analysis*, Geymonat, G. (ed.). C.I.M.E., 793–840.

Sweldens, W. and Piessens, R. Asymptotic error expansions for wavelet approximations of smooth functions. (To appear.)

Vilenkin, N.Ja. (1968). *Special functions and theory of group representations*. AMS, Providence, RI.

Walter, G.G. (1992). Wavelets and generalized functions. In: *Wavelets: a tutorial in theory and applications*, Chui, C.K. (ed.). Academic Press, San Diego, 51–70.

Walter, G.G. (1993). Wavelet subspaces with an oversampling property. *Indag. Math. (N.S.)*, **4**, 499–507.

Whittaker, E.T. (1915). On the functions which are represented by the expansion of the interpolation theory. *Proc. Roy. Soc. Edinburgh*, **35**, 181–194.

Wickerhauser, M.V. (1991). Codage et compression du signal et de l'image par les ondelettes et les paquets d'ondelettes. In: *Problèmes non linéaires appliqueés, ondelettes et paquets d'ondes*. INRIA, 31–99.

Xia, X.-G. and Zhang, Z. (1993). On sampling theorem, wavelets, and wavelet trans-forms. *IEEE Trans. Signal Proc.*, **41**, 3524–3535.

Chapter 8

Beatson, R.K. and Light, W.A. (1992). Quasi-interpolation in the absence of polyno-mial reproduction. In: *Numerical methods of approximation theory*, D. Braess and L.L. Schumaker (eds.). Birkhäuser-Verlag, Basel, 21–39.

de Boor, C. and Ron, A. (1992). Fourier analysis of the approximation power of principal shift-invariant spaces. *Constr. Approx.*, **8**, 427–462.

Buhmann, M.D. (1988). Convergence of univariate quasi-interpolation using multi-quadrics. *IMA J. Numer. Anal.*, **8**, 365–383.

Buhmann, M.D. (1990). Multivariate cardinal-interpolation with radial-basis functions. *Constr. Approx.*, **6**, 225–255.

Buhmann, M.D. (1993). New developments in the theory of radial basis function interpolation. In: *Multivariate approximation: from CAGD to wavelets*, K. Jetter and F.I. Utreras (eds.). World Scientific, Singapore, 35–75.

Buhmann, M.D. and Dyn, N. (1993). Spectral convergence of multiquadric interpolation. *Proc. Edinburgh Math. Soc.*, **36**, 319–333.

Buhmann, M.D., Dyn, N., and Levin, D. (1995). On quasi-interpolation with radial basis functions with scattered centers. *Constr. Approx.*, **11**, 239–254.

Buhmann, M.D. and Ron, A. (1994). Radial basis functions: L^p-approximation orders with scattered centers. In: *Wavelets, images and surface fitting*, P.J. Lauren, A. Le Mehaute, and L.L. Schumaker (eds.). A.K. Peters, Wellesley, MA, 93–122.

Duchon, J. (1976). Interpolation des fonctions de deux variables suivant le principe de la flexion des plaques minces. *RAIRO, Analyse Numeriques*, **10**, 5–12.

Duchon, J. (1977). Splines minimizing rotation-invariate semi-norms in Sobolev spaces. In: *Constructive theory of functions of several variables*, W. Schempp and K. Zeller (eds.). Springer-Verlag, Berlin, 85–100.

Dyn, N. (1987). Interpolation of scattered data by radial functions. In: *Topics in multivariate approximation*, C.K. Chui, L.L. Schumaker and F.I. Utreras (eds.). Academic Press, New York, 47–61.

Dyn, N. (1989). Interpolation and approximation by radial and related functions. In: *Approximation theory VI*, C.K. Chui, L.L. Schumaker and J.D. Ward (eds.). Academic Press, New York, 211–234.

Dyn, N., Jackson, I.R.H., Levin, D., and Ron, A. (1992). On multivariate approximation by integer translates of a basis function. *Israel J. Math.*, **78**, 95–130.

Dyn, N., Levin, D., and Rippa, S. (1986). Numerical procedures for surface fitting of scattered data by radial functions. *SIAM J. Scien. Stat Comp.*, **7**, 639–659.

Dyn, N. and Ron, A. (1995). Radial basis function approximation: from gridded centres to scattered centres. *Proc. London Math. Soc.*, **71**, 76–108.

Franke, R. (1982). Scattered data interpolation: tests of some methods, *Math. Comp.*, **38**, 181–200.

Gelfand, I.M. and Shilov, G.E. (1964). *Generalized functions*, vol. 1. Academic Press, New York.

Guo, K., Hu, S., and Sun, X. (1993). Conditionally positive definite functions and Laplace–Stieltjes integrals. *J. Approx. Theory*, **74**, 249–265.

Hardy, R.L. (1971). Multiquadric equations of topography and other irregular surfaces. *J. Geophysical Res.*, **76**, 1905–1915.

Hardy, R.L. (1990). Theory and applications of the multiquadric-biharmonic method. *Comput. Math. Applic.*, **19**, 163–208.

Jackson, I.R.H. (1989). An order of convergence for some radial basis functions. *IMA J. Numer. Anal.*, **9**, 567–587.

Madych, W.R. and Nelson, S.A. (1990). Multivariate interpolation and conditionally positive definite functions II. *Math. Comp.*, **54**, 211–230.

Madych, W.R. and Nelson, S.A. (1992). Bounds on multivariate polynomials and exponential error estimates for multiquadric interpolation. *J. Approx. Theory.*, **70**, 94–114.

Meinguet, J. (1979). Multivariate interpolation at arbitrary points made simple. *Z. Angew. Math. Phys.*, **30**, 292–304.

Micchelli, C.A. (1986). Interpolation of scattered data: distance matrices and conditionally positive definite functions. *Constr. Approx.*, **1**, 11–22.

Powell, M.J.D. (1992). The theory of radial basis function approximation in 1990. In: *Advances in numerical analysis II: wavelets, subdivision, and radial functions*, W.A. Light (ed.). Clarendon Press, Oxford, 105–210.

Ron, A. (1992). The L_2-approximation orders of principal shift-invariant spaces generated by a radial basis function. In: *Numerical methods of approximation theory*, D. Braess and L.L. Schumaker (eds.). Birkhäuser-Verlag, Basel, 245–268.

Schaback, R. and Wu, Z. (1993). Local error estimates for radial basis function interpolation of scattered data. *IMA J. Numer. Anal.*, **13**, 13–27.

Strang, G. and Fix, G. (1973). A Fourier analysis of the finite element variational method. In: *Constructive aspects of functional analysis*, vol. 1, G. Geymonet (ed.). Edizioni Cremonese, Roma, 793–840.

Wu, Z. (1994). Multivariate compactly supported positive definite functions (preprint).

Chapter 9

Balakrishnan, A.V. (1957). A note on the sampling principle for continuous signals. *IRE Trans. Information Theory*, **IT-3**, 143–146.

Belyaev, Yu.K. (1959). Analytical random processes. *Teor. Verojat. Primenen*, **IV**, no. 4, 437–444 (Russian).

Beutler, F.E. (1961). Sampling theorems and bases in a Hilbert space. *Inform. Contr.*, **4**, 97–117.

Butzer, P.L., Splettstösser, W., and Stens, R.L. (1988). The sampling theorem and linear prediction in signal analysis. *Jahresber. Deutsch. Math.- Verein*, **90**, 1–70.

Cambanis, S. and Masry, E. (1976). Zakai's class of bandlimited functions and processes: its characterization and properties. *SIAM J. Appl. Math.*, **30**, no. 1, 10–21.

Chang, D.K. and Rao, M.M. (1983). Bimeasures and sampling theorems for weakly harmonizable processes. *Stoch. Anal. Appl.*, **1**, no. 1, 21–55.

Fikhtengol'c, G.M. (1969). *A course of differential and integral calculus*, III. Fifth edition, Nauka, Moscow (Russian).

Gaposhkin, V.F. (1977). A theorem on the almost sure convergence of the sequences of measurable functions and its application to the sequences of stochastic integrals. *Mat. Sbornik (N.S.)*, **104(146)**, no. 1(9), 3–21 (Russian).

Gulyás, O. (1970). On the truncation error in the sampling theorem. *Proc. Coll. Microwave Comm.*, vol. I (held at Budapest), April 21–24, 1970, Bognár, G. (ed.), ST-13/1-ST-13/5 (Russian).

Harkevich, A.A. (1955). *The questions of the general communication theory* Gostehizdat, Moscow (Russian).

Higgins, J.R. (1991). Sampling theorems and the contour integral method. *Applic. Anal.*, **4**, no. 1, 155–171.

Houdré, C. (1995). Reconstruction of band limited processes from irregular samples. *Ann. Probab.*, **23**, 674–696.

Houdré, C. (1994). Wavelets, probability, and statistics: Some bridges. In: *Wavelets: mathematics and applications*, Benedetto, J.J. and Frazier, M.W. (eds.). CRC Press, Boca Raton, 365–399.

Jerri, A.J. (1977). The Shannon sampling theorem—its various extensions and applications: a tutorial review. *Proc. IEEE*, **65**, 1565–1596.

Karamata, J. (1949). *Theory and Practice of Stieltjes integral.* Special edition of the Serbian Academy of Sciences **CLIV**, Mathematical Institute, Belgrade (Serbian).

Klesov, O.I. (1983). On the conditions of the signal restoration by discrete samples. *Doklady Akad. Nauk. Ukraine SSR, Ser.A*, **11**, 15–17 (Russian).

Klesov, O.I. (1984). On the almost sure convergence of the multiple Kotel'nikov Shannon series. *Probl. Peredachi Inform.*, **XX**, no. 3, 79–93 (Russian).

Klesov, O.I. (1985). The restoration of a Gaussian random field with finite spectrum by readings on a lattice. *Kibernetika*, **4**, 41–46 (Russian).

Kolmogorov, A.N. (1956). On the Shannon theory of information transmission in the case of continuous signals. *IRE Trans. Inform. Theory*, **IT-2**, no. 4, 102–108.

Lee, A.J. (1976a). On bandlimited stochastic processes. *SIAM J. Appl. Math*, **30**, no. 2, 269–277.

Lee, A.J. (1976b). Characterization of bandlimited functions and processes, *Inform. Control*, **31**, 258–271.

Lee, A.J. (1977). Approximate interpolation and the sampling theorem, *SIAM J. Appl. Math.*, **32**, no. 4, 731–744.

Lee, A.J. (1978). Sampling theorems for nonstationary random processes. *Trans. Amer. Math. Soc.*, **242**, 225–241.

Leonenko, N.N. (1993). Personal communication to the author.

Linden, D.A. and Abramson, N.M. (1960). A generalization of the sampling theorem. *Inform. Control*, **3**, 26–31; (see correction to equation (40), (1961), *ibid.* **4**, 95–96).

Lloyd, S.P. (1959). A sampling theorem for stationary (wide sense) stochastic processes. *Trans. Amer. Math. Soc.*, **92**, 1–12.

Oswald, J. (1951). Signaux aléatoires à spectre limité. *Cables et Transmissions*, **5**, no. 2, 158–177.

Parzen, E. (1956). A simple proof and some extensions of the sampling theorem. *Technical report* 7, Stanford University, California, 1–9.

Parzen, E. (1984). *Time series analysis of irregularly observed data.* Lect. Notes in Stat., vol. 25, Springer-Verlag, New York.

Piranashvili, Z.A. (1967). On the problem of interpolation of random processes. *Teor. Verojat. Primenen*, **XII**, no. 4, 708–717 (Russian).

Pogány, T. (1989). An approach to the sampling theorem for continuous time processes. *Austral. J. Stat.*, **31**, 427–432.

Pogány, T. (1991). Almost sure sampling reconstruction of non-band-limited homogeneous random fields. *Publ. Inst. Math. Belgrade (N.S.)*, **49(63)**, 221–232.

Pogány, T. (1994). Some irregular-derivative sampling theorems. In *Proc. functional analysis*, vol. IV, Butković, D. et al. (ed.). Various Publication Series **43**, Mathematical Institute, Aarhus.

Pogány, T. (1995). On the irregular-derivative sampling with uniformly dense sample points for bandlimited functions and processes. In: *Exploring stochastic laws, Festschrift in Honour of the 70th birthday of Academician Vladimir Semenovich Koroljuk*, Skorokhod, A.V., Boroskhikh, Yu.V. (eds.). VSP, Utrecht, The Netherlands, 395–408.

Pogány, T. and Peruničić, P. (1991). On the multidimensional sampling theorem. *Glasnik Mat. Ser.*, **III**. (in print).

Pogány, T. and Peruničić, P. (1992). On the spectral representation of the sampling cardinal series expansion of weakly stationary stochastic processes. *Stochastica*, **XIII**, no. 1, 89–100.

Pogány, T. and Peruničić (1995). On the sampling theorem for homogeneous random fields. *Theor. Imovirnost ta Matem. Statyst.*, **N53**(Ukraine); 142–148 see also *Theor. Prob. and Math. Statist.*, **N53**(USA). 153–159.

Rao, M.M. (1967). Inference in stochastic processes III. *Wahrscheinlichkeitstheorie und verw. Gebiete*, **8**, 49–72.

Rao, M.M. (1989a). A view of harmonizable processes. In: *Statistical data and inference*, Dodge, Y. (ed.), Elsevier Science Publishers, Amsterdam, 597–615.

Rao, M.M. (1989b). Harmonizable signal extraction, filtering and sampling. In: *Topics in non-Gaussian signal processing*, Wegman, E.J. et al. (ed.), Springer-Verlag, New York, 98–117.

Rao, M.M. (1991). Sampling and prediction for harmonizable isotropic random fields. *J. Comb. Inf. System Sci.*, **16**, no. 2–3, 207–220.

Seip, K. (1990a). A note on sampling of bandlimited stochastic processes. *IEEE Trans. Inform. Theory*, **IT-36**, no. 5, 1186.

Seip, K. (1990b). On certain irregular sampling formulas for bandlimited functions. *The Univ. Trondheim Preprints Mathematics*, no. 4, 1–16.

Skorokhod, A.V. (1990). Personal communication to the author.

Ville, J. (1953). Signaux analytiques à spectre borne II. *Cables et Transmission*, **7**, no. 1, 44-53.

Wong, E. (1971). *Stochastic processes in information and dynamical systems*. McGraw-Hill, New York.

Zakai, M. (1965). Band-limited functions and the sampling theorem. *Inform. Control*, **25**, 143–158.

Yadrenko, M.I. (1980). *Spectral theory of random fields*. Višča Škola, Kiev (Russian).

Yaglom, A.M. (1949). On the problem of linear interpolation of stationary random sequences and processes. *Usp. Mat. Nauk*, **4**, no. 4, 173–178 (Russian).

Yaglom, A.M. (1955). *The correlation theory of continuous processes and fields with contributions to the statistical extrapolations of time series and to the turbulence theory*. Doctoral Thesis, Moscow (Russian).

Yaglom, A.M. (1962). *An introduction to the theory of stationary random functions*. Dover Publications Inc., New York.

Chapter 10

Bachman, G. (1964). *Elements of abstract harmonic analysis*. Academic Press, New York.

Beaty, M.G. (1994). Multichannel sampling for multiband signals. *Signal Processing*, **36**, 133–138.

Beaty, M.G. and Higgins, J.R. (1994). Aliasing and Poisson summation in the sampling theory of Paley–Wiener spaces. *Journal of Fourier Analysis and its Applications*, **1**, 67–85.

Beaty, M.G., Dodson, M.M., and Higgins, J.R. (1994). Approximating Paley–Wiener functions by smoothed step functions. *J. Approx. Th.*, **78**, 433–445.

Brown, J.L. Jr. (1967). On the error in reconstructing a non-bandlimited function by means of the bandpass sampling theorem. *J. Math. Anal. Appl.*, **18**, 75–84. Erratum (1968) *ibid*. **21**, 699.

Brown, J.L. Jr. (1981). Multichannel sampling of lowpass signals. *IEEE Trans. Circuits and Systems*, **CAS-21**, 101–106.

Butzer, P.L., Splettstößer, W., and Stens, R.L. (1988). The sampling theorem and linear prediction in signal analysis. *Jber. d. Dt. Math.-Verein.*, **3**, 1–70.

Dodson, M.M. and Silva, A.M. (1985). Fourier analysis and the sampling theorem. *Proc. Royal Irish Acad.*, **85A**, 81–108.

Dodson, M.M., Silva, A.M., and Souček, V. (1986). A note on Whittaker's cardinal series in harmonic analysis. *Proc. Edinb. Math. Soc.*, **29**, 349–357.

Faridani, A. (1994). A generalised sampling theorem for locally compact abelian groups. *Math. Computation*, **63**, 307–327.

Hewitt, E. and Ross, K.A. (1963). *Abstract harmonic analysis*, vol. 1. Springer-Verlag, New York.

Hewitt, E. and Ross, K.A. (1970). *Abstract harmonic analysis*, vol. 2. Springer-Verlag, New York.

Higgins, J.R. (1985). Five short stories about the cardinal series. *Bull. Amer. Math. Soc.*, **12**, 45–89.

Higgins, J.R. (1991). Another look at aliasing in Paley–Wiener spaces. *Proc. 13th IMACS World Congress on Computation and Applied Math.*, 256–257.

Higgins, J.R. (1996). *Sampling theory in Fourier and signal analysis: Foundations*. Clarendon Press, Oxford.

Katznelson, Y. (1976). *An introduction to harmonic analysis*. Dover, New York.

Kluvánek, I. (1965). Sampling theorem in abstract harmonic analysis. *Mat.-Fyz. Casopis Sloven. Akad. Vied.*, **15**, 43–48.

Loomis, L.H. (1953). *An Introduction to abstract harmonic analysis*. Van Nostrand, New York.

Montgomery, D. and Zippin, L. (1955). *Topological transformation groups*. Interscience, New York.

Nachbin, L. (1965). *The Haar integral*. Van Nostrand, New York.

Papoulis, A. (1977). Generalised sampling expansion. *IEEE Trans. Circuits and Systems*, **CAS-24**, 652–654.

Rudin, W. (1962). *Fourier analysis on groups*. John Wiley, New York.

Rudin, W. (1977). *Functional analysis*. McGraw-Hill, New York.

Rudin, W. (1986). *Real and complex analysis*. McGraw-Hill, New York.

Stanković, R.S. and Stanković, M.S. (1984). Sampling expansions for complex valued functions on finite Abelian groups. *AUTOMATIKA*, **25**, 147–150.

Titchmarsh, E.C. (1962). *Introduction to the theory of Fourier integrals* (2nd edn). Clarendon Press, Oxford.

AUTHOR INDEX

SUBJECT INDEX

Printed in the United Kingdom
by Lightning Source UK Ltd.
112261UKS00001B/61-63